An Introduction to Aircraft Performance

An Introduction to Aircraft Performance

Mario Asselin
Royal Military College of Canada
Kingston, Ontario
Canada

EDUCATION SERIES
J. S. Przemieniecki
Series Editor-in-Chief
Air Force Institute of Technology
Wright-Patterson Air Force Base, Ohio

Published by
American Institute of Aeronuatics and Astronuatics, Inc.
1801 Alexander Bell Drive, Reston, VA 20191

American Institute of Aeronautics and Astronautics, Inc., Reston, Virginia

Library of Congress Cataloging-in-Publication Data

Asselin, Mario, 1965–
    An Introduction to Aircraft Performance / Mario Asselin,
      p. cm.—(AIAA education series)
    Includes bibliographical references and index.
    ISBN 1-56347-221-X (alk.paper)
    1. Airplanes—Performance. I. Title. II. Series.
    TL671.4.A87   1997
    629.132—dc21   97-308
                                    CIP

Data and information appearing in this book are for informational purposes only. AIAA is not responsible for any injury or damage resulting from use or reliance, nor does AIAA warrant that use or reliance will be free from privately owned rights.

A mon épouse, Sylvie Saint-Georges

Foreword

An Introduction to Aircraft Performance by Mario Asselin is an important addition to the AIAA Education Series. It provides a needed text for undergraduate courses in which the student can be introduced to all the important elements of aeronautical engineering: aerodynamics, propulsion, performance, stability and control, and elements of aircraft design. Professor Asselin, currently with the Royal Military College (RMC) of Canada, had extensive prior experience with the Canadian Forces working in engineering design as well as maintenance, including the CF-188 (Canadian version of the F/A-18 Hornet). The material presented in this text was used extensively in courses offered at RMC.

The Education Series of textbooks and monographs embraces a broad spectrum of theory and application of different disciplines in aeronautics and astronautics, including aerospace design practice. The series also includes texts on defense science, engineering, and management. It includes over 50 titles, which serve as teaching texts and reference materials for practicing engineers, scientists, and managers. This recent addition to the series will be a valuable text for undergraduate courses in aeronautical engineering programs. It complements several other texts on aircraft performance previously published in this series.

J. S. Przemieniecki
Editor-in-Chief
AIAA Education Series

Table of Contents

Preface

How fast can this aircraft go? What altitude can it reach? Can this aircraft get to the designated patrol area and remain on station for at least an hour? Can this aircraft carry 300 passengers from New York to London? These are all aircraft performance-related questions that could drive the requirements of a particular aircraft design, whether military or civilian.

This book was started out of curiosity. As I was completing research on aircraft icing, focusing mostly on aerothermodynamics, I felt that the results by themselves were incomplete. I needed to know how ice accretion on the aircraft would affect it, apart from the obvious aerodynamic penalties of increased drag and decreased maximum lift coefficient and stall angle. This curiosity led me to consult several texts for information on aircraft performance. Then I was required to evaluate different Ground Proximity Warning Systems (GPWS) as part of my duties as project manager for the integration of a GPWS in the CF-188 (the Canadian version of the F/A-18 Hornet). This required a good understanding of the performance of the aircraft during a dive recovery.

The investigation of aircraft performance is now a passion for me. As I broaden my knowledge of aerospace engineering, I can relate the various fields (such as propulsion, structures, stability, and control) to performance. Now, as an instructor, I make certain that students are aware of the implications on performance when modifying a parameter on the aircraft.

This book is formatted for teaching graduate and undergraduate courses in aircraft performance. The format of the book was set following discussions with Dr. Lemieux, of the Royal Military College (RMC) of Canada, on the possibility of providing a study guide for the aircraft performance course being offered at RMC. This book provides a gradual approach to the subject of aircraft performance. Initial estimates of performance include many assumptions to get simplified equations from which the effects of several factors affecting performance can be evaluated. Specialized subjects that I had the chance of personally investigating are included to provide a broader view of the field of aircraft performance. I am currently working on software to complement this book and give the students a better understanding of the field.

This book should serve as a day-to-day reference for aircraft consultants, aircraft accident investigators, government officials, and engineering students.

Acknowledgment

I would like to thank my wife, Sylvie, for her understanding as I spent many manhours away from her while I was completing this textbook. Without her support, this work could have taken a lot longer to finish. Sincere thanks to Guy

Lemieux of the Royal Military College of Canada in Kingston for taking the time to review and comment on this textbook. He took many hundreds of hours of his own time to review the first and second drafts of the textbook, including the subjects that do not form part of the core material for the course MEE467—Aircraft Performance. This high level of effort on his part, with the continuing research and updating on my part, has greatly improved the readability of the book. The magnificent painting on the preceding page was done by a Canadian aviation artist, Don Connolly of Sydenham, Ontario. Finally, I would like to thank my parents!

Mario Asselin
April 1997

1
Aircraft Aerodynamics

1.1 Introduction

T HE aim of this chapter is to identify the various aerodynamic parameters influencing aircraft performance. This chapter is not intended to provide an in-depth analysis of each parameter but rather assumes that the reader has sufficient knowledge of aerodynamics to understand the following discussion. This chapter defines and classifies the different aerodynamic sources that affect the performance of an aircraft. The reader is encouraged to consult specialized references in the matter of aerodynamics while reading this chapter.

An aircraft in flight (see Fig. 1.1), at a given velocity, will be subjected to an aerodynamic force F created by the distribution of pressure and shear stress over its entire wetted surface (surface in contact with the moving air). Intuitively, one may expect the aerodynamic force to depend on the aircraft velocity V_∞, the surrounding air density ρ_∞, the size of the aircraft expressed in terms of a reference area (usually the gross wing surface S), the viscosity of the air μ_∞, and the shape and orientation of the aircraft with respect to the incoming airflow, as well as on the compressibility of the air as generally expressed in terms of the velocity of sound a_∞. Thus, for a given aircraft at a given orientation, the aerodynamic force can be expressed as

$$F = f(V_\infty, \rho_\infty, S, \mu_\infty, a_\infty, \text{aircraft shape, orientation})$$

In fact, if a dimensional analysis is performed, the aerodynamic force can be expressed as

$$F = \tfrac{1}{2}\rho_\infty V_\infty^2 S C_F \tag{1.1}$$

where

$$C_F = f(Re, M_\infty, \text{aircraft shape, orientation})$$

and where C_F is a nondimensional force coefficient, which accounts for the aircraft shape (and configuration), the aircraft orientation with respect to the incoming airflow, the airflow compressibility (through the airflow Mach number), and the air viscosity (through the Reynolds number). Equation (1.1) represents a simple-to-use expression for the aerodynamic force yet it contains all of the information required to study aircraft performance.

The aerodynamic force F vector can be broken down into components assigned to a reference system of axes. The system of axes to be used throughout this book (unless specified otherwise) will be one that is dependent on the airflow orientation, that is, the aircraft flight path. The component of F perpendicular to the flight path is defined as the lift force L, whereas the component of F parallel to the flight path is the drag force D (see Fig. 1.2).

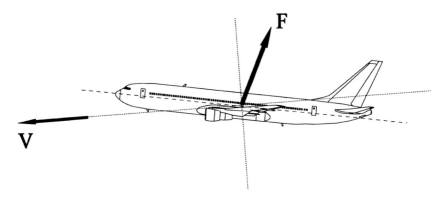

Fig. 1.1   Aerodynamic force acting on an aircraft in flight.

These two components are mathematically defined in the same way as the aerodynamic force F. Thus,

$$L = \tfrac{1}{2}\rho_\infty V_\infty^2 S C_L$$
$$D = \tfrac{1}{2}\rho_\infty V_\infty^2 S C_D$$

$$(1.2)$$

where

$$C_L = f_1(Re, M_\infty, \text{aircraft shape, orientation})$$

$$C_D = f_2(Re, M_\infty, \text{aircraft shape, orientation})$$

It is then necessary to define the environment in which the aircraft operates to quantify the value of the aerodynamic force components acting on the aircraft. The air density is a function of ambient air pressure and temperature. A standard atmosphere will be used throughout this book so that the evolution of the air density with altitude can be quantified. It is suggested that the reader consult "Appendix B: Atmosphere" of this book.

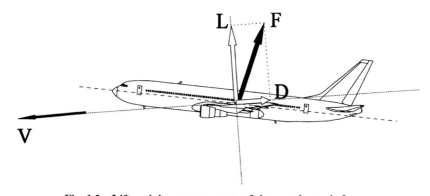

Fig. 1.2   Lift and drag components of the aerodynamic force.

1.2 Aircraft Shape and Orientation Effects

As mentioned previously, the aerodynamic force is created by the distribution of pressure and shear stress over the entire wetted surface of the aircraft. This involves a complex flow distribution, which can be best understood by analyzing simple elements first. One may suspect that the majority of the lift is produced by the wing, and so the following analysis will focus primarily on the wing and its elements. The reader is encouraged to read "Appendix A: Aircraft Nomenclature" before proceeding with this section.

As the air flows around an object, the relative velocity of the layer immediately adjacent to the surface must be zero because of the viscosity of the air. Ludwig Prandtl introduced the boundary-layer concept in 1904, which stipulates that there exists a thin region near the surface where the effects of friction are very important and outside of this region the flow can be considered irrotational. Because the boundary layer is thin, there is a rapid change of velocity from zero at the surface to the velocity just outside the boundary layer. This change in velocity is usually referred to as the velocity profile of the boundary layer (see Fig. 1.3). The slippage of one air layer relative to the next creates a shear stress (force per unit surface), which is tangential to the surface and in the direction of the air flow. This shear stress is equal to the air viscosity coefficient times the velocity gradient at the surface in the direction perpendicular to it,

$$\tau = \mu \left[\frac{du}{dn} \right]_{n=0} \tag{1.3}$$

Because friction is a dissipative force, as the fluid moves along the surface of the object it will lose more and more energy. This will slow down more and more fluid, which leads to an increase in boundary-layer thickness. The resulting retarding force is equal to the integration of the local shear stress over the entire wetted surface of the object. This loss of momentum is called skin-friction drag.

The flow within the boundary layer can be of two types. The first one, called laminar flow, consists of layers of air sliding one over the other in a regular fashion without mixing. The second one is called turbulent flow and consists of particles of air that move in a random and irregular fashion with no clear individual path. In specifying the velocity profile within a turbulent boundary layer, one must look at the mean velocity distribution measured over a relatively long period of time. There is usually a transition region between these two types of boundary-layer flow. It

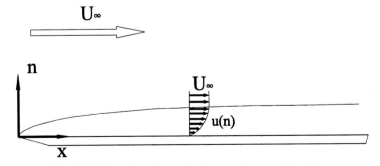

Fig. 1.3   Boundary-layer development over a flat plate.

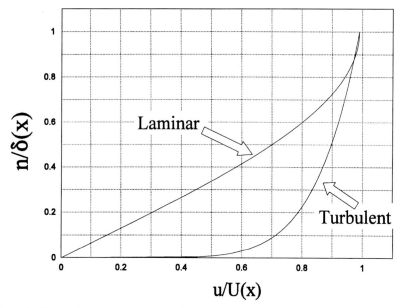

Fig. 1.4 Normalized velocity profiles within a boundary layer, comparison between laminar and turbulent flow.

was observed by Osborne Reynolds in 1883 that a fluid flow changes from laminar to turbulent at approximately the same value of the dimensionless ratio $(\rho u \ell / \mu)$ where ℓ is a characteristic length for the object in the flow. This ratio is now called the Reynolds number, and it is the governing parameter for viscous flows.

The turbulent boundary layer is thicker than the laminar one. The breakup of the streamlines means that there is air mixing and thus energy exchange between the high-energy flow of the outer boundary-layer region and the low-energy flow near the surface. The velocity profile for the turbulent boundary layer will, thus, be fuller, and the velocity gradient at the surface will be larger, which leads to larger skin-friction drag in a turbulent boundary layer than in a laminar one. Thus, to reduce skin friction drag, one must promote laminar flow (see Fig. 1.4).

Contrary to the shear stress, which acts tangentially to the surface, pressure acts perpendicularly to the surface. From Bernoulli's equation for incompressible flows, one may expect that the flow of air about an object will create a pressure distribution around it. In the region of decreasing pressure, the velocity will increase and the boundary layer will tend to grow relatively slowly because it is being pulled by the suction in front. In the region of increasing pressure (called the region of adverse pressure), the velocity will decrease and the boundary layer will thicken more rapidly. Because the boundary layer has less energy, it will slow down faster than the surrounding air flow. If the adverse pressure gradient is strong enough, the flow within the boundary layer may come to a stop and even be reversed. The boundary layer is said to be separated when there exists a region of reverse flow (see Fig. 1.5). The separation point is located at the point where the shear stress is zero, that is, where the velocity gradient normal to the surface equals zero [see Eq. (1.3)]. The consequence of flow separation is that the pressure on the rearward surface of an object is less than the pressure on the front part. This

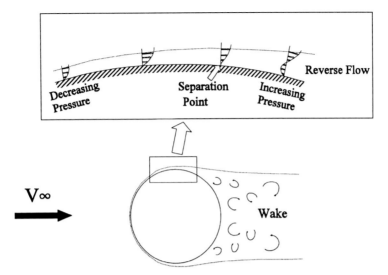

Fig. 1.5 Boundary-layer flow separation about a cylinder.

creates a drag force called form drag or pressure drag due to flow separation or even wake drag. The wake of an object is the region of low-energy air behind the object caused by friction and flow separation.

Form drag is roughly proportional to the size of the wake. The farther upstream the boundary layer separates from the surface, the larger the wake will be. Because a laminar boundary layer has less energy near the surface (see Fig. 1.4), it will separate sooner than a turbulent boundary layer, thus increasing form drag. But a turbulent boundary layer will create more skin-friction drag than a laminar boundary layer. Profile drag is the sum of both skin-friction drag and form drag. Profile drag can be minimized by carefully shaping an object to achieve the best compromise between skin-friction drag and form drag.

For a specific aircraft shape, the aerodynamic forces just described depend on the aircraft orientation. The angle of attack α is defined as the angle between the incoming flow and a reference line in the plane of symmetry of the aircraft [the longitudinal axis (three-dimensional flow) or the chord line for a profile (two-dimensional flow)]. The angle of attack (AoA) α will be positive if the nose of the aircraft is above the velocity vector. The sideslip angle β is defined as the angle between the incoming flow and a reference line in the water plane. The sideslip angle is positive if the nose is to the left of the velocity vector (as shown in Fig. 1.6).

In general, for low AoAs, the lift coefficient of an airfoil increases linearly with an increase in α (see Fig. 1.7). Above a certain value, the point of separation of the boundary layer starts to move upstream on the upper surface of the wing, decreasing the slope of the lift coefficient curve up to an AoA where the boundary layer separates completely, and the lift coefficient starts decreasing with further increases in AoA. The angle at which the lift coefficient is maximum is called the stall AoA (α_{stall}). Beyond that angle the airfoil is said to be stalled. The loss in lift coefficient at stall is greatly influenced by the shape of the leading edge of the wing. A sharp leading edge generally creates a large and sudden decrease in lift coefficient because the boundary layer tends to separate at the leading edge

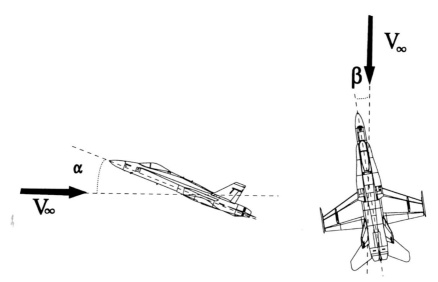

Fig. 1.6   Definitions of aerodynamic angles.

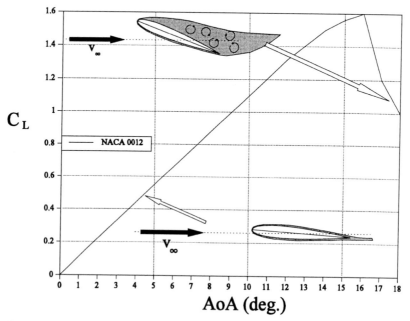

AoA (deg.)

Fig. 1.7   Lift coefficient as a function of AoA for a NACA 0012 ($Re = 6 \times 10^6$).

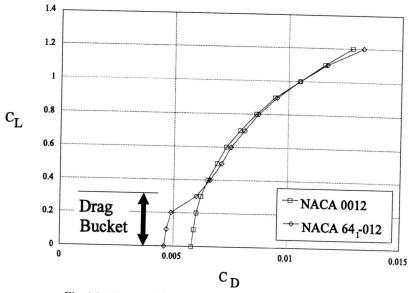

Fig. 1.8   Drag coefficient for two airfoils at $Re = 6 \times 10^6$.

and propagate rapidly toward the trailing edge, whereas for a large leading-edge radius, the boundary-layer separation will start at the trailing edge and propagate slowly toward the leading edge as the AoA is increased. The AoA for zero lift (α_0) depends on the camber of the airfoil. Symmetrical airfoils will have an α_0 equal to zero, whereas airfoils with a positive camber will have an α_0 smaller than zero, thus developing lift at zero AoA.

At low AoAs, the drag of an airfoil is mainly skin-friction drag as most of the boundary layer remains attached to the surface. As the AoA increases, the boundary layer starts to separate at the trailing edge, increasing the airfoil drag slightly at first, then very rapidly. This means that at low AoA, the drag coefficient will remain essentially constant, being only affected by the percentage of the surface being covered by turbulent flow. Some airfoils are designed to promote laminar flow for a range of low-lift coefficients, but above a given lift coefficient value the boundary layer transitions to turbulent flow and the advantage this airfoil had over other conventional airfoils is lost. The lift coefficient range over which this occurs is often referred to as a drag bucket due to the shape of the drag polar (see Fig. 1.8).

1.3   Effects of the Reynolds Number (Viscosity)

The Reynolds number Re has a major influence on the airflow around the airfoil. It represents the ratio of the inertial forces to the viscous forces of the airflow. The Reynolds number for an aircraft at a given flight state is

$$Re = \rho_\infty V_\infty \bar{c} / \mu_\infty \qquad (1.4)$$

where $\bar{c}$ is the mean aerodynamic chord (MAC) of the aircraft.

The greater the Reynolds number, the more energy the boundary layer has to go around the top of the airfoil. Remember that in an ideal flow there are no viscous

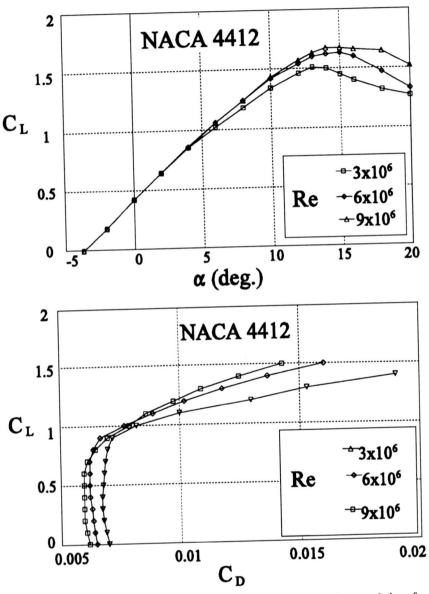

Fig. 1.9   Effects of the Reynolds number on the lift and drag characteristics of a NACA 4412.

forces and, thus, the Reynolds number is equal to infinity. There is also no flow separation and no stall. For real flow, no matter how high the Reynolds number is, there is always viscosity, which will bring flow separation at a certain AoA. But as the Reynolds number increases, flow separation can be delayed, resulting in a lift coefficient increase as shown in Fig. 1.9.

A high Reynolds number will reduce form drag by causing the boundary-layer laminar flow to transition earlier to turbulent flow, thus delaying boundary-layer flow separation. On the other hand, a higher Reynolds number will increase skin-friction drag by having a greater portion of the surface covered by turbulent flow. The actual behavior of the drag coefficient at low AoA due to the change in Reynolds number is highly dependent on the shape of the airfoil.

A typical Reynolds number range for a Boeing 767 class aircraft with a MAC of about 20 ft would be between 1.7×10^7 (high altitude and low speed) and 1.15×10^8 (low altitude and high speed). For a F-20 size fighter, with a MAC of about 7.5 ft, that range would be between 8.5×10^6 (high and slow) and 5.3×10^7 (low and fast). The aerodynamic characteristic of an aircraft wing will be given for a Reynolds number within its range of operation.

1.4 Effects of the Mach Number: Compressibility

The Mach number M is defined as the ratio of the air velocity V to the local velocity of sound a [see Eq. (1.5)]. The flight Mach number is the ratio of the aircraft velocity to the velocity of sound of the air that the aircraft is flying through. The airflow is said to be subsonic if its Mach number is everywhere smaller than unity. It is supersonic if the velocity is greater than Mach 1.0,

$$M = V/a \qquad (1.5)$$

As the air goes around an airfoil it will be accelerated up to a maximum local velocity. At a given flight Mach number (dependant on airfoil thickness and AoA) the point of maximum local velocity will reach Mach 1.0 even if the flight Mach number is less than one. The flight Mach number at which this happens is called the critical Mach number M_{crit}. A further increase in velocity will result in the formation of a region of supersonic flow on the top surface of the airfoil; the flow is now said to be transonic.

As the supersonic region expands, a normal shock wave will be created at the back of the supersonic region. As the air goes through a normal shock wave, there will be a sudden increase in pressure and a reduction in airflow velocity to subsonic conditions. This compression will increase air pressure and temperature behind the shock wave resulting in a loss of energy. If the shock is strong enough, the large pressure rise can produce flow separation (Fig. 1.10). The loss of energy

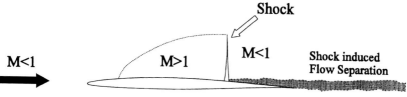

Fig. 1.10 Transonic flow and shock induced flow separation.

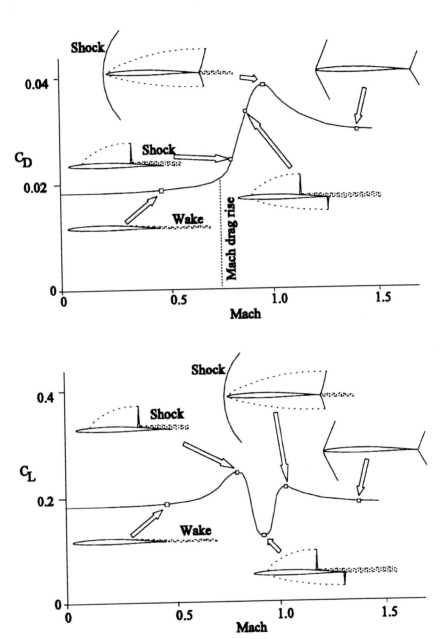

Fig. 1.11   Effect of Mach number on the lift coefficient for a given AoA and on the drag coefficient for a given lift coefficient.

combined with the shock induced flow separation is called wave drag. The flight Mach number at which wave drag starts is called the drag rise Mach number M_{DR}. Flow separation also leads to loss of lift. Figure 1.11 illustrates typical variation in lift and drag coefficients with Mach number.

As the flight Mach number reaches 1.0, a normal shock wave develops in front of the airfoil's leading edge, which further increases wave drag. At higher Mach numbers, the shock waves of the leading and trailing edges will become more and more oblique. The air going through an oblique shock wave will still be compressed, but the pressure and temperature rise will not be as great as that of a normal shock wave. The flow velocity is still reduced but remains supersonic.

Wave drag can be minimized by using thin airfoils, thereby delaying the occurrence of shock waves and reducing their intensity and effects. The use of specially designed supercritical airfoils also reduces wave drag by bringing the drag rise Mach number closer to 1.0 and by reducing the strength of the normal shock wave on the top surface of the airfoil.

The wave drag of a complete aircraft can be minimized by carefully shaping the volume distribution of the aircraft. Figure 1.12 illustrates the cross-sectional area distribution of a fighter type aircraft compared to that of the Sears–Haack area distribution, which yields the smallest wave drag possible.

1.5 Effects of Wing Planform

The flow around a wing is the extension in the third dimension of the flow about an airfoil. The lift is still created by the average pressure over the upper surface being less than the average pressure under the lower surface. However, the pressure on both surfaces must be equal at the wing tips where they both meet. This will introduce pressure gradients along the wing span on both surfaces. There will be a resulting flow of air from the middle of the wing toward the wing tips under the lower surface and one in the opposite direction over the upper surface. A vortex sheet is thus formed behind the wing.

This vortex sheet (Fig. 1.13) will induce a velocity, called downwash, normal to the freestream velocity and directed toward the ground. This downwash, which varies along the span, modifies the local velocity V_e both in direction and in magnitude for a particular wing section. The apparent lift L_e, which is perpendicular to the local velocity, will no longer be perpendicular to the freestream velocity. This lift force has a component normal to the freestream velocity (which is the required lift) and one parallel to V_∞. This latter component is a drag force and is called vortex drag D_v (see Fig. 1.14). It is due solely to the generation of lift and not to friction. The effective AoA α_e for a particular wing section will be equal to the geometric AoA α minus the induced AoA α_i,

$$\alpha_e = \alpha - \alpha_i$$
$$\alpha_i = \arctan(\omega/V_\infty) \approx \omega/V_\infty \tag{1.6}$$

Because the aircraft still requires the same lift to stay in the air, the geometric AoA has to be increased so as to have the wing generate the required lift. This effectively reduces the lift coefficient slope of the wing. This does not, however, change the value of the zero lift AoA because there would then be no induced velocity and, therefore, no induced AoA.

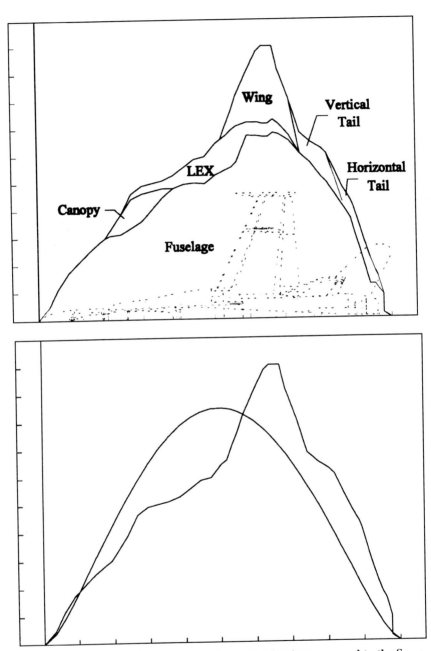

Fig. 1.12   Fighter aircraft cross-sectional area distribution compared to the Sears–Haack area distribution of the same volume, all curves are normalized.

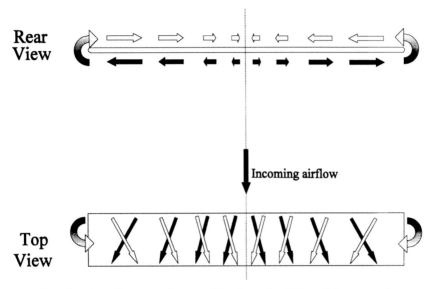

Fig. 1.13   Creation of a vortex sheet resulting from the differential pressure between the upper and lower surfaces.

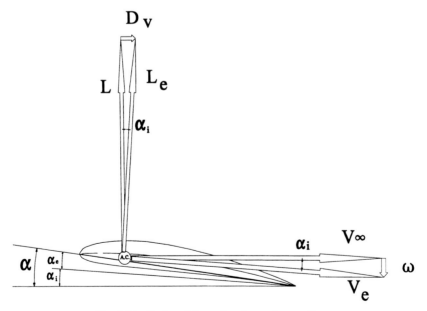

Fig. 1.14   Induced flow at a given wing section.

$(\pi/4)\, b$

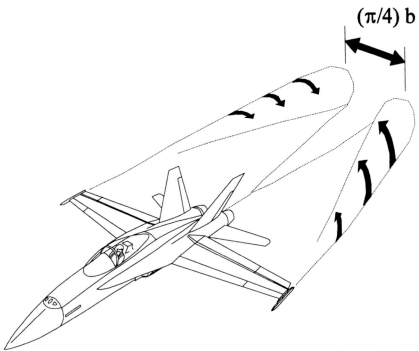

Fig. 1.15   Vortex sheet will tend to roll up with cores separated by a distance of about $(\pi/4)b$.

The vortex sheet is unstable and will tend to roll up into two large counter-rotating vortices at a certain distance behind the wing. The core of these vortices will be separated by a distance of approximately $\pi/4$ times the span (78.5% of the span). For this reason, and because the vortices within the vortex sheet are strongest at the wing tip, these two large vortices are often referred to as wing tip vortices (see Fig. 1.15).

Vortex drag can be reduced by either increasing the separation between the vortex cores (by increasing the wing span), by dissipating the wing tip vortex, or by designing the wing so as to have a wing span lift distribution that is of elliptical shape. In the first case, one can increase the aspect ratio of the wing (Sec. 1.5.1), and in the second case, one may use winglets (Sec. 1.5.2) or other wing tip vortex control devices.

1.5.1 Aspect Ratio (AR)

The aspect ratio (AR) has a major influence on vortex drag and on the lifting ability of a wing at a given AoA. At subsonic speeds the profile drag of a wing is proportional to its area, whereas the drag due to lift is proportional to the span loading squared $[(W/b)^2]$. Thus if the ratio of span squared to wing area, AR, can be increased, the lift-dependent drag can be reduced for no increase in profile drag. A large AR will yield a high lift-to-drag ratio, which is beneficial in all aspects of aircraft performance (see Appendix C). For example, the high-lift curve slope conferred by high AR is useful on takeoff and landing, when the usable AoA is restricted by the demands of ground clearance and pilot vision. The aircraft induced drag will include all drag due to the production of lift, including trim drag (Sec. 1.7.3). The aircraft induced drag coefficient can be written as

$$C_{D_i} = K C_L^2 \qquad K = 1/\pi AR e \tag{1.7}$$

The variable e is called the span efficiency factor or the Oswald number. It accounts for the shape of the wing planform, the wing sections used, geometric twist, etc. The AR will affect the wing lift coefficient slope in the following way:

$$\frac{dC_L}{d\alpha} = a = a_\infty \bigg/ \left(1 + \frac{a_\infty}{\pi AR} \right) \tag{1.8}$$

where a_∞ is the lift slope of an infinitely long wing. This lift slope has a value of approximately (2π) for all wing sections, which means that the lift slope of a wing of AR of 3.5 would be approximately 4.

A high AR also has its drawbacks. For a given wing area, there is a weight penalty because the higher bending moments arising from the increased span have to be countered by more massive root sections due to their shorter root chord. Wing weight can be very sensitive to AR beyond a certain value. On the F-16, for example, the start-of-combat weight increased only slightly as AR increased from 3.0 to 3.5, whereas from 3.5 to 4.0 the weight increase was tripled. An AR of 3.5 appears to be the upper limit for combat aircraft.

1.5.2 Winglets

Because the vortices shed by the wing are strongest at the tips, the addition of wing tip surfaces can reduce and diffuse the strength of these vortices, thus reducing the overall vortex drag of the aircraft. Winglets are small, nearly vertical aerodynamic surfaces mounted at the wing tips. They must be placed behind the region of lowest pressure of the wing to avoid the increased velocities over the inner surface of the winglets being superimposed to the high velocities over the forward region of the wing upper surface. Therefore, increase in transonic compressibility effects can be minimized. Furthermore, the winglets operate in the circulation flowfield around the wing tip. This wing tip circulation flow tends to produce large sidewash (toward inboard) on the winglet even at low airplane AoA. Because the resultant side force is approximately perpendicular to the local airflow, it produces a forward thrust component, which reduces the wing's vortex drag. The magnitude of the thrust produced depends on the strength of wing tip circulation, which, in turn, is a function of the rate of change in wing tip loading (see Fig. 1.16).

Special care must be taken while fitting winglets to a wing tip so as not to create more parasite drag than what can be saved in induced drag. A smooth fairing should be used at the juncture of the winglet and wing tip. Also, a tradeoff must be done between winglet span (which increases its efficiency) and aircraft weight penalty due to the winglet weight and increased wing root weight (larger aerodynamic bending moments). An ideally shaped winglet can reduce cruise drag by about 3–6%.

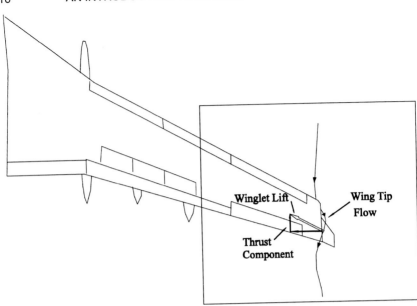

Winglet Lift

Wing Tip Flow

Thrust Component

Fig. 1.16   Winglet flow.

Other devices can be used at the wing tip to reduce vortex drag. The addition of tip tanks or of tip-mounted missiles can produce the same effect as a slight increase in AR, as well as reduce the strength of the vortices at the tips. Wing tip spanwise blowing can also increase the effective AR of the wing, but this requires air ducts (extra weight) and a source of high-pressure air (engine bleed air), which may alleviate any advantage this system may offer.

1.5.3   Wing Design for High Speed

The formation of shock waves at high-subsonic speeds creates a sharp increase in drag coefficient, as shown in Fig. 1.11. This increase in drag results in what is commonly known as the sound barrier. It was seen in Sec. 1.4 that these effects can be reduced by using thinner wing sections. By shaping the wing planform, the effects of compressibility can be reduced and delayed to higher Mach numbers.

A swept wing will postpone the development of regions of supersonic flow terminating in shock waves. The flow most affecting the aerodynamic behavior of the wing is the one perpendicular to the suction peak of the local wing sections (approximately the same angle as the wing leading-edge sweep), thus the velocity at the leading edge of a swept wing $[V \cos(\Lambda)]$ will be smaller than the freestream velocity. This leads to a larger freestream Mach number at which the first sign of supersonic flow appears. Thus, for two wings of the same relative thickness to chord ratio, the swept wing will display a reduction and delay in peak wave drag, as shown in Fig. 1.17. This feature can be very beneficial to an aircraft's acceleration and climb performance in the transonic regime. On the other hand, there will be a larger spanwise pressure gradient resulting in a larger outflow along the top surface of the wing. Aerodynamic penalties of swept wings are lower lift

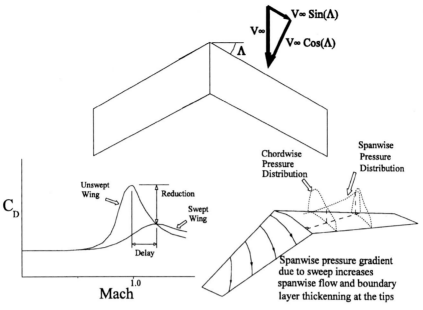

Fig. 1.17   Sweep effects on the flow about a wing.

slopes, as approximated in Eq. (1.9). There will also be a higher induced drag for a given AR and wing tip stall tendencies due to the spanwise outflow

$$\frac{dC_L}{d\alpha} = \pi AR \left/ \left(1 + \sqrt{1 + \left(\frac{AR}{2\cos(\Lambda)} \right)^2} \right) \right.$$ (1.9)

A variable sweep wing aircraft combines good field (takeoff and landing) performance with high-supersonic speeds performance by continuously changing the wing sweep angle. The large AR and low sweep in the low-speed regime provide high wing-lift slope and high-lift coefficient; low AR and high sweep provide better transonic and supersonic performance. Also, increasing the wing sweep will actually reduce the relative thickness of the wing with respect to the freestream Mach number, as shown in Fig. 1.18. In general, a variable sweep wing will have good performance throughout the design flight envelope by optimizing the sweep to provide high lift-to-drag ratio. The main drawback of variable sweep wing, other than the effects of sweep angle, is its higher weight and mechanical complexity.

Another wing design used for high-speed flights is the delta wing (Fig. 1.19). Such wings combine the effects of large sweep, thin wing sections (typically between 3 and 5%), low AR, and good area ruling required for high-speed flights. The large wing surface also means that a delta wing aircraft will generally have a lower wing loading (W/S) than a wing of similar AR. Delta wings also have thick wing root due to their long root chord. This provides more internal volume for fuel and landing gear. For example, the Avro Arrow (AR = 1.61, S = 1550 ft²) had a wing root that could house the landing gear even though the wing root relative

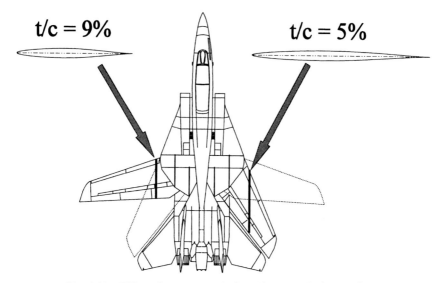

Fig. 1.18   Effect of sweep on relative thickness of wing sections.

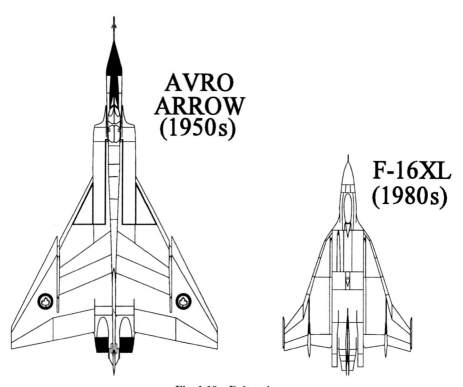

Fig. 1.19   Delta wing.

thickness was only 3.5% because the wing root chord was 35.5 ft long. But the low AR and large sweep of the delta wing has a detrimental effect on its lift coefficient slope. Also a tailless delta wing aircraft cannot use trailing-edge flaps because it cannot trim out the large nose-down pitching moments of such devices. This means that such aircraft will usually have poor airfield performance. The high-span loading of a delta wing also results in high-vortex drag during subsonic flights.

1.6   High-Lift Coefficient Devices

While designing an aircraft, designers are faced with conflicting requirements such as the need for high wing loading and low drag for efficient cruise, while low wing loading and high lift coefficient are required for good field performance. In fact, the takeoff and landing distance are both proportional to the takeoff and landing velocity squared, as will be seen in Chapter 7. Because takeoff and landing represent a very small portion of the intended mission of any aircraft, the wing is usually optimized to provide the high lift-to-drag ratio required for efficient cruise. The lift characteristics and maximum lift coefficient of the wing are thus set.

There exist many types of retractable devices that can modify the lift characteristics of the wing so as to provide better field performance. They can be grouped into one, or a combination, of the following categories: 1) camber modifier, 2) energy adder/boundary-layer controller, 3) wing surface modifier, and 4) powered lift.

Camber modifiers, when deflected, change the pressure distribution over the airfoil by altering the camber (Fig. 1.20). Trailing-edge types (plain trailing-edge flap, split flap) increase the value of the lift coefficient for a given AoA. The

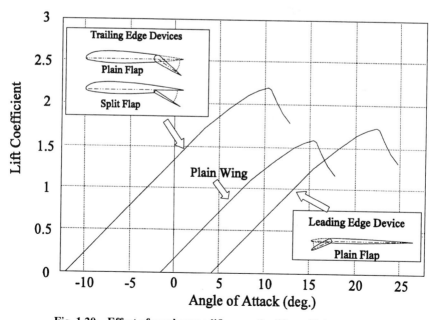

Fig. 1.20   Effect of camber modifiers on the lift coefficient curve.

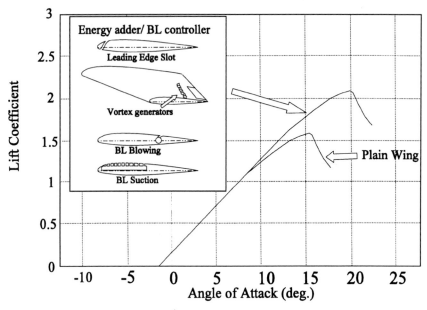

Fig. 1.21   Effect of energy adders/BL controllers on the lift coefficient curve.

maximum lift coefficient is also increased but the stall AoA is lowered due to an earlier boundary-layer separation caused by the higher suction peak at the leading edge. Plain leading-edge flaps, on the other hand, reduce the peak suction at the leading edge, thus delaying stall to a larger AoA. Leading-edge flaps do increase the maximum lift coefficient of the airfoil and delay the onset of stall to higher AoAs, but they also reduce the available lift at low AoAs.

Energy adders/boundary-layer controllers are used to prevent the separation of the boundary layer (BL) by removing it or by injecting high-energy air into it (Fig. 1.21). They delay the separation of the BL, thus increasing the maximum lift coefficient and stall AoA but they do not affect lift at low AoAs. These devices can either be passive (slot, vortex generators) or active (blowing, suction).

Wing surface modifiers actually increase the surface of the wing. Because the lift characteristics of the wing are still calculated using the referenced surface, the resulting effect is to increase the slope of the lift coefficient. There is no commonly used retractable device that only increases the surface of the wing.

The effects on the lift curve of the three categories mentioned thus far can be combined to provide even greater lift. Single-slotted flaps combine the effects of a camber modifier with those of an energy adder. Krueger flaps, which act as air dams forcing air above the upper surface, combine the effects of a camber modifier and a surface modifier. Many other devices combine all three categories to provide high lift coefficients. Some of these devices are shown in Fig. 1.22.

High lift coefficient devices do provide the lift required to enable effective field performance but there are drawbacks to having such systems. Most systems, when deflected, create a lot of drag. Even when retracted, high lift coefficient devices are responsible for part of the roughness drag due to the discontinuities over the wing surface, as shown in Fig. 1.23. As well, the more complicated the

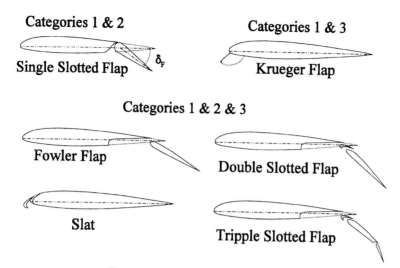

Categories 1 & 2
Single Slotted Flap δ_F

Categories 1 & 3
Krueger Flap

Categories 1 & 2 & 3

Fowler Flap

Double Slotted Flap

Slat

Tripple Slotted Flap

Fig. 1.22 High lift coefficient devices.

system is, the heavier it also is, which is a burden when not in use such as during cruise. Designers must perform tradeoff studies to determine the right compromise between high lift coefficient and minimal weight/drag increase.

Finally, powered lift consists in using the propeller wash or the jet engine exhaust flow to locally increase the lift coefficient of the wing. Two systems that were tested in the late 1970s are shown in Fig. 1.24. The upper surface blowing, tested on the YC-14 and quiet short-haul research aircraft (QSRA), could turn the jet exhaust stream by as much as 70 deg with less than 10% loss in thrust by using the coanda effect. The externally blown flap, tested on the YC-15 and used on the C-17, actually sends the jet exhaust stream to be deflected directly onto the flap.

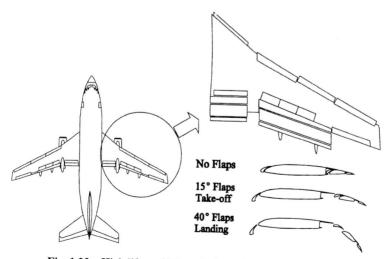

No Flaps

15° Flaps
Take-off

40° Flaps
Landing

Fig. 1.23 High lift coefficient devices of the Boeing 737-100.

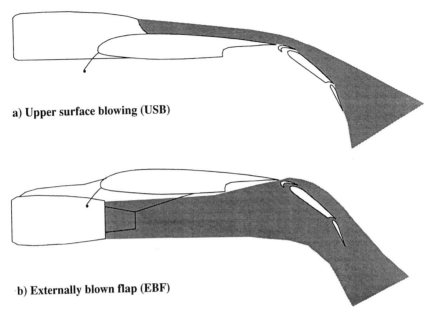

a) Upper surface blowing (USB)

b) Externally blown flap (EBF)

Fig. 1.24   Two powered-lift systems.

In either case, significant increase in lift coefficient was achieved. Under extreme testing conditions, the QSRA demonstrated a lift coefficient as high as 11.4. The lift produced by any powered-lift system is, of course, highly dependent on the throttle setting.

1.7   Drag Components

To obtain a simple-to-use drag equation for the prediction of aircraft performance, the sources of drag must be regrouped. The main causes of aircraft drag were introduced in the preceding sections of this chapter. Three additional sources of drag will now be described, and they have been identified in the drag tree (Fig. 1.25). These drag sources are: excrescent drag, interference drag, and trim drag. In the end, the total drag coefficient will have two main components: the parasite drag and the induced drag. The former is independent of lift, whereas the latter is lift dependent.

The equation describing the drag coefficient commonly takes one of the following two forms:

$$C_D = C_{D_0} + K C_L^2$$
$$C_D = C_{D_{\min}} + K \left(C_L - C_{L_0} \right)^2 \tag{1.10}$$

depending on whether the minimum drag coefficient occurs when the lift coefficient is zero or not.

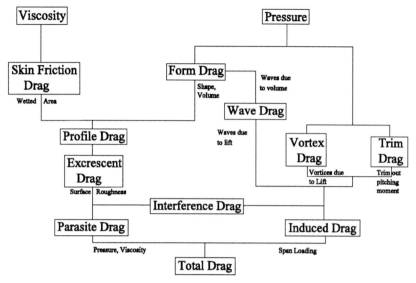

Fig. 1.25 Drag tree for a complete aircraft.

1.7.1 Excrescence Drag

Excrescence drag (also called roughness drag) is due to local flow separation and vortex formation. It is a mix of tangential forces and normal forces, the latter being dominant. Whereas profile drag includes the skin-friction drag and the form drag of a smooth surface, excrescence drag accounts for the imperfection of the aircraft skin such as produced by surface discontinuities (panel joints, engine inlet contour, and gaps around doors, windows, and control surfaces), rough surface finish, pressurization leaks, antennas, misrigged controls, protrusions and recesses, mass transfer in and out of the aircraft, etc. Some of these excrescences are unavoidable due to the manufacturing techniques and specifications used to build the aircraft, which are in part driven by the need to keep the manufacturing cost as low as possible.

Not all excrescences are detrimental to the overall aerodynamic characteristics of the aircraft. Vortex generators, although they produce extra drag, are beneficial because they delay separation of the BL by re-energizing it, thus delaying stall. As well, at high speeds, carefully placed vortex generators help break up shock waves over the wing, thus reducing the wave drag.

A military aircraft can be expected to suffer some kind of battle damage during a war. Aircraft battle damage repair philosophy is to do just enough repair to be able to return the aircraft to service for one more mission. The patches made to a damaged aircraft will usually not correspond to the original design but they will follow strict guidelines to ensure that the aircraft can fly another mission. Excrescence drag then becomes an important part of total aircraft drag. To minimize this, attention to the patch's shape and orientation must be included in the guidelines.

Roughness drag represents about 3.5% of the total drag of the Lockheed C-5A military transport at cruise condition. Excrescence drag can increase the direct operating cost of an aircraft. An example of this[1] is the drag produced by bad pressurized area seal leak along windows and doors of the Boeing 727, which increased the annual rate of fuel consumption by about 71,000 kg per aircraft (roughly U.S. $18,000 in 1992, see Chapter 11, Sec. 11.4). A 1-deg sideslip caused by a misrigged control on a Boeing 727 resulted in 108,000 kg per year of wasted fuel (roughly U.S. $27,700). The last two items can be corrected at the maintenance level and could save a lot of money in reducing fuel consumption. One forward facing, 3-mm-high sheet metal joint along the entire diameter of the fuselage (diameter of 5.73 m) of an Airbus type fuselage that accounted for half the drag count at cruise condition is an example of excrescence drag that originated during manufacturing.[2]

It should be remembered that excrescence drag results from a complex combination of interaction phenomena such as the pressure force on the excrescence itself and the changes in the local surface shear forces forward and aft of the excrescence, as well as the effect on the evolution of the BL downstream and the possibility of flow separation. Excrescence drag can be minimized by careful design and manufacturing, and it can be kept low by good maintenance practices.

1.7.2 Interference Drag

Interference drag occurs when the drag of the assembled components of an aircraft is greater than the sum of the drag of the individual components. This is due to the flow interaction between the various assembled components, which can cause flow separation, vortices, and even shock waves. Such interference can also affect induced drag by modifying the lift distribution along the span.

The closer two objects are to one another, the faster the flow between them will be. This effect is greatly magnified as the Mach number approaches or exceeds unity. The interference between stores on military aircraft can be large enough to limit the maximum speed of a supersonic aircraft to speeds below Mach 1.0 due to drag increase or buffeting.

Interference drag can be minimized by using fairings at the junction of two elements or by increasing the distance between two elements. In Fig. 1.26, the fairing used on the F-16 (aircraft on the right) between the wing and the fuselage not only reduces the interference at the junction but also reduces the wetted surface of the aircraft, thus reducing the skin friction drag as well.

Another example of aircraft shaping to reduce the interference effects is the tailoring of the wing tip tanks on the F-5 (Fig. 1.27). Not only was the aircraft fuselage area ruled to achieve maximum gain in wave drag reduction at around Mach 1.15 but so were the wing tip tanks. Compared to the original, more traditional shape of wing tip tanks, these tanks reduced the cruise drag of the aircraft by about 23 drag counts.

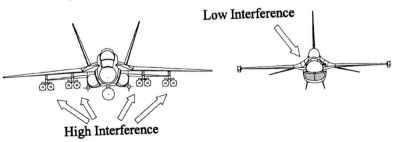

Fig. 1.26   Interference drag.

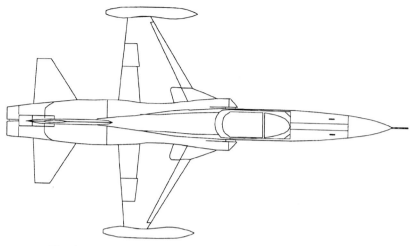

Fig. 1.27   Application of area rule on F-5 fuselage and tip tanks.

1.7.3   Trim Drag

Trim drag is the additional vortex drag created by the wing and tail of the aircraft while trimming out the pitching moment. Even the best designed aircraft will not have all four basic force vectors act through a common centroid. The weight vector is said to act through the c.g., while the lift and drag act through the aerodynamic center and the thrust will generally act along the centerline of the engine. On conventional designs, with the tail at the back, the elevator will have to create a download, as shown in Fig. 1.28, to counteract the moments created by these basic forces acting through their respective centers plus the aerodynamic pitching moment of the wing.

$$L = W + L_T$$

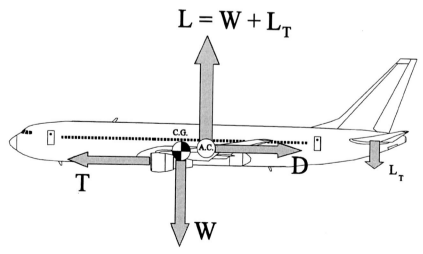

Fig. 1.28   Trim drag created to trim out the pitching moment.

This download is the equivalent to a lift vector pointing downwards, thus creating vortex drag, which is part of trim drag. Trim drag does not include the tail profile drag, which is accounted for in the overall aircraft parasite drag but it does include the vortex drag resulting from the additional lift required of the wing to counter the download produced by the tail. Typically, trim drag may vary from 0.5 to 5% of total aircraft cruise drag, depending on aircraft type and c.g. location.

Trim drag can be minimized if the elevator is located forward of the c.g. (canard configuration) because an upward force is now needed to trim out the pitching moment instead of a downward force. This layout is especially useful for supersonic aircraft because their aerodynamic centers move back as the aircraft goes from subsonic to supersonic flight. Another way of reducing trim drag is by displacing the c.g. in flight (by transferring fuel fore and aft, for example) to keep the c.g. close to the aerodynamic center throughout the flight envelope of the aircraft. The Concorde and the SR-71 use such a system.

1.8   Basic Aerodynamic Assumptions for Aircraft Performance

The discipline of aircraft performance deals with the long-term behavior of the aircraft motion. For this, the aircraft can be considered to be a point-mass on which the basic forces of lift, weight, drag, and thrust act, and the only aerodynamic data required are the drag polar and the thrust available. Then, assuming that the aircraft can overcome any disturbances, only three degrees of freedom (three linear displacements, no moments considered) can be used to study aircraft performance. Chapter 10, "Stability and Control," will deal with the reactions of the aircraft to disturbances and how the design to cope with stability and control can affect

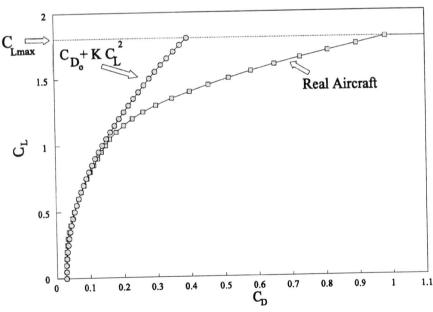

Fig. 1.29   Drag polar, Eq. (1.11), compared to real aircraft.

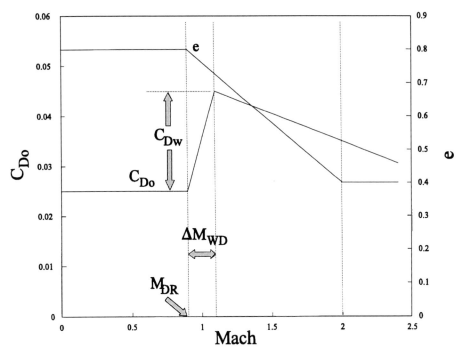

Fig. 1.30 Evolution of the minimum drag coefficient and of the Oswald coefficient with Mach number.

the performance of an aircraft. The following form of the drag polar will be used throughout:

$$C_D = C_{D_0} + K C_L^2 \qquad \text{where } K = 1/\pi \text{AR}e \qquad (1.11)$$

Figure 1.29 compares Eq. (1.11) with the actual drag polar of a fighter type aircraft with high-AoA capability. In general, Eq. (1.11) provides a reasonable curve fit for lift coefficients varying from 0 to 1 and Mach numbers lower than 4 for wings of AR greater than 3 and wing sweep less than 30 deg in trimmed or untrimmed conditions. To be consistent in our calculation of the performance of any aircraft throughout its full regime of velocity and altitude, the following approximations will be used throughout:

1) The value of C_{D_0} and K will be constant in the subsonic regime (below Mach drag rise M_{DR}).

2) In the transonic regime the value of the minimum drag coefficient will vary linearly up to a maximum value at the end of the regime ($M_{\text{DR}} + \Delta M_{\text{WD}}$).

3) In the supersonic regime the value of the minimum drag coefficient will decrease linearly so that at Mach 2 it has a value equal to the subsonic one plus half of the maximum increase (i.e., $C_{D_0} = C_{D_{0,\text{sub}}} + \frac{1}{2} C_{Dw}$).

4) The value of the Oswald coefficient will decrease linearly from its subsonic value at M_{DR} to about half its value at Mach 2 from where it will remain constant thereafter (Fig. 1.30).

2
Aircraft Propulsion

2.1 Introduction

T HE aim of this chapter is to provide insight to the many aircraft propulsion systems. An aircraft engine tree is provided in Fig. 2.1 to illustrate the many types of aircraft propulsion systems available today.

This chapter covers the basic theory governing the field of aircraft propulsion. It shows, for example, how propellers convert rotational energy into propulsive energy. It also describes how the many propulsion systems are affected by atmospheric parameters and flight conditions. All this is aimed at illustrating how these propulsive systems react at different regimes to ultimately see how they affect aircraft performance.

2.2 Propeller

All propulsive systems produce thrust by imparting a change of momentum to a mass of air or propulsive fluid. Propellers operate by producing a relatively small change in velocity to a relatively large mass of air. Propellers have two or more blades (Fig. 2.2). These blades are made up of airfoils (also called blade elements) with a twist distribution that gives the optimal AoA at some design condition.

2.2.1 Rankine–Froude Momentum Theory

Starting with Newton's second law, which states that the force is equal to the change in momentum per unit time, the Rankine–Froude momentum theory assumes that the propeller disk may be replaced physically by an actuator disk, which acts as a pure energy supplier. The actuator can be viewed as having an infinite number of infinitely thin propeller blades and has a frontal area A equal to the area covered by a rotating propeller blades. Further assumptions are 1) the disk offers no resistance to the air passing through it, 2) the velocity through the disk (V_P) is constant (continuity), 3) the air mass receives energy in the form of differential pressure ($p_3 - p_2$) uniformly distributed across the disk surface, and 4) the air is assumed to be a perfect incompressible fluid. Here the flow is assumed irrotational in front of and behind the disk but not through it. Figure 2.3 illustrates an actuator disk with the stream tube of air flow through it.

Note that the pressure far from the disk, on either side of it, will equal the atmospheric pressure ($p_1 = p_\infty = p_4$). It should also be noted that the high-velocity region behind the actuator disk is called the slipstream or wake of the propeller. The mass flow through the disk (and within the stream tube) is

$$\dot{m} = \rho A V_P \tag{2.1}$$

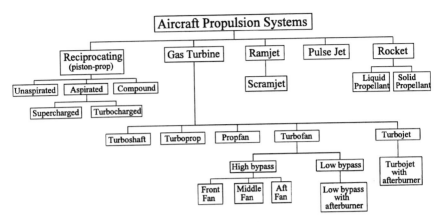

Fig. 2.1   Aircraft engine tree.

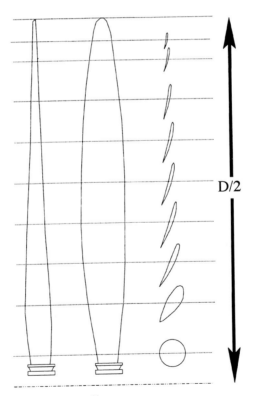

Hub center line

Fig. 2.2   Typical propeller blade.

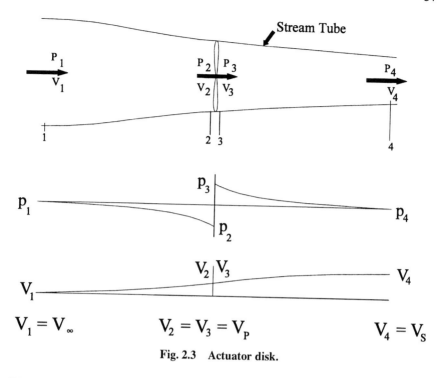

$$V_1 = V_\infty \qquad V_2 = V_3 = V_P \qquad V_4 = V_S$$

Fig. 2.3 Actuator disk.

The thrust produced by the disk is found from Newton's second law,

$$T = m\frac{dV}{dt} = \dot{m}\Delta V = \rho A V_P (V_S - V_\infty) \qquad (2.2)$$

But the thrust is also equal to the differential pressure on either side of the disk times its surface, therefore,

$$T = A(p_3 - p_2) \qquad (2.3)$$

Using Bernoulli's equation for the flow on either side of the disk, but not through it (energy input at the disk), gives

$$p_\infty + \tfrac{1}{2}\rho V_\infty^2 = p_2 + \tfrac{1}{2}\rho V_P^2$$
$$p_3 + \tfrac{1}{2}\rho V_P^2 = p_\infty + \tfrac{1}{2}\rho V_S^2 \qquad (2.4)$$

Therefore,

$$p_3 - p_2 = \tfrac{1}{2}\rho\left(V_S^2 - V_\infty^2\right) \qquad (2.5)$$

Combining Eqs. (2.2), (2.3), and (2.5) yields

$$V_P = \tfrac{1}{2}(V_S + V_\infty) \qquad (2.6)$$

Thus,

$$T = \tfrac{1}{2}\rho A\left(V_S^2 - V_\infty^2\right) \tag{2.7}$$

This development indicates that the flow velocity through the actuator disk is the average of the flow velocities far upstream and far downstream. In fact, half of the acceleration of the air mass occurs upstream of the propeller and the other half occurs downstream. The velocity at the propeller disk is, thus, the freestream velocity plus a speed increase (ω), whereas the slipstream velocity is equal to the freestream velocity plus two times ω,

$$V_P = V_\infty + \omega \qquad V_S = V_\infty + 2\omega$$
$$\therefore \quad T = \rho A(V_\infty + \omega)2\omega = 2\dot{m}\omega \tag{2.8}$$

From this last equation, the speed ω can be written in terms of freestream velocity and thrust,

$$\omega = \frac{1}{2}\left[-V_\infty + \sqrt{V_\infty^2 + \frac{2T}{\rho A}}\right] \tag{2.9}$$

The ideal efficiency of the actuator disk can be obtained by comparing the power input to the power output. The power output (or useful work) is equal to the thrust generated by the disk multiplied by the velocity of the actuator disk through the air (freestream velocity),

$$P_{\text{out}} = T V_\infty \tag{2.10}$$

whereas the power input is the thrust generated by the disk multiplied by the airflow velocity through the disk,

$$P_{\text{in}} = T V_P \tag{2.11}$$

The propeller efficiency is defined as the ratio of the power output to the power input,

$$\eta_i = \frac{P_{\text{out}}}{P_{\text{in}}} = \frac{T V_\infty}{T V_P} = \frac{V_\infty}{\tfrac{1}{2}(V_S + V_\infty)} = \frac{V_\infty}{V_\infty + \omega} \tag{2.12}$$

Therefore,

$$\eta_i = \frac{2}{(V_S/V_\infty) + 1} = \frac{1}{1 + (\omega/V_\infty)} \tag{2.13}$$

This is the ideal efficiency or theoretical upper limit efficiency. It is equal to zero for zero forward velocity and approaches 1.0 as ω moves toward zero. Note that as the velocity differential nears zero so does the thrust [Eqs. (2.7) and (2.8)]. The ideal value cannot be reached in practice because of losses such as energy lost in slipstream rotation, losses due to nonuniform thrust loading, blade interference losses, propeller profile drag losses, and losses due to increased drag and changes in flow in the compressible range are not accounted for.

2.2.2 Real Propeller Performance

To determine the exact performance of a propeller, one must analyze the aerodynamics of the propeller blades, in particular, the aerodynamics of the propeller blade elements. Each element is an airfoil that is subjected to a rotational speed equal to the number of revolution per unit time (n) times the radial location of the element (r). As well, the element is subjected to the aircraft forward motion V_∞. For maximum propeller efficiency, each blade elements should be set at an AoA α for maximum lift-to-drag ratio. The AoA is a function of the blade element geometric pitch angle β and of the effective pitch angle θ as defined in Fig. 2.4. The rotational speed of the blade elements is a function of their radial location, but the aircraft forward motion is the same for each blade element, which means that the pitch must be varied from the propeller hub to the tip so as to maintain the proper AoA for each blade element. The lift and drag of a blade element are perpendicular and parallel to the relative wind, respectively. For propellers, these forces are usually converted to thrust and torque, which are parallel to the aircraft forward motion and propeller blade rotational speed, respectively.

The advance ratio J is the dimensionless ratio of the propeller forward speed to the propeller rotational speed, expressed as

$$J = V_\infty/nD \tag{2.14}$$

where V_∞ is the forward speed, n is the rotational speed expressed in revolution per unit time, and D is the propeller diameter. The design advance ratio is the value at which the propeller efficiency is maximum, that is, the value at which each blade element will be at an AoA for maximum lift-to-drag ratio. From Fig. 2.4 and Eq. (2.14), it can be seen that for a small advance ratio each blade element will be at a larger than optimum AoA, which will reduce its efficiency. In fact, for very low-advance ratio values (such as during takeoff), the propeller blades may be partially or totally stalled. For high-advance ratios, the AoA of each blade element will be smaller than optimum. Furthermore, if the advance ratio becomes

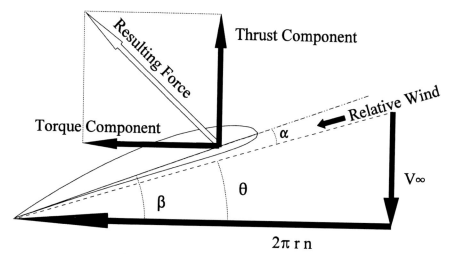

Fig. 2.4   Blade element aerodynamics.

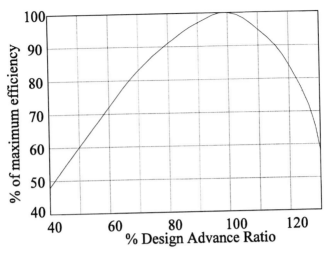

Fig. 2.5   Typical effect of deviating from the design advance ratio on the propeller efficiency.

very large such as during a dive, the propeller blades may even produce negative thrust, i.e., act as a windmill. Thus, an aircraft with its propeller stopped may go faster in a vertical dive than with its propeller turning. Figure 2.5 illustrates the effect on the propeller efficiency of deviating from the design advance ratio.

It is clear from the graph of Fig. 2.5 that deviating from the design advance ratio will result in a loss of propeller efficiency. A fixed pitch propeller, one that cannot alter its geometric pitch, must therefore be well matched to the engine to be used and the flight conditions at which optimum performance are required. In general, a fine pitch is required for low-advance ratios such as during takeoff or in a climb where the airspeed is relatively small compared to the engine revolutions per minute (rpm) whereas for cruise conditions and top speed, a larger pitch is required. Variable pitch propellers, either adjustable on the ground or controllable in the air, usually offer at least two different pitch settings, one fine and one coarse, so as to maximize the propeller efficiency according to the desired performance.

Constant-speed propellers automatically change the propeller pitch so as to maintain proper torque on the engine shaft such that the engine rpm remains approximately constant, close to its optimum value. This maximizes the propeller efficiency over a wider range of advance ratio values as shown in Fig. 2.6. It also maximizes the engine performance by allowing it to remain at an rpm for maximum efficiency.

The thrust T produced by the propeller and the torque τ and power P required to maintain a given rpm are defined as follows:

$$T = \rho n^2 D^4 C_T$$

$$\tau = \rho n^2 D^5 C_\tau \qquad (2.15)$$

$$P = \rho n^3 D^5 C_P$$

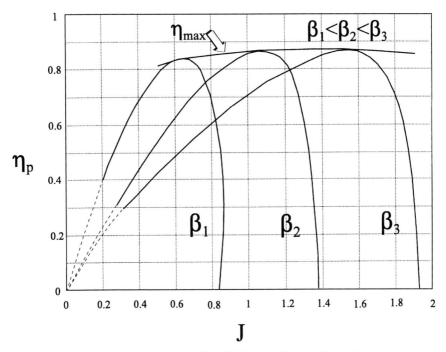

Fig. 2.6   Variation in maximum propeller efficiency with propeller pitch and advance ratio.

From this, the real propeller efficiency can be written as

$$\eta_p = \frac{TV_\infty}{P} = \frac{TV_\infty}{2\pi n\tau} \qquad (2.16)$$

Thus,

$$\eta_p = \frac{C_T}{2\pi C_\tau}\frac{V_\infty}{nD} = \frac{C_T}{C_P}J \qquad (2.17)$$

where

$$C_P = 2\pi C_\tau \qquad (2.18)$$

Aircraft engines that require a propeller for propulsion are generally rated in terms of power developed such as horsepower or kilowatt. From Eq. (2.16), it can be seen that for a constant power input (P) to the propeller from the engine and a constant propeller efficiency, the thrust generated by the propeller is proportional to the inverse of the velocity. This means that the thrust would go to infinity as the velocity nears zero. The actual propeller thrust from 0 to approximately 50 kn typically varies from a maximum static thrust value to the forward-flight value calculated by Eq. (2.16) as shown in Fig. 2.7.

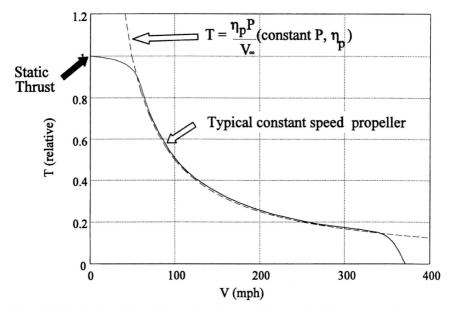

Fig. 2.7   Typical variation of propeller thrust with velocity (thrust normalized to static thrust).

In general, the larger the propeller diameter is, the greater the propeller efficiency. But the larger the propeller diameter is, the higher the propeller tip speed for a given airspeed and propeller rotational speed. The propeller tip speed is equal to

$$V_{\text{tip}_{\text{helical}}} = \sqrt{(\pi n D)^2 + V_\infty^2} \qquad (2.19)$$

At high tip speed, compressibility effects are encountered and shock waves may be present. The presence of shock waves will reduce propeller blade elements' lift and increase their drag, resulting in a loss of efficiency. Metal propellers, which have thin blade elements, are generally limited to tip speeds of about Mach 0.85 (around 950 ft/s at sea level), whereas wooden propellers with their thicker profiles are generally limited to tip speeds of Mach 0.75 (around 850 ft/s at sea level). Recent propeller developments have seen the introduction of propellers with swept tips to maintain propeller efficiency to higher tip speed values. Propfans, which were first tested in the 1980s, are highly swept thin blades that can maintain high propeller efficiency for flight Mach numbers up to 0.85.

2.3   Reciprocating Piston Engines (Piston-Prop)

The first aircraft engines were the reciprocating piston engines [or internal combustion engines (IC)] driving a propeller. In fact, this combination, which came to be known as the piston-prop engine, was the main propulsion system for the first 50 years of aviation history. A German inventor, Nikolaus August Otto, was the first to design a successful four-stroke IC engine in 1876. However, it was

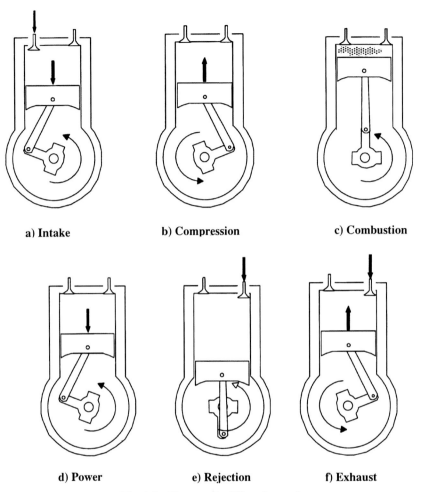

a) Intake   b) Compression   c) Combustion

d) Power   e) Rejection   f) Exhaust

Fig. 2.8   Four-stroke IC engine cycle.

not until more powerful and lighter weight engines were designed that they could be used on aircraft. An IC engine converts the translational back and forth motion (reciprocating) of a piston into a rotational motion of a crankshaft. This rotational motion is used to turn a propeller to produce thrust (Fig. 2.8).

2.3.1 Power Cycle

The power developed by an IC (four-stroke cycle) engine can be determined by the power cycle, also called the Otto cycle, using a pressure/volume (p–$\forall$) diagram, as shown in Fig. 2.9.

In the first part of the cycle (a–b) a fuel–air mixture is drawn in the cylinder at constant pressure. Then the intake valve closes and the mixture is compressed isentropically (b–c). At maximum compression (c) the mixture is ignited, and

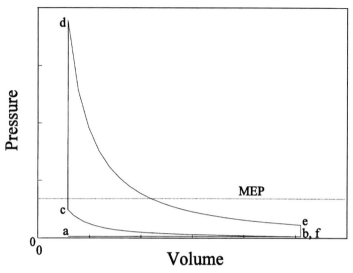

Fig. 2.9   Ideal Otto cycle.

combustion occurs at constant volume (*c–d*). The charge then expands isentropically and transmits power to the piston (*d–e*). The exhaust valve opens at the full
expansion point, quickly rejecting the exhaust gas to intake pressure (at constant
volume) (*e–f*), then the piston goes up again to further reject exhaust gas (*f–a*)
(at constant pressure).

   The compression and expansion processes are isentropical, as already mentioned. The thermodynamic equation for isentropic compression (and expansion) is

$$p = c \forall^{\gamma} \tag{2.20}$$

where c is a constant and γ is the ratio of specific heats of the mixture, which
varies between 1.2 and 1.35 depending on the fuel–air ratio (*F/A*). A value of 1.3
will be used to standardize the discussion on IC engine performance. Here, $\forall$ is
the volume.

   Using Fig. 2.9, it can be seen that the work W done by the piston to compress
the gas or by the gas, on the piston, during the power stroke is

$$dW = p \, d\forall \tag{2.21}$$

where $d\forall$ is the change of volume. The total work done during one cycle is

$$W = \int_{a}^{b} p \, d\forall + \int_{b}^{c} p \, d\forall + \int_{c}^{d} p \, d\forall + \int_{d}^{e} p \, d\forall + \int_{e}^{f} p \, d\forall + \int_{f}^{a} p \, d\forall$$

step   1   2   3   4   5   6

   For step 3 (*c–d*) and step 5 (*e–f*), the change of volume is zero; thus, the integral
is zero. Step 1 (*a–b*) and step 6 (*f–a*) are both at the same pressure but their change

of volume has opposite signs; thus, they cancel out. This leaves the following equation for work done during one ideal cycle for an internal combustion engine (remembering that $\forall_b = \forall_e$ and $\forall_c = \forall_d$):

$$W = \int_b^c p\,d\forall + \int_d^e p\,d\forall = \int_{\forall_b}^{\forall_c} p\,d\forall + \int_{\forall_c}^{\forall_d} p\,d\forall$$

Thus, the total work done is represented by the area bounded by the complete cycle on the p–$\forall$ diagram and is equal to

$$W = \int_{\forall_c}^{\forall_b} (p_{\text{power}} - p_{\text{comp}})\,d\forall = W_{\text{power}} - W_{\text{comp}} \qquad (2.22)$$

for an ideal Otto cycle. The net power produced by this cycle is the total work done per unit time. Here, since the piston must do two strokes (one up and one down) for each engine shaft revolution and a complete cycle is four strokes, the number of complete cycles per shaft revolution is one-half. The number of shaft revolutions per unit time is usually expressed in revolutions per minute (rpm). Now if the engine has N cylinders, the indicated power (IP) is

$$\text{IP} = \frac{1}{2}\left(\frac{\text{rpm}}{60}\right) NW \qquad (2.23)$$

Note that the rpm is divided by 60 to get the number of shaft revolutions per second. Because the usual unit used to express power is horsepower, Eq. (2.23) can be rewritten in terms of indicated horsepower (IHP),

$$\text{IHP} = \text{IP}/K \qquad (2.24)$$

where

$$K = 746\,\frac{J}{\text{s\,HP}} = 550\,\frac{\text{ft-lb}}{\text{s\,HP}}$$

The IHP is often expressed in terms of engine parameters as follows:

$$\text{IHP} = \left[\text{MEP} \times \frac{\pi b^2}{4} \times \frac{\ell}{12} \times \frac{\text{rpm}}{2} \times N\right] \Big/ 33{,}000 \qquad (2.25)$$

where the mean effective pressure (MEP) is in units of pounds per square inch acting on the piston head during a complete cycle. The area of the piston head is $\frac{1}{4}\pi$ times the cylinder bore b squared (the bore is the cylinder's head diameter in inches). Here, ℓ is the length of the stroke of a piston (in inches). These units are the standard of industry for IC engines. The grouping ($\frac{1}{4}\pi b^2 \ell N$) is the engine displacement d (in cubic inches) and is another value given by IC engine manufacturers. The indicated horsepower is, thus,

$$\text{IHP} = \frac{\text{MEP} \times d \times \text{rpm}}{792{,}000} \qquad (2.26)$$

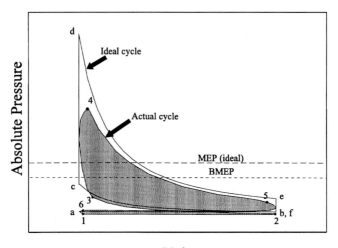

Volume

Fig. 2.10   Actual engine cycle compared to ideal Otto cycle.

An actual engine, however, does not deliver such power due to frictional losses that are dissipated in the form of heat (Fig. 2.10). Other factors affecting the engine cycle are losses because of the negative work required to exhaust fumes and intake fresh mixture, the ignition occurring slightly before the piston reaches the top of its stroke, and the minimum time required to discharge the piston. The actual power delivered to the shaft is the brake horsepower (BHP). It is called BHP because of the way it is usually measured using a brake system to measure torque τ (in foot-pound) and monitoring the rpm. The BHP is then

$$BHP = \frac{2\pi \, \text{rpm} \, \tau}{33,000} \tag{2.27}$$

The BHP is also expressed as

$$BHP = \eta_{\text{mech}} IHP \tag{2.28}$$

Table 2.1   Small IC engine

Engine	Textron Lycoming IO-360-A
Number of cylinders	4
Max horsepower	200 at 2,700 rpm
Displacement	361 in.^{3}
Stroke	4.375 in.
Compression ratio	8.7
Weight	293 lb

where η_{mech} is the mechanical efficiency of the engine. The brake MEP (BMEP) is then

$$\text{BMEP} = \eta_{\text{mech}}\text{MEP} \tag{2.29}$$

The thermal input (here in British thermal units per hour) to the cycle ($\dot{Q}$) is then a function of the fuel flow (here in pounds mass per hour) and of the heating value (HV) of the fuel (in British thermal units per pounds mass per hour)

$$\dot{Q} = \dot{W}_f\text{HV} \tag{2.30}$$

For average gasoline with an HV of 18,500 Btu/lbm the approximate heat release per pound of air is (from Ref. 3) shown in Table 2.2.

The brake thermal efficiency η_{th} of the engine is defined as the ratio of the rate of production of mechanical power to the total energy consumption. It is a measure of the efficiency of the IC engine to convert the fuel latent energy into mechanical power to turn the propeller.

$$\eta_{\text{th}} = \frac{P}{\dot{Q}} = \left[2545\frac{\text{Btu}}{\text{h}\cdot\text{HP}}\right]\frac{\text{BHP}}{\dot{W}_f\,\text{HV}} \tag{2.31}$$

The brake horsepower specific fuel consumption (BSFC) is defined as the ratio of the fuel flow to the BHP. It is a measure of the efficiency to produce power for a given fuel flow rate. Its usual units are pounds mass-fuel per hour per BHP,

$$\text{BSFC} = \frac{\dot{W}_f}{\text{BHP}} \tag{2.32}$$

The overall efficiency of the piston-prop combination is

$$\eta = \eta_{\text{th}}\eta_p \tag{2.33}$$

where η_p is the propeller efficiency as defined previously. Thus, the power available to propel to aircraft is

$$P_a = \eta\dot{Q} \tag{2.34}$$

Table 2.2   Approximate heat release

F/A	Btu/lb air	Condition
0.06	1,200	Lean
0.066	1,310	Chemically correct
0.08	1,520	Max heat release (best power)
0.10	1,400	Overrich

2.3.2 Factors Influencing Performance

Heat release per pound of air.   The horsepower developed by an IC engine is proportional to the heat release in the piston [Eq. (2.26)]. When the heat release in the piston is high, the temperature and, hence, the mixture pressure (MEP) are high. The heat release per pound of air is a function of the fuel HV, [Eq. (2.30)] and of the fuel–air ratio (*F/A*), Table 2.2. If the *F/A* is too high, the mixture may be too rich, and the combustion may not be complete. If it is too lean, the heat release will be lower and, hence, the power available will be lower.

Maximum permissible rpm.   The horsepower produced by an IC engine is directly proportional to the engine rpm [Eq. (2.27)]. The engine rpm is usually limited by such factors as engine designed life (reliability of the engine) or propeller tip speed.

Charge per stroke and altitude effects.   The mass of air introduced into the cylinders of an IC engine controls the quantity of heat released for a given *F/A*. This quantity of air is a function of the intake pressure [called manifold absolute pressure (MAP)]. For complete burning of the fuel, there must be a sufficient quantity of air. At higher altitudes, the air is less dense; thus, for a given throttle setting the power output will be reduced.

To maintain the power output of the engine at higher altitudes, an IC engine can be supercharged. This is accomplished by using a centrifugal compressor driven by either the crankshaft (gear-driven supercharger) or by a turbine (turbocharger), which will increase the MAP and hence power output. The power used to operate gear-driven superchargers is drawn from the crankshaft (usually 6–10% required). The turbocharger uses the exhaust gas to drive a turbine to increase the MAP. Finally, if the turbine is connected to the engine shaft in addition to driving the compressor, the arrangement is called compounding. Compound engines were used on the Lockheed L-1049C and DC-7C passenger transport but, due to their inherent complexity, they were set aside in favor of the more powerful and lighter gas turbine engines. Superchargers will usually maintain the IC engine power output up to a critical altitude, after which the power output will decrease with altitude increase.

2.4   Gas Turbine Engines

Gas turbine engines are the most widely used propulsion systems for commercial and military aircraft. Unlike the piston engine, which must handle intermittent flow, gas turbine engines deal with a continuous flow and are able to process a much larger mass flow rate for a given engine weight. The gas cycle for turbine engines can be best illustrated on a temperature/entropy diagram (see Fig. 2.11). The freestream air is introduced into the gas turbine through the diffuser up to the compressor face (0 → 1 → 2). From there it is compressed by the latter (2 → 3). Energy is then added to the compressed air in the form of heat by burning fuel (3 → 4). Part of the energy of the gas mixture is removed by the turbines (4 → 5) to power the compressor. The remaining energy (5 → 6) may be used completely to power a transmission shaft (turboshaft), or completely to develop thrust by

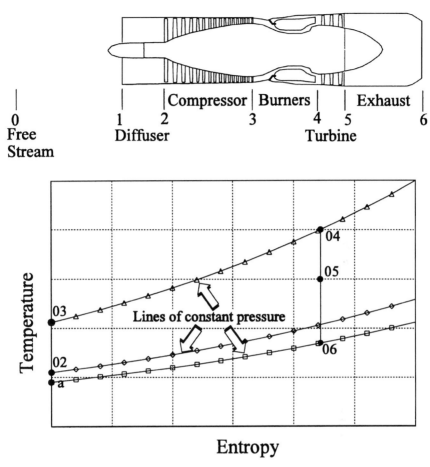

Entropy

Fig. 2.11 Brayton cycle for a single-shaft turbojet engine.

expansion and acceleration of the hot gas in a nozzle (turbojet), or something in between [turboprop, propfan, turbofan (see Fig. 2.1)].

2.4.1 Thrust Equation

One way to use the high-temperature, high-pressure gas mixture at the turbine exit is to expand and accelerate it to ambient pressure by using a nozzle, thus transforming heat energy into kinetic energy. Such is the case for turbojet engines (see Fig. 2.1).

To determine the thrust produced by such a system, consider a control volume (∀) about a turbojet engine installed on a static test stand (Fig. 2.12). This control volume was chosen because the sides and front surfaces are far enough from engine to consider the flow velocity through them to be negligible ($V \approx 0$) and the pressure at these surfaces is approximately equal to the ambient pressure ($p \approx p_\infty$). The back part of the control volume was chosen because the velocity

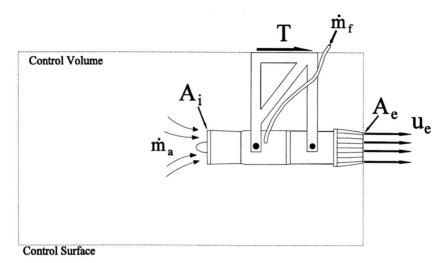

Fig. 2.12   Static engine test installation.

at the jet engine exhaust (u_e) is fairly uniform and the area (A_e) is known, whereas farther downstream it would be harder to determine both the size and velocity of the exhaust flow.

From the continuity equation,

$$\frac{\mathrm{d}}{\mathrm{d}t} \int_{\mathsf{V}} \rho \, \mathrm{d}\mathsf{V} + \int_{A} (\rho V \cdot \bar{n}) \mathrm{d}A = 0 \qquad (2.35)$$

and assuming steady flow ($\mathrm{d}/\mathrm{d}t = 0$), it is found that

$$\dot{m}_a + \dot{m}_f = \dot{m}_e = \rho_e u_e A_e \qquad (2.36)$$

where $\dot{m}_a$ is the mass flow rate of the air going into the engine, $\dot{m}_f$ is the mass flow rate of the fuel going into the engine, and $\dot{m}_e$ is the mass flow rate of the burned gas mixture coming out of the engine. From the momentum equation,

$$\frac{\mathrm{d}}{\mathrm{d}t} \int_{\mathsf{V}} \rho V \mathrm{d}\mathsf{V} + \int_{\mathsf{V}} V (\rho V \cdot \bar{n}) \mathrm{d}A = \sum F \qquad (2.37)$$

and still assuming steady flow, Eq. (2.37) reduces to: the change of momentum of the fluid going through the control volume is equal to the summation of the external forces acting on the control surface,

$$\int_{A} V (\rho V \cdot \bar{n}) \mathrm{d}A = \sum F \qquad (2.38)$$

The summation of external forces on the control surface is

$$\sum F = [T + A_e (p_\infty - p_e)] \bar{i} + 0\bar{j} + 0\bar{k}$$

Because the velocity of the fluid through the control surface is negligible except at A_e, then the change of momentum of the fluid going through the control volume can be expressed as

$$\int_A V(\rho V \cdot \bar{n})\,dA = \int_A (u_x \bar{i} + u_y \bar{j} + u_z \bar{k})(\rho V \cdot \bar{n})\,dA = \int_A \rho u_x^2\,dA = \rho_e u_e^2 A_e$$

Therefore,

$$\rho_e u_e^2 A_e = T + A_e(p_\infty - p_e)$$

which is rewritten to give the thrust equation for a static engine installation,

$$T = \dot{m}_e u_e + A_e(p_e - p_\infty) \tag{2.39}$$

This is the engine uninstalled gross thrust because this equation does not account for the engine forward motion or installation effects on the engine thrust. If the effect of forward motion is included, the thrust equation becomes

$$T = \dot{m}_a [(1 + f)u_e - V_\infty] + (p_e - p_\infty)A_e \tag{2.40}$$

where f is the fuel–air mass ratio,

$$f = \dot{m}_f / \dot{m}_a \tag{2.41}$$

This ratio is generally very small for any air-breathing engines, i.e., $f \ll 1$. Equation (2.40) is the engine uninstalled net thrust. The installation effects on the thrust equation will be discussed in Sec. 2.5. The term $[(p_e - p_\infty)A_e]$ is usually zero since the flow is accelerated to ambient pressure. In a convergent nozzle, the flow is accelerated to a maximum velocity of Mach 1.0 (for the nozzle operating temperature) at the exit, thus limiting the mass flow through the engine (choked flow). For a convergent–divergent nozzle, the exhaust jet flow may be supersonic and may now be overexpanded or underexpanded, but the term $[(p_e - p_\infty)A_e]$ is still much lower than the change of momentum term. For these reasons, the term $[(p_e - p_\infty)A_e]$ is often neglected in the thrust equation

$$T = \dot{m}_a [(1 + f)u_e - V_\infty] \tag{2.42}$$

A jet engine is very sensitive to operating parameters, as one could guess by simply observing Eq. (2.42). These parameters include: engine rpm, size of nozzle area, fuel flow rate, turbine inlet temperature, quantity of air bled from the engine compressor for use in other systems in the aircraft or the power extracted to drive accessories, the aircraft velocity, and the temperature, pressure and humidity of the air, which affects the density and thus the air mass flow.

Engine rpm effect. The engine rpm greatly influence the thrust produced by the engine as it determines the air mass flow rate being pumped by the compressor and, from Eq. (2.42), it can be seen that the thrust is directly proportional to the air mass flow rate. Since the compressor has fixed blade shapes, it will be more

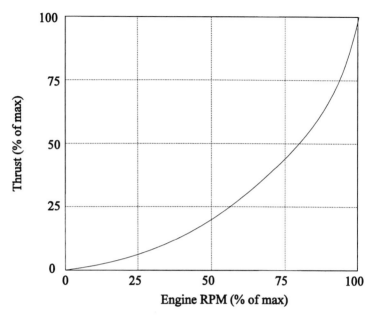

Fig. 2.13   Typical effect of RPM on engine thrust.

efficient in pumping air at a given rpm range, usually close to the maximum rpm of the engine. The variation in rpm does not yield a linear variation in thrust, as can be seen in Fig. 2.13. An engine rpm is expressed in terms of the maximum rpm for the engine: idle thrust being around 30%, whereas cruise thrust could be at 90%, and maximum continuous thrust is at 100% rpm. In some cases, an engine may be able to provide more than 100% continuous thrust for a very short period; this is called the maximum takeoff thrust, and it may damage the engine if used for a prolonged period.

Fuel flow rate and turbine inlet temperature.   Fuel flow determines the amount of energy transferred to the flow. As can be seen in the temperature–entropy $(T–s)$ diagram (Fig. 2.14), the more fuel is burned in the gas turbine combustion chamber (from point 03 to 04), the more heat energy will be available to the flow to be transformed into kinetic energy (point 05 to 06). But the total temperature at the turbine inlet (T_{04}) must be limited by the thermal/mechanical limits of the turbine blades. Equation (2.43) provides a means of estimating the fuel–air mass ratio for a given temperature rise in the combustion chamber,

$$f = c_p/(T_{04} - T_{03})/HV \tag{2.43}$$

Table 2.3 provides some information on commonly used gas turbine fuel by military (JP 4) and civilian (Jet A) aircraft. Typical values of fuel–air mass ratio is usually $f \leq 0.05$.

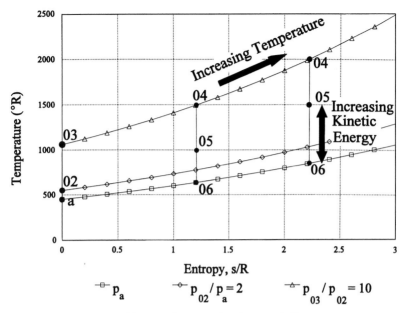

Fig. 2.14   *T–s* diagram of an isentropic flow.

Nozzle area and temperature.   As mentioned previously, the nozzle area determines the maximum exhaust mass flow rate, the limiting value being a velocity of Mach 1.0 at exhaust temperature for convergent nozzle. At this condition, the nozzle is said to be choked. Another factor that may limit the kinetic energy available at the nozzle is the maximum operating temperature of the nozzle. To maintain an acceptable nozzle temperature under the various combinations of aircraft speed and altitude, the fuel flow may have to be reduced, thus reducing the available thrust. This is usually done automatically by the fuel control system.

Bleed air and power extraction.   Most gas turbine equipped aircraft require pressurization, heating, and ventilation flow. This airflow is bled off from a given section of the compressor, which provides adequate pressure for the environmental control system. This results in a lower pressure in the burner and a lower mass flow at the nozzle, and, thus, a reduction in thrust for a given fuel flow. Other systems, such as the hydraulic system and the electrical system, require a power source. This power is usually extracted from the engine by using a shaft connected to the

Table 2.3   Commonly used gas turbine fuels

	JP 4	Jet A
Fuel density	760 kg/m³, 47.4 lbm/ft³	810 kg/m³, 50.6 lbm/ft³
HV	43,400 kJ/kg; 18,658 Btu/lbm	43,400 kJ/kg; 18,658 Btu/lbm
f stoichiometric	0.0673	0.0678

turbine. This results in less energy in the exhaust flow that can be used for thrust, or for a given thrust, a higher fuel consumption.

Aircraft speed effect. From Eq. (2.42), it can be seen that an increase in forward velocity results in a direct reduction in thrust for a given exhaust jet velocity. This loss in thrust can be partially offset by the increase in air mass flow rate due to what is called the ram pressure rise. The actual air mass flow rate increase into the engine is directly proportional to the rise in air density due to the compression of the air flow as it is slowed down from the aircraft forward velocity to the air velocity at the compressor face.

Atmospheric conditions effects. Because the thrust is directly proportional to the air mass flow rate, a variation in pressure, temperature, and/or humidity level will change the air density, thus affecting the air mass flow rate. An engine will produce more thrust under dry, cold, sea level conditions than under humid, hot, high-altitude conditions. Engine manufacturers generally rate an engine thrust in terms of standard atmosphere, sea level conditions.

2.4.2 Engine Efficiency

Specific fuel consumption. The engine thrust specific fuel consumption (TSFC) SFC_T is defined as the ratio of the fuel mass flow to the engine thrust

$$SFC_T = \dot{m}_f/T \tag{2.44}$$

which is the efficiency to produce thrust from a given fuel mass flow rate. The lower the value of the specific fuel consumption (SFC), the more fuel efficient the engine is in producing the required thrust. SFC will be affected by altitude, generally decreasing slightly with an increase in altitude, up to the tropopause and then increasing with an increase in altitude. SFC can be rewritten using Eqs. (2.41) and (2.42) to the following form, which shows the effect of forward speed on SFC for a given f:

$$SFC_T = \frac{f}{(1+f)u_e - V_\infty} = \frac{f}{u_e(1 + f - V_\infty/u_e)} \tag{2.45}$$

Propulsive efficiency. Propulsive efficiency η_{pr} is defined as the ratio of the thrust power ($T V_\infty$) over the rate of production of kinetic energy. Knowing that the fuel–air mass ratio is very small ($f \ll 1$), then

$$\eta_{\mathrm{pr}} = \frac{T V_\infty}{\dot{m}_a\left[(1+f)\left(u_e^2/2\right) - \left(V_\infty^2/2\right)\right]} \approx \frac{2(V_\infty/u_e)}{1 + (V_\infty/u_e)} \tag{2.46}$$

The propulsive efficiency, thus, will be maximum when the ratio of the aircraft velocity to the gas turbine exhaust jet velocity is unity. Unfortunately, as can be seen in Eq. (2.42), this also means that the thrust will be zero. A compromise between efficiency and thrust must then be achieved.

Thermal efficiency. Thermal efficiency η_{th} is defined as the ratio of the rate of production of kinetic energy to the total energy consumption rate,

$$\eta_{th} = \frac{\dot{m}_a\left[(1+f)\left(u_e^2/2\right) - \left(V_\infty^2/2\right)\right]}{\dot{m}_f HV} = \frac{\left[(1+f)\left(u_e^2/2\right) - (V_\infty^2/2)\right]}{fHV} \qquad (2.47)$$

which is a measure of the efficiency of the gas turbine to convert the fuel latent energy into kinetic energy to produce thrust.

Total efficiency. Total efficiency is defined as the product of the propulsive efficiency and the thermal efficiency

$$\eta = \eta_{pr}\eta_{th} = \frac{TV_\infty}{\dot{m}_f HV} = \frac{((1+f)u_e - V_\infty)\,V_\infty}{fHV} \qquad (2.48)$$

2.4.3 Turbofan Engine

A turbojet engine is relatively inefficient at low velocities due to the high-exhaust jet velocity. To increase the propulsive efficiency, while maintaining thrust, one must increase the air mass flow rate and reduce the exhaust jet velocity [Eq. (2.42)]. The way to accomplish this is to take part of the energy left in the flow to drive a fan to take on a larger air mass flow rate. This results in less heat energy available to be transformed into kinetic energy at the nozzle and, thus, a lower exhaust jet velocity.

The turbofan engine does just that by adding a fan with a cold airflow duct (air that does not flow through the combustion chamber). The fan is usually driven by its own turbine located immediately behind the compressor turbine. The cold airflow can either be mixed with the hot gas flow (air that does go through the combustion chamber) before being accelerated through a common nozzle (but after the turbines) or both flows can have their separate nozzles. In any case, there results a significantly lower average velocity exhaust jet flow, which increases the propulsive efficiency of the engine. The thrust equation for a turbofan engine is

$$T = \dot{m}_{a_{hot}}\left[(1+f)u_{e_{hot}} - V_\infty\right] + \dot{m}_{a_{cold}}\left(u_{e_{cold}} - V_\infty\right) \qquad (2.49)$$

where

$$f = \dot{m}_f/\dot{m}_{a_{hot}} \qquad (2.50)$$

The bypass ratio is defined as the ratio of the cold air mass flow rate to the air mass flow rate that goes through the combustion chamber (hot flow),

$$\beta = \dot{m}_{a_{cold}}/\dot{m}_{a_{hot}} \qquad (2.51)$$

Thus, the thrust equation can be rewritten as

$$T = \dot{m}_{a_{hot}}\left[(1+f)u_{e_{hot}} + \beta u_{e_{cold}} - (1+\beta)V_\infty\right] \qquad (2.52)$$

Table 2.4   Engine comparison

Characteristics	J79 (military)	CJ805-3	CJ805-23
Weight, lbm	3,670	2,815	3,760
Thrust, lbf	10,900 (dry)/17,900 (A/B)	11,200	16,100
Hot flow, lbm/s	170	171	171
Cold flow, lbm/s	0	0	251
Bypass ratio	0	0	1.65
TSFC, lbm/h/lbf	0.84 (dry)/1.96 (A/B)	0.81	0.53
Fuel flow, lbm/h	9,156 (dry)/35,084 (A/B)	9,072	8,533

In general, turbofan engines accelerate a larger mass of air (per unit of fuel) to a smaller average exhaust velocity than a turbojet, resulting in a lower fuel consumption per unit thrust and a better propulsive efficiency.

Table 2.4 provides an example of how the same basic gas generator from General Electric was adapted to different use (also see Fig. 2.15). In its military form, known as the J79 afterburning turbojet, it was used in the McDonnell Douglas F-4 Phantom, Lockheed F-104 Starfighter, and a myriad of other aircraft designed in the 1950s and 1960s. In the civilian form, it was used either as a turbojet (CJ805-3) or a turbofan (CJ805-23). This particular engine, although a somewhat older design, is used in this example because in all three cases the flow is the same through the basic gas generator, i.e., the fan, the nozzles, and the afterburner section are all located behind the basic gas generator. This enables a direct comparison of the effects of adding a fan or afterburner to the basic gas generator. This gas generator is made up of a 17-stage axial-flow compressor with the inlet guide vanes and the first 6 stages of stator vanes are variable. The compressor has a 13:1 compression ratio and is driven by a 3-stage turbine.

To better appreciate the size of the mass flow rate of turbofan and turbojet engines, imagine this: the time required to empty the air from a typical classroom of 20 by 30 by 10 ft high (about 460 lbm of air under standard sea level conditions) is about 2.7 s for a J79 turbojet engine, 1.1 s for a CJ805-23 turbofan of 1.65 bypass ratio, and 0.3 s for the JT9D high-bypass ($\beta = 5.25$, engine of Boeing 747-100) turbofan.

In the case of the JT9D, even if its hot flow is only about 40% higher than the CJ805-3 turbojet (240 lbm/s compared to 171 lbm/s under static maximum thrust), the JT9D develops 302% of the thrust produced by the CJ805-3 due to its very large secondary cold flow of 1260 lbm/s. The average velocity of the cold flow under static maximum sea level thrust is about 885 ft/s and the hot flow velocity is about 1190 ft/s.

The main advantage of the turbofan engine is that it has a lower SFC than a turbojet engine of the same thrust. Current generation fighters have low-bypass turbofans instead of the turbojets used for the fighters of the 1950s and 1960s. The General Electric F404 was designed to replace the J79, (Fig. 2.16). It is a low-bypass turbofan ($\beta = 0.34$) with a higher compression ratio (25:1) and slightly lower mass flow rate (142 lbm/s). It develops approximately the same thrust (11,000 lb dry and 18,000 lb with afterburner) and has about the same fuel consumption ($SFC_T = 0.8$ lbm/h/lbf dry and 1.92 lbm/h/lbf afterburner). Its main

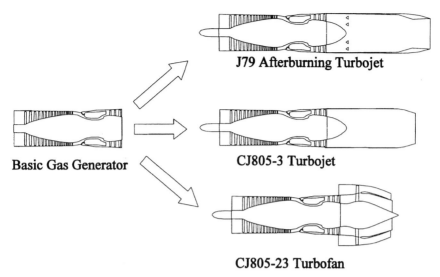

J79 Afterburning Turbojet

Basic Gas Generator

CJ805-3 Turbojet

CJ805-23 Turbofan

Fig. 2.15   Adaptation of a basic gas generator to various propulsive systems.

advantage is that it weighs 2,200 lb compared to the 3,670 lb of the J79; thus, it has a greater thrust-to-weight ratio. The F404 is used on such aircraft as the F/A-18, X-29, X-31, F-20, SAAB Grippen, and F-117A (with no afterburner).

2.4.4 Turboprop, Turboshaft, and Propfan

Turboprop engines use the same principle as the turbofan engine but they extract even more energy to drive a propeller, through a reduction gear box, to accelerate an even larger air mass flow rate. Turboprop engines (Fig. 2.17), because they drive propellers, are generally rated in terms of horsepower. The engine BHP is the horsepower measured at the turbine shaft. The shaft horsepower (SHP) is the horsepower available to turn the propeller, and it is slightly less than the BHP as it accounts for the mechanical losses in the reduction gear box. The equivalent shaft

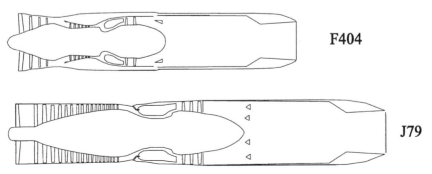

F404

J79

Fig. 2.16   F404 compared to J79.

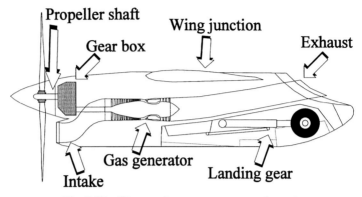

Fig. 2.17   Given turboprop components layout.

horsepower (ESHP) is the SHP plus the thrust horsepower of the exhaust jet at a given velocity

$$\mathrm{ESHP} = \mathrm{SHP} + (T V_\infty / \eta_p) \qquad (2.53)$$

where η_p is the propeller efficiency. About 85% of the propulsive force of a turboprop comes from the propeller, while the rest is provided by the exhaust jet flow. Turboprop engines have an even lower SFC than turbofan engines, but the efficiency of the propellers drops sharply as they encounter compressibility effects. For this reason, turboprop engines are efficient propulsion systems in the 300–500 mph range.

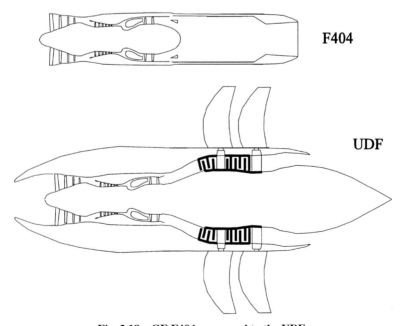

Fig. 2.18   GE F404 compared to the UDF.

Table 2.5   SFC ranges

Turbojet:	Turbofan ($\beta < 1$):	Turbofan ($\beta > 1$):
0.8–1.15 lbm/h/lbf (dry)	0.7–0.9 lbm/h/lbf	0.3–0.65 lbm/h/lbf
1.77–2.5 lbm/h/lbf (A/B)		
Propfan:	Turboprop:	Piston-prop:
0.24–0.3 lbm/h/lbf	0.45–0.78 lbm/h/ESHP	0.4–0.5 lbm/h/SHP

A turboshaft engine removes practically all of the energy left in the flow, after the turbine that drives the compressor, to drive a shaft that can be connected to a rotor or other system. These engines are rated exclusively in horsepower because the energy left at the exit of the engine is practically negligible.

A propfan is a compromise between the turboprop and the turbofan. The gas generator drives high-speed propellers and does provide some thrust from the exhaust jet flow. These propellers are designed to remain efficient up to higher Mach numbers, closer to the Mach 0.8 flight speed of most modern airliners. The bypass ratio of such an engine can be as high as 36:1 compared to the current trend for high-bypass turbofan engines of 15:1. In the late 1980s, engine manufacturers proceeded in testing several types of propfans. The General Electric candidate, called the unductedfan (UDF) had an F404 gas generator and two rows of high-speed counter-rotating propellers (eight blades per row) with a diameter of 10 ft. General Electric dispensed with the reduction gear box in this configuration by having two 6-stage intermeshing counter-rotating turbine spools to drive the propellers. The UDF demonstrated a takeoff thrust of 25,000 lb (compared to 11,000 lb for a non-augmented F404) and a SFC_T, of 0.24 lbm/h/lbf (compared to the 0.8 lbm/h/lbf for the low-bypass F404) (Fig. 2.18). The UDF has a bypass ratio of 25:1.

Table 2.5 provides typical SFC ranges for different types of propulsion systems under static (no forward motion) conditions.

2.4.5 Afterburner

An afterburner installation consists in adding heat energy into the gas flow after the last turbine stage (see Fig. 2.16, F404 and J79 engines). This is accomplished by injecting fuel directly into the hot flow and igniting it. There results a large increase in heat energy that can be converted into kinetic energy. The drawback to such a concept is the large increase in fuel consumption and SFC (see Tables 2.4 and 2.5), which limits an aircraft's range. For example, the mass fuel flow of the J79 (Table 2.4) is 9156 lbm/h in at maximum military (no afterburner) setting and 35,084 lbm/h with afterburner, an increase of 283%, although the SFC_T increased only by 133%. Afterburners are usually used sporadically in such situations as takeoff, rapid climbs, or during combat.

2.5 Installed vs Uninstalled Propulsion System

The preceding sections of this chapter addressed the uninstalled performance of different types of engines. No comments were made on the effects on engine and aircraft performance of how and where an engine is installed. In this section,

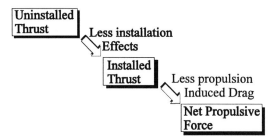

Fig. 2.19   Typical thrust/drag bookkeeping.

there is a brief discussion on such effects. The concept of thrust/drag (power/drag) bookkeeping is also introduced (Fig. 2.19).

2.5.1   Propeller Driven Aircraft

The most obvious problem with propeller driven aircraft is the propeller stream tube (Fig. 2.3). For high efficiency, the flow of air to the propeller disk must be as smooth as possible. Thus, one would install the propeller in a tractor configuration with the engine nacelle behind the propeller disk. In this case, however, the propeller slipstream with its increased turbulence would increase the drag on the aircraft components touched by that slipstream. For many single-engine aircraft, this means that the entire fuselage, the empennage, and part of the wing would be affected. The relative importance of the extra drag due to the slipstream is a function of the aircraft velocity and throttle setting. Because this extra drag will vary with throttle setting, most manufacturers account for this extra drag as a reduction in power available, either as reduced propeller efficiency or reduced SHP. This is the basis of power/drag bookkeeping where the aircraft manufacturer must ensure that the drag and power values are accounted for totally and only once to predict the performance of an aircraft. The slipstream of a tractor configuration may also influence the aircraft lift in creating a region of high velocity at relatively constant flow angle with respect to the wing.

One way to reduce the extra drag produced by the propeller slipstream is to place the propeller disk behind the aircraft in a pusher configuration (Fig. 2.20). In this case, the flow on the aircraft will be undisturbed by the propeller turbulent slipstream, and no extra drag will be created due to throttle setting. In fact, the propeller may even reduce the drag on the back end of a short stubby fuselage by preventing early separation of the BL due to its sucking effect (remember that

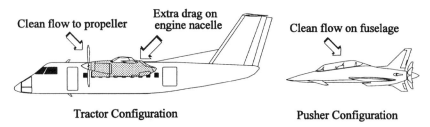

Fig. 2.20   Influence of engine location on aircraft/engine performance.

half of the acceleration of the airflow occurs in front of the propeller disk). The disadvantage of this configuration is that the airflow at the propeller disk will not be undisturbed, which will result in a loss of propeller efficiency and thus in power available to propel the aircraft.

Another factor to consider in the power/drag bookkeeping is the effect of the exhaust jet flow, which is itself throttle setting dependent, on the aircraft aerodynamics. Once again, this must be accounted for in some manner and usually is included in the aircraft power available.

2.5.2 Turbojet and Turbofan Equipped Aircraft

The thrust produced by turbojets and turbofans is directly proportional to the engine mass flow as seen in Eq. (2.42) (turbojet) and (2.52) (turbofan). The engine will take in only the mass flow it requires to operate. At low-forward speed and high-engine mass flow (see Fig. 2.21), the capture area (A_∞) will be larger than the intake area (A_i) resulting in a relatively smaller airflow around the engine. At high-forward speed and low-engine mass flow the capture area will be smaller than the intake area resulting in an increase in airflow (spilled air) around the engine.

The engine exhaust flow also influences the airflow around the aircraft, especially if it is located ahead of other components such as the wing or tailplane (see

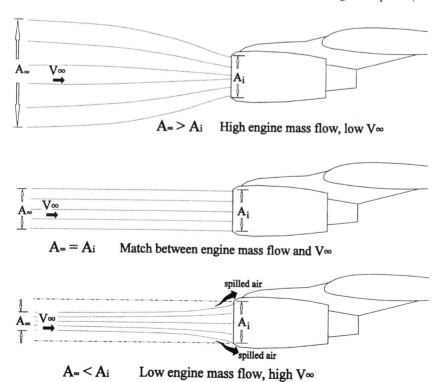

$A_\infty > A_i$ High engine mass flow, low $V\infty$

$A_\infty = A_i$ Match between engine mass flow and $V\infty$

$A_\infty < A_i$ Low engine mass flow, high $V\infty$

Fig. 2.21   Airflow to the engine intake.

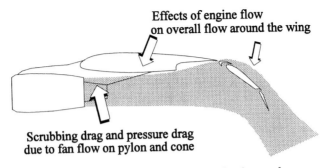

Fig. 2.22   Subsonic engine installation in a pod.

Fig. 2.22). Here again, the general flow will be affected by the engine mass flow, which is itself a function of the throttle setting.

Whereas pure subsonic or low-supersonic aircraft are usually fitted with short intakes to maximize the pressure recovery, supersonic aircraft must be fitted with long intakes of variable geometry (Fig. 2.22). This ensures that the flow can be deccelerated from the supersonic value to a low-subsonic value (closer to Mach 0.4) before entering the engine compressor (see Fig. 2.23). Clarence Kelly Johnson, head of the famous Lockheed Skunk Works for most of the 1950s, 1960s, and 1970s, quoted in a paper[4] that the SR-71 Blackbird mighty engines were no more than flow inducers while flying in high-speed cruise (Mach 3+) and that the aircraft's nacelles were actually pushing the SR-71 (83% of the installed thrust).

The design of the air intake is very important, as can be shown in the following example. When the U.S. Air Force revived the B-1 project under the Reagan

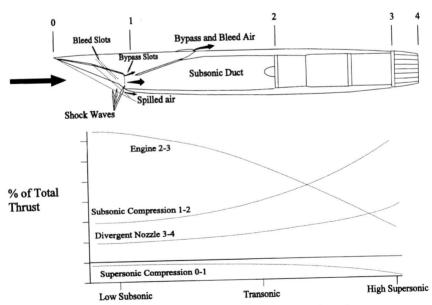

Fig. 2.23   Engine installation for a supersonic aircraft.

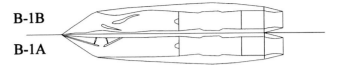

B-1B

B-1A

Fig. 2.24   Difference between the B-1A and B-1B air intakes.

administration, the emphasis was put on reduced radar signature rather than speed. To reduce the radar signature of the engine nacelles, Rockwell changed the design of the air intake by removing the variable geometry ramps and inserting flow deflectors (Fig. 2.24) that shielded the compressor faces completely from any direct radar energy. The drawback was a reduction in maximum speed attainable from Mach 2.2+ for the B-1A to Mach 1.25 for the B-1B even though both aircraft were equipped with the same engine. But this later installation resulted in a reduction in frontal radar cross section for the aircraft, which increases the aircraft survival chances, as discussed in Chapter 11, Sec. 11.3.

The exhaust nozzle will also have a large influence on the performance of an engine (Fig. 2.25). Usually, subsonic aircraft will be fitted with a simple fixed geometry convergent nozzle. This type of nozzle will accelerate the flow from the high-pressure and low-velocity conditions after the last turbine stage to atmospheric pressure and high-velocity conditions. The exit plane will be sized so as to give maximum velocity (Mach 1.0 for the exhaust flow temperature) at full throttle. Supersonic aircraft, on the other hand, need much higher exhaust velocities [thrust being proportional to the velocity difference between the exhaust flow and flight

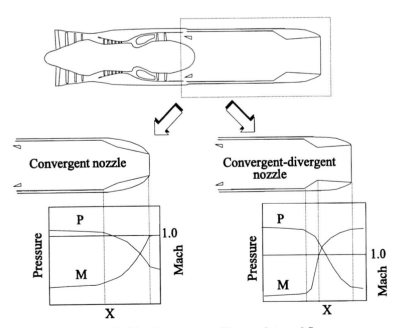

Convergent nozzle

Convergent-divergent nozzle

Fig. 2.25   Nozzle geometry effects on internal flow.

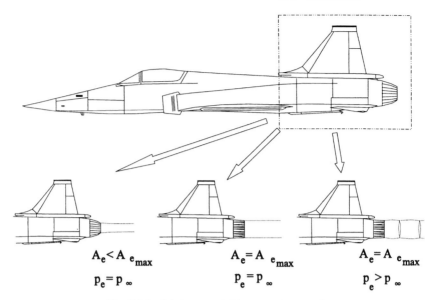

$$A_e < A_{e_{max}}$$
$$p_e = p_\infty$$

$$A_e = A_{e_{max}}$$
$$p_e = p_\infty$$

$$A_e = A_{e_{max}}$$
$$p_e > p_\infty$$

Fig. 2.26   Iris type variable geometry nozzle.

velocity, Eq. (2.42)] and are usually fitted with a variable geometry convergent–divergent nozzle. This variable geometry nozzle is able to better process the wide range of flow conditions of the engines installed on supersonic aircraft. The flow will be accelerated from the high-pressure, low-velocity conditions after the turbine to Mach 1.0 at the throat (smallest cross-sectional area) to a given pressure at the exit plane, which may be equal to or greater than atmospheric; Eq. (2.40) must then be used.

The external shape of the nozzle will obviously have an effect on the external flow around the aircraft. A fixed geometry nozzle will have little effect on the aerodynamics of the aircraft, whereas a variable geometry nozzle, especially the iris type used on most modern combat aircraft, will actually change the shape of the aft end of the aircraft (Fig. 2.26). This variation in shape will affect the aerodynamics and must be accounted for as a thrust variation in the bookkeeping.

2.6   Basic Propulsion Assumptions for Aircraft Performance

To study aircraft performance, some basic assumptions on the propulsion system behavior with altitude, velocity, and throttle setting must be made. As a first assumption, no correction on the propulsion system performance will be made due to its installation.

2.6.1   Piston-Prop Performance

All propeller driven aircraft will be grouped within this category. The engines for this class of aircraft are rated in terms of power. The power developed by these engines will be assumed to be independent of airspeed. It will vary from idle power to maximum power by throttle movement and, for a given throttle setting, it will vary with altitude, in feet, as follows:

$$\frac{P}{P_{SL}} = \begin{cases} \sigma^{0.765} & h \leq 36,089 \\ 1.331\sigma & h > 36,089 \end{cases} \tag{2.54}$$

For supercharged piston engines and flat power rated turboprop, the sea level power will be maintained up to a critical altitude and then decrease with altitude as follows:

$$\frac{P}{P_{SL}} = \begin{cases} 1.0 & h \leq h_{crit} \\ \left(\dfrac{\sigma}{\sigma_{crit}}\right)^{0.765} & h_{crit} < h \leq 36,089 \\ \dfrac{1.331\sigma}{\sigma_{crit}^{0.765}} & h > 36,089 \end{cases} \tag{2.55}$$

These equations provide the power available to turn the propeller (shaft power), which is not the power available for flight. The power available for flight will be the shaft power times the propeller efficiency. The propeller efficiency is assumed independent of throttle setting and altitude but it will vary with velocity and Mach number. For the study of aircraft performance, it will be assumed that the propeller efficiency is constant from Mach 0.1 to Mach drag rise, at which point it will decrease sharply. Below Mach 0.1, the thrust produced by the propeller, for a given throttle setting, will be assumed to be constant and equal to the Mach 0.1 value using the following equations:

$$T = \left(\frac{\eta_p P}{V_\infty}\right)_{Mach=0.1} \qquad M \leq 0.1 \tag{2.56}$$
$$P_a = \eta_p P \qquad M > 0.1$$

The power SFC (PSFC) will be assumed constant throughout.

2.6.2 Turbojet Performance

All jet propelled aircraft will be grouped within this category. It was determined experimentally that the variation of thrust produced with altitude for a turbojet is roughly as follows:

$$\frac{T}{T^*} = \left(\frac{\sigma}{\sigma^*}\right)^x \approx \left(\frac{\sigma}{0.297}\right)^x \tag{2.57}$$

Where the asterisk represents the conditions at the tropopause (36,089 ft). The value of x is shown in Table 2.6. Throughout our discussion on aircraft performance, an x value of 0.7 will be used (middle range) for altitudes lower than 36,089 ft. This gives a thrust value at 36,089 ft of

$$T^* = 0.4275 T_{SL}$$

Table 2.6   Thrust ratio exponent values

$h \leq 36,089$ ft (troposphere)	0.5 (turbojet) $\leq x \leq 0.9$ (turbofan)
$h > 36,089$ ft (stratosphere)	$x = 1$

so that the preceding equation becomes

$$\frac{T}{T_{\text{SL}}} = \begin{cases} \sigma^{0.7} & h \leq 36,089 \\ 1.439\sigma & h > 36,089 \end{cases} \tag{2.58}$$

For a turbojet, it will be assumed that the thrust will not vary with velocity. Furthermore, the TSFC will be assumed constant.

3
Level Flight

3.1 Introduction

E VERY aircraft is designed for a specific mission profile and is optimized to fulfill that mission. In evaluating the detailed performance of a given aircraft, a good knowledge of its aerodynamics and propulsion system behavior with altitude, velocity, and Mach number is required, but a good approximation can nevertheless be obtained with the assumptions stated at the end of Chapter 1, "Aircraft Aerodynamics," and Chapter 2, "Aircraft Propulsion."

An aircraft is designed to meet specific performance requirements, which are normally defined in a statement of operational requirements (SOR). For an aircraft to be airworthy, it must also meet safety related performance specifications. Most countries have a set of regulations (or use other countries' regulations) that define clearly what performance must be met by a certain type of aircraft for a given mission. These regulations were written with the experience of time and are revised regularly. They provide guidelines, which will not only ensure that an aircraft design has a satisfactory performance, but that it also has acceptable flying qualities in the event of one or more in-flight failures of crucial components. The U.S. Federal Aviation Administration (FAA) is the organization responsible for issuing regulations for civil aircraft design and operation, air traffic control, and other aviation-related activities in that country. The Federal Aviation Regulation (FAR) part 21 applies to the experimental (homebuilt) category of aircraft. FAR 23 applies to the design of small aircraft with less than nine passengers (excluding the pilot) and weighing less than 12,500 lb for the normal, utility, and acrobatic categories. FAR 23 also applies to the commuter category with less than 19 passengers and weighing less than 19,000 lb. FAR 25 applies to all aircraft carrying more than 19 passengers with no weight restrictions. The U.S. Air Force uses the MIL-C-005011B as its standard, whereas the U.S. Navy and Marine Corps use the AS-5263. In Canada, the Ministry of Transportation (MOT) is responsible for issuing such regulations, and in Europe there is the European Joint Aviation Authorities.

By using typical aircraft, a better understanding of aircraft performance can be achieved. Three aircraft are defined. Aircraft A is of the large commercial transport aircraft type with design characteristics approaching those of the Boeing 767; it will be used to illustrate most of the points to be studied. Aircraft B is of the turboprop category used to represent the light commuter type with characteristics representative of those of the DeHavilland DHC 8. Aircraft C is of the fighter type, used to represent high-performance aircraft; its characteristics approximate those of the Northrop F-20 Tigershark (see Table 3.1 and Fig. 3.1).

3.2 Level Flight Basic Equations

Although not the initial phase of flight, the level flight will probably be the longest phase of any mission. From it the equations for minimum/maximum

Table 3.1   Typical aircraft characteristics

Characteristic	Aircraft A	Aircraft B	Aircraft C
b, ft	156.08	85.00	26.67
S, ft^2	3,080	585	200
Incidence, deg	4.25	0	0
Γ, deg at $c/4$	31.5	3	25
AR	7.9	12.35	3.55
e	0.80	0.80	0.80
C_{D_0}	0.018	0.020	0.025
$C_{L_{max}}$ (no flap)	1.65	1.5	1.9
M_{DR}	0.85	0.7	0.9
ΔM_{DW}	0.4	0.5	0.2
C_{Dw}	0.1	0.1	0.02
Empty weight, lb	178,000	22,600	13,150
Fuel weight, lb	112,725	5,678	5,050
Maximum TO weight, lb	300,000	34,500	18,540 (clean)
Thrust, lb/power, hp	$2 \times 50{,}000$ lb	2×2000 hp	11,000/18,000 lb
TSFC, lb/h/lb	0.65	——	0.80/1.95
PSFC, lb/h/hp	——	0.485	——
h/b/min	0.079	0.13	0.18
Maximum load factor (struc)	$+4/-1$	$+4/-1$	$+9/-3$
Other		Propeller efficiency is assumed constant after takeoff at a value of $\eta_p = 0.8$	This aircraft has an afterburning engine, thus the two thrust and TSFC values. This aircraft can be configured to carry external stores, which will increase its drag. This aircraft has high AoA capability, $C_{L_{max}}$ reached at 30 deg.

velocity, best range, and endurance can be derived. The forces acting on the aircraft during steady level flight are thrust opposing drag and lift opposing weight (Fig. 3.2).

In the present analysis, the thrust angle (α_T) will be considered to be negligible and the effect of α_T will be discussed in Chapter 9. To satisfy steady-state conditions there must exist an equilibrium of forces

$$T - D = 0 \tag{3.1}$$

$$L - W = 0 \tag{3.2}$$

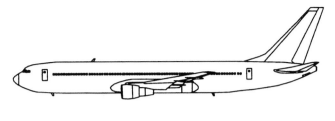

a) Aircraft A

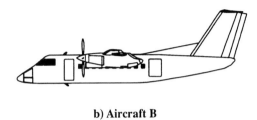

b) Aircraft B

c) Aircraft C

Fig. 3.1   Typical aircraft.

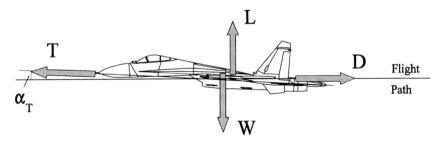

Fig. 3.2   Forces acting on an aircraft in level flight.

where

$$L = \tfrac{1}{2}\rho_{\text{SL}}\sigma V^2 S C_L$$
$$D = \tfrac{1}{2}\rho_{\text{SL}}\sigma V^2 S C_D \tag{3.3}$$

From the preceding equations, it can be seen that for a given thrust setting and aircraft weight (thrust-to-weight ratio) the aircraft will fly at a given lift-to-drag ratio

$$\frac{T}{W} = \frac{1}{(L/D)} \tag{3.4}$$

The thrust required for steady level flight can thus be defined as a function of the lift-to-drag ratio (also called aerodynamic efficiency E) and the aircraft weight,

$$T_r = \frac{W}{(L/D)} = \frac{W}{E} \tag{3.5}$$

The minimum thrust required will, therefore, occur when the aircraft is flying at its maximum value of lift-to-drag ratio. This value of the maximum aerodynamic efficiency (E_m) can be determined by differentiating the lift-to-drag ratio with respect to the lift coefficient and setting the result equal to zero.

$$\left[\frac{\partial E}{\partial C_L}\right]_{E_m} = \frac{\partial}{\partial C_L}\left[\frac{C_L}{C_{D_0} + K C_L^2}\right]_{E_m} = 0$$

$$\frac{\left[C_{D_0} + K C_L^2\right] - C_L[0 + 2K C_L]}{\left[C_{D_0} + K C_L^2\right]^2} = 0$$

$$C_{D_0} - K C_L^2 = 0$$

Thus,

$$C_{L_{E_m}} = \sqrt{\frac{C_{D_0}}{K}} \qquad C_{D_{E_m}} = 2 C_{D_0} \tag{3.6}$$

$$E_m = \frac{1}{2\sqrt{C_{D_0} K}} \tag{3.7}$$

The following thrust-required curve (Fig. 3.3), for aircraft A (at 90% maximum weight), as a function of velocity for sea-level conditions, was created using the preceding equations. The parasite drag and induced drag curves have also been included, as well as the maximum thrust-available curve. The minimum thrust required occurs at a velocity where the induced drag is equal to the parasite drag. Below that velocity, the dominant drag component is the induced drag (lift dependent), whereas above that velocity, the parasite drag is dominant.

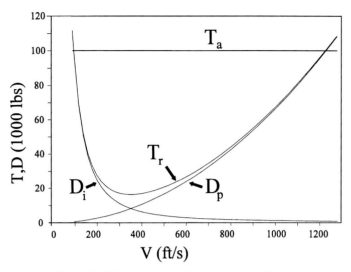

Fig. 3.3   Thrust required for steady level flight.

The lift-to-drag ratio and lift coefficient curves have been superimposed on the thrust-required curve in Fig. 3.4. Because flying at a constant lift coefficient is the same as flying at a constant AoA while in the subsonic regime, the lift coefficient curve will yield the corresponding AoA at a given velocity. Some modern aircraft are now equipped with air data computers and AoA probes, which determine the AoA of the aircraft in flight. The lift coefficient cannot be measured in-flight; it can only be approximated from the air density measurements, the aircraft velocity, and an approximation of the aircraft weight at a given point in time or by knowing the value of the AoA and the variation of the lift curve with Mach number.

As the weight of an aircraft decreases during the flight, it can be seen that the thrust required [Eq. (3.5) and Fig. 3.5] to fly at a given velocity and altitude decreases as well. Furthermore, the velocity at which minimum thrust required occurs will also decrease

$$V_{E_m} = \sqrt{\frac{2W}{\rho_{SL} \sigma S}} \sqrt[4]{\frac{K}{C_{D_0}}} \tag{3.8}$$

As can be seen in Fig. 3.5 the weight variation has negligible effects on the thrust-required curve in the high-velocity domain. This is because of the induced drag, that is, the drag due to lift (thus weight), which is quite small at high speeds. Combining Eqs. (1.11) and (3.1–3.3) the effects of aircraft weight on the thrust-required equation can be clearly seen

$$T_r = \tfrac{1}{2} \rho_{SL} \sigma V^2 S C_{D_0} + \frac{2KW^2}{\tfrac{1}{2} \rho_{SL} \sigma V^2 S} \tag{3.9}$$

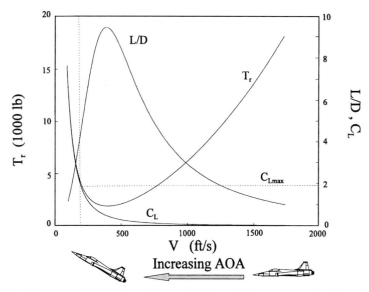

Fig. 3.4   Thrust-required, lift-to-drag, and lift coefficient curves for aircraft C at sea level.

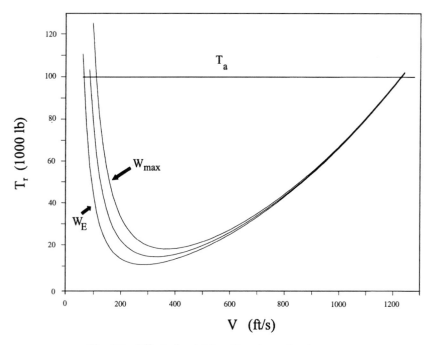

Fig. 3.5   Effect of weight on thrust-required curve.

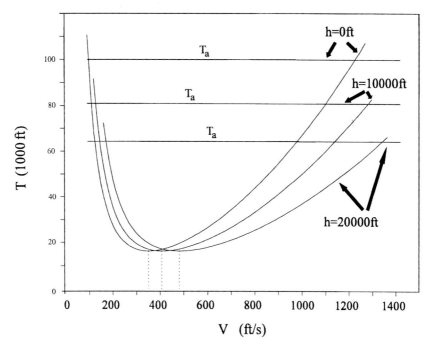

Fig. 3.6   Effect of altitude on thrust-required and thrust-available curves.

This last equation indicates that the effects of weight on thrust required can be significant at low velocities and become negligible at high velocities. As the velocity increases, the parasite drag term will become more important.

For a given weight, the minimum thrust required is independent of altitude since the aircraft will be operating at E_m [see Eq. (3.5) and Fig. 3.6], but the velocity at which minimum thrust required will occur will increase as the altitude increases [Eq. (3.8)].

An aircraft's ceiling is the maximum altitude it can reach while flying at a given velocity and throttle setting. The absolute ceiling of an aircraft is the highest altitude that aircraft can maintain under steady-state conditions while at maximum throttle setting. Inspection of Fig. 3.6 reveals that the absolute ceiling is reached when maximum thrust available at altitude is equal to the minimum thrust required of the aircraft, thus flying at maximum lift-to-drag conditions. The ceiling density ratio can be found by using Eq. (2.58).

$$\sigma_{c,\text{abs}} = \left(\frac{1}{E_m (T_{a,\text{max}}/W)_{\text{SL}}} \right)^{(1/0.7)} \quad \text{for} \quad h < 36,089$$

$$\sigma_{c,\text{abs}} = \frac{1}{1.439 E_m (T_{a,\text{max}}/W)_{\text{SL}}} \quad \text{for} \quad h \geq 36,089$$

(3.10)

A reduction in the minimum drag coefficient by the addition of systems to control the BL or by decreasing the number of excrescences such as landing gear, flaps, external stores, antennas, etc., will decrease the thrust required to fly at a

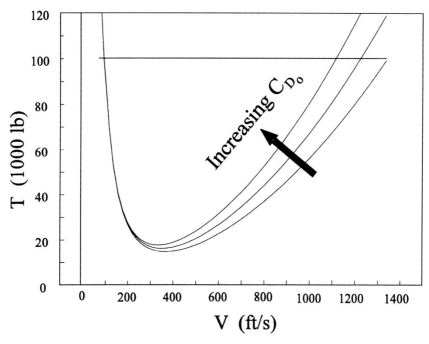

Fig. 3.7   Effect of parasite drag variation on T_r curve.

given velocity, especially in the high-velocity range. On the other hand, a reduction in the induced drag coefficient, by increasing the AR or the Oswald coefficient, will also decrease the thrust required to fly at a given velocity with the greatest impact in the low-velocity range. This can be seen in Figs. 3.7 and 3.8. Observe the influence on the minimum drag velocity.

At the extreme left of the thrust-required curve (low-velocity range), although the aircraft may still have enough thrust available to counter drag, it may not produce enough lift to sustain level flight, i.e., the aircraft would have to fly at a lift coefficient higher than $C_{L_{max}}$, in which case the aircraft would stall. The velocity at which $C_{L_{max}}$ occurs is called the stall velocity

$$V_{\text{stall}} = \sqrt{\frac{2W}{\rho_{\text{SL}} \sigma S C_{L_{max}}}} \tag{3.11}$$

This is called the power-off stall velocity where the thrust from the engine does not contribute in anyway to the lift. If the angle of the thrust vector with respect to the velocity vector (α_T) is known at $C_{L_{max}}$, it is then possible to determine the power-on stall velocity

$$V_{\text{stall}} = \sqrt{\frac{2[W - T \sin(\alpha_{T,\text{stall}})]}{\rho_{\text{SL}} \sigma S C_{L_{max}}}} \tag{3.12}$$

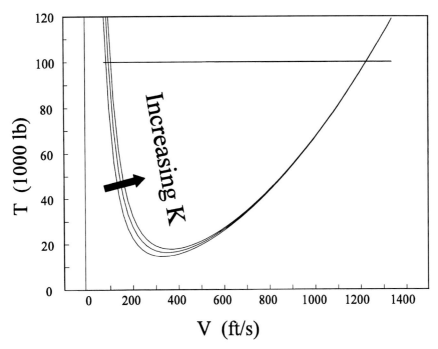

Fig. 3.8   Effect of induced drag variation on T_r curve.

This aspect will be discussed in more detail in Chapter 9. The stalling velocity of the aircraft will be considered as being equal to the power-off value thus avoiding, for the time being, such complications as the influence of the flow of the propeller slipstream over the wing or the thrust AoA (α_T).

Superimposing the stall limit on the thrust-required curve, two distinct regions of possible steady-state level flight can be identified: one below the velocity for minimum thrust required and one above. In the region below the velocity for minimum drag, a decrease in velocity will require an increase of thrust to maintain altitude. This is the opposite of what would normally be expected because an aircraft's normal operating speed regime is the region above the minimum thrust-required velocity where a reduction in velocity requires a reduction in thrust to remain at the same altitude. The region below minimum drag velocity is usually called the region of reverse-throttle command. It is also often referred to as flying on the back side of the thrust curve or the region of slow flight (Fig. 3.9).

The region of slow flight is bounded on one side by the stall velocity [Eq. (3.11)] and on the other side by the minimum drag velocity [Eq. (3.8)]. Its extent is determined by a combination of aircraft minimum drag coefficient, induced drag coefficient, and maximum lift coefficient. An increase in wing sweep, which would result in a reduction in Oswald coefficient, or a decrease in AR will tend to extend the region of slow flight. The use of speed brakes and the lowering of the landing gear will increase the minimum drag coefficient and thus will reduce the region of slow flight, as shown in Fig. 3.10. The application of flaps will increase drag and

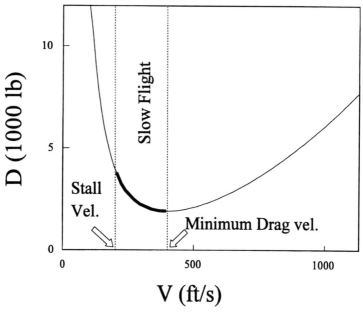

Fig. 3.9   Slow flight velocity regime for aircraft C in a clean configuration at $W = 18,500$ **lb.**

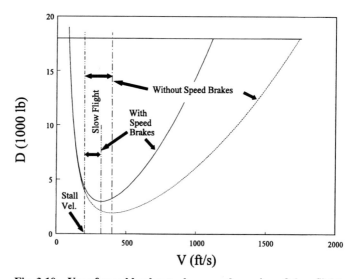

Fig. 3.10   Use of speed brakes to decrease the region of slow flight.

will reduce V_{stall}, which means that both the upper and lower limits of slow flight will be lowered.

Points to keep in mind while flying in a slow flight region is that the throttle controls altitude while the elevator (change in AoA) controls speed. This again will be explained in more detail in Chapter 9.

The curves shown up to now apply equally well to turbojet aircraft and piston-prop aircraft, although they are more suited to the turbojet type because of the assumption that the thrust available is independent of velocity. For a piston-prop aircraft, the power-required curve is usually used because it is generally assumed that the power available is independent of velocity. Because the power required is, in fact, the thrust required multiplied by the velocity, the mentioned factors influencing the drag curve apply directly to the power-required curve (Fig. 3.10)

$$P_r = DV = \left(W\frac{C_D}{C_L}\right)\sqrt{\frac{2W}{\rho_{\text{SL}}\sigma S C_L}} = \sqrt{\frac{2W^3}{\rho_{\text{SL}}\sigma S}}\left(\frac{C_D}{C_L^{3/2}}\right) \tag{3.13}$$

Thus, the minimum power required (subscript MP) occurs when the ratio $(C_L^{3/2}/C_D)$ is maximum. Differentiating this ratio with respect to the lift coefficient, it is found that the maximum occurs when the induced drag coefficient (KC_L^2) is equal to three times the minimum drag coefficient (C_{D_0}):

$$\left(KC_L^2\right)_{\text{MP}} = 3C_{D_0} \tag{3.14}$$

and

$$C_{L_{\text{MP}}} = \sqrt{3}C_{L_{E_m}} \quad C_{D_{\text{MP}}} = 4C_{D_0}$$
$$E_{\text{MP}} = \sqrt{3/4}E_m \tag{3.15}$$

Steady-state flights at minimum power required occur at a velocity below the minimum drag velocity (in the reverse command region).

The power required equation can be rewritten in the following way:

$$P_r = DV = \frac{1}{2}\rho_{\text{SL}}\sigma V^3 S C_{D_0} + \frac{2KW^2}{\rho_{\text{SL}}\sigma S V} \tag{3.16}$$

This last equation reveals that the zero-lift power required increases as the cube of the velocity. This is one of the important factors that limit the maximum velocity of a piston-prop aircraft (propeller efficiency degradation being another) because of the relatively constant power available. Remember that for a turbojet aircraft, the thrust available is assumed constant while the zero-lift thrust required increases as the square of the velocity.

Before going any further, a discussion on power and thrust curves is required. It was demonstrated that minimum thrust required occurs while flying at maximum lift-to-drag ratio conditions (Fig. 3.4). This point also corresponds to the point of maximum excess thrust for a jet aircraft whose thrust-available curve is assumed constant. For a propeller driven aircraft, where constant power is assumed, the

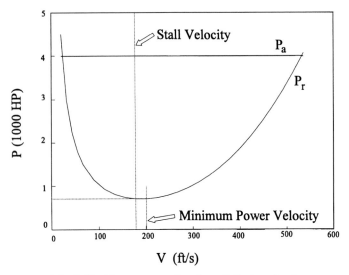

V (ft/s)

Fig. 3.11   Power curves for aircraft B at sea level.

maximum excess thrust will occur at a lower velocity than the minimum drag velocity, as shown in Fig. 3.12. It generally occurs very close to, and at times below, the stall velocity. It will be demonstrated in a later chapter that large excess thrust is important when large angles of climb are desired.

Using the power curves, it was noticed that the minimum power required occurs at the maximum value of the $(C_L^{3/2}/C_D)$ ratio. This point also corresponds to the maximum excess power velocity for a propeller driven aircraft (Fig. 3.11) when the power available is assumed constant. A jet aircraft power-available curve looks quite different from that of a prop aircraft. Assuming that thrust available is independent of velocity, the power-available (T_aV) curve is directly proportional

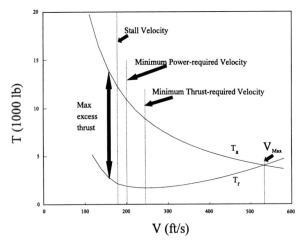

V (ft/s)

Fig. 3.12   Thrust curves for aircraft B.

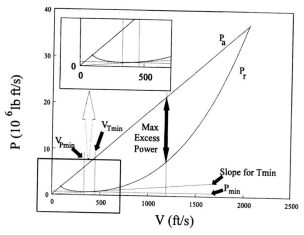

Fig. 3.13 Power curves for a jet aircraft.

to the velocity (as seen in Fig. 3.13). The power-required $(T_r V)$ curve, on the other hand, is independent of the propulsion system.

From here on, the thrust curve will be used for turbojet aircraft and the power curve for piston-prop aircraft unless specified otherwise.

3.3 Flight Envelope

The maximum and minimum velocities attainable by an aircraft in steady level flight, for a given weight, altitude, and throttle setting, are obtained at the intersection of the thrust-required curve and the thrust-available curve:

$$T_a = T_r = \tfrac{1}{2}\rho_{\mathrm{SL}}\sigma V^2 S C_D$$

$$T_a = \tfrac{1}{2}\rho_{\mathrm{SL}}\sigma V^2 S\left(C_{D_0} + K C_L^2\right)$$

$$T_a = \tfrac{1}{2}\rho_{\mathrm{SL}}\sigma V^2 S\left[C_{D_0} + K\left(\frac{W}{\tfrac{1}{2}\rho_{\mathrm{SL}}\sigma V^2 S}\right)^2\right]$$

$$T_a = \tfrac{1}{2}\rho_{\mathrm{SL}}\sigma V^2 S C_{D_0} + \frac{K W^2}{\tfrac{1}{2}\rho_{\mathrm{SL}}\sigma V^2 S}$$

$$\tfrac{1}{2}\rho_{\mathrm{SL}}\sigma S C_{D_0} V^4 - T_a V^2 + \frac{K W^2}{\tfrac{1}{2}\rho_{\mathrm{SL}}\sigma S} = 0$$

$$V^2 = \frac{T_a \pm \sqrt{T_a^2 - 4 C_{D_0} K W^2}}{\rho_{\mathrm{SL}}\sigma S C_{D_0}}$$

$$\therefore \ V = \sqrt{\frac{(T_a/S)}{\rho_{\mathrm{SL}}\sigma C_{D_0}}\left[1 \pm \sqrt{1 - \frac{4 K C_{D_0}}{(T_a/W)^2}}\right]} \tag{3.17}$$

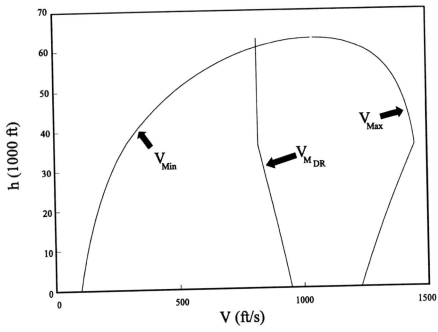

Fig. 3.14   Flight envelope of aircraft A using Eq. (3.17); no compressibility effects.

The minimum velocity is calculated by using the minus sign in this last equation; the maximum velocity is calculated by using the plus sign. The absolute minimum and maximum velocities are obtained by setting full throttle. The magnitude of the maximum and minimum velocities will change with altitude since the thrust-required and thrust-available curves change with altitude. A graph of V_{min} and V_{max} as a function of altitude (Fig. 3.14) defines what is called the flight envelope of an aircraft. The level flight steady-state envelope of an aircraft defines the limits of that aircraft both in terms of possible steady-state velocity or Mach number and altitude. By using Eqs. (3.17) and (2.58) the basic flight envelope of aircraft A ($W = 270,000$ lb) can be drawn up. These velocities are theoretical limits, which may not be achievable in steady level flight (Fig. 3.15).

The maximum velocity of the aircraft can be limited by compressibility effects that were not considered in Eq. (3.17). As explained earlier, C_{D_0} and K are assumed constant up to the Mach drag rise M_{DR} beyond which point both coefficients begin to be affected by compressibility effects such as transonic flow and the resulting presence of shock waves, which in turn have an effect on parasite and induced drag coefficients (see Sec. 1.8 of Chapter 1). Knowing the variation of both coefficients with Mach number, an iterative process can be used to determine the maximum velocity for a given altitude. Similarly, the theoretical minimum velocity will not be achievable if it is lower than the aircraft's stall velocity. The modified flight envelope that results is shown in Fig. 3.16.

There are other limitations to an aircraft flight envelope beyond those imposed by the aerodynamic and thrust-available ones. The aircraft's structural design and construction will impose a maximum dynamic pressure and Mach number,

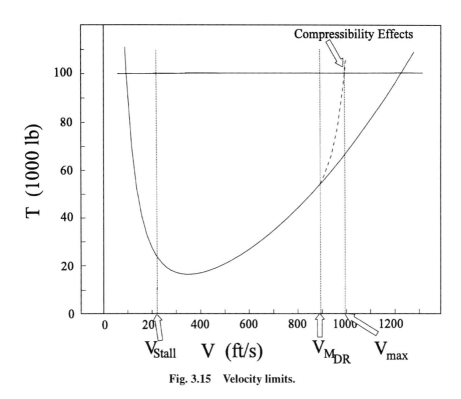

Fig. 3.15   Velocity limits.

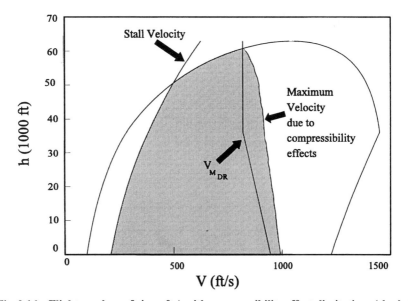

Fig. 3.16   Flight envelope of aircraft A with compressibility effects limitations (shaded region).

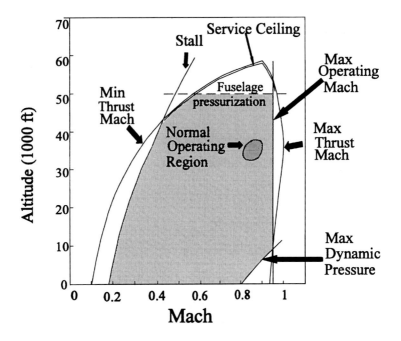

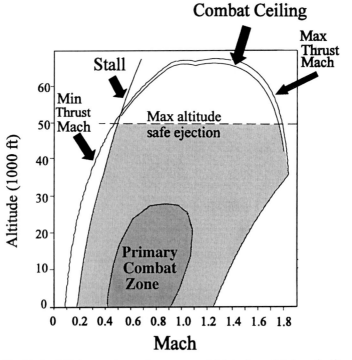

Fig. 3.17 Typical flight envelope of a commercial jet aircraft (aircraft A) and of a fighter (aircraft C).

may it be in level flight or in a dive. The velocity and altitude may also be limited by the engine's flight envelope, which is defined by its ability to develop thrust without surging, stalling, or flaming out. Typical aircraft flight envelopes are represented by the shaded areas in Figs. 3.17 and 3.18. Note the dip in the flight envelope near Mach 1 for the fighter. This is due to the transonic drag increase.

The power available for a piston-prop aircraft is a function of the SHP developed by the engine and the propeller efficiency (η_p). The former is essentially independent of the aircraft's airspeed but varies with altitude as per Eq. (2.54) and the throttle setting. The latter varies with engine rpm, airspeed, and Mach number. The propeller efficiency is normally assumed to remain constant throughout the normal level flight envelope of the piston-prop aircraft as long as no compressibility effects are encountered (below M_{DR} as a reference).

The power-to-weight ratio of an aircraft is defined as

$$P_a/W = \eta_p \text{SHP}/W \tag{3.18}$$

Because it is generally easier to use units of feet-pounds per second in calculations involving power, a coefficient may be used to convert the horsepower units, which are so frequently used by manufacturers. This coefficient has a value of 550 ft-lb/s/hp. Aircraft manufacturers usually provide the inverse of this ratio (W/SHP), which is called the power loading, as an aircraft design specification.

Using Eq. (3.15),

$$V_{\text{MP}} = \sqrt{\frac{2W}{\rho_{\text{SL}}\sigma S}}\sqrt[4]{\frac{K}{3C_{D_0}}} = \sqrt[4]{\frac{1}{3}}V_{E_m} \approx 0.76V_{E_m} \tag{3.19}$$

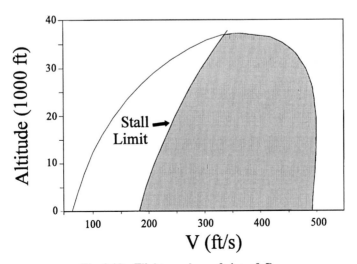

Fig. 3.18   Flight envelope of aircraft B.

and, finally, from Eqs. (3.13) and (3.15)

$$\left[\frac{P_r}{W}\right]_{\min} = \frac{V_{MP}}{\sqrt{3/4}E_m} \tag{3.20}$$

The minimum power required increases with altitude because V_{MP} increases with altitude (remember that the minimum thrust required does not change with altitude). This is another limiting factor for piston-prop aircraft, which explains why propeller-driven aircraft do not generally fly at altitudes as high as those of a jet aircraft.

For a piston-prop aircraft, the absolute ceiling will occur at an altitude where the aircraft minimum power required is equal to the aircraft maximum power available at that altitude,

$$P_{a_{\max}} = P_{r_{MP}} \qquad\qquad P_{a_{\max}} = \frac{W V_{MP}}{E_{MP}}$$

Thus, for an unsupercharged piston-prop engine, using Eq. (2.54) the density ratio at the absolute ceiling is, for h in feet

$$\sigma_{c_{abs}} = \begin{cases} \left[\left(\dfrac{W V_{MP_{SL}}}{\eta_p P_{A_{SL_{max}}} E_{MP}}\right)\dfrac{1}{\sqrt{\sigma_{c_{abs}}}}\right]^{(1/0.765)} & h \le 36{,}089 \\[2em] \left(\dfrac{W V_{MP_{SL}}}{1.331\eta_p P_{A_{SL_{max}}} E_{MP}}\right)\dfrac{1}{\sqrt{\sigma_{c_{abs}}}} & h > 36{,}089 \end{cases} \tag{3.21}$$

The value of $\sigma_{c,abs}$ can now be found by successive iterations.

3.4 Range

An aircraft's range is the distance an aircraft can travel on a given load of fuel and for a given set of flight parameters. With the growth of the Pacific rim market, long range is becoming a driving factor during the design of new aircraft. Bombardier, for example, is offering the Global Express to executives worldwide to enable them to reach major trading centers from almost anywhere in the world, and that without having to land midway to refuel (Fig. 3.19). Long range has always been predominant during the design of bomber, with the Convair B-36 Peacemaker becoming the first true intercontinental bomber. This section covers in detail the range aspect of aircraft performance and those parameters (such as airspeed, altitude, wind speed, etc.) that affect it.

Knowing that velocity is the rate of change of distance over time and assuming that the decrease in aircraft fuel weight with time is directly proportional to the thrust of the aircraft's engines, the following equations are obtained:

$$\frac{\partial X}{\partial t} = V \tag{3.22}$$

$$\frac{\partial W}{\partial t} = -SFC_T T \tag{3.23}$$

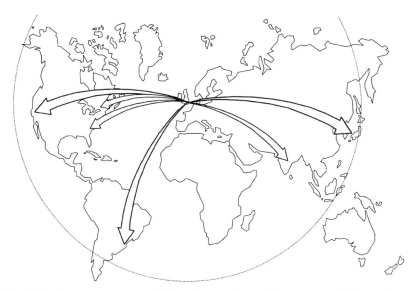

Fig. 3.19   Long-range aircraft bring the major trading centers within reach in a single flight.

The specific instantaneous range [Eq. (3.24)] is obtained by combining the preceding equations and Eq. (3.5). Its value can be maximized while flying at low-weight conditions, having engines with a low SFC_T (a function of engine design), and by optimizing the product of the velocity and lift-to-drag ratio. The latter is itself a function of velocity, altitude, weight, and the aircraft's aerodynamic characteristics

$$-\frac{\partial X}{\partial W} = \frac{V}{SFC_T\,T} = \frac{V E}{SFC_T W} \tag{3.24}$$

Two quantities, which will be useful throughout the discussion of range, will now be introduced. The first quantity is the fuel-weight fraction (ζ), which is the ratio of the fuel weight used during a mission leg to the total aircraft weight, including fuel, at the beginning of that leg. The second quantity is the mass ratio (MR), which is the ratio of the total aircraft weight at the beginning of a mission leg (W_1) to the total aircraft weight at the end of that leg (W_2).

$$\zeta = \frac{\Delta W_f}{W_1} = \frac{W_1 - W_2}{W_1}$$

$$\mathrm{MR} = \frac{W_1}{W_2} = \frac{W_1}{W_1 - \Delta W_f} = \frac{1}{1 - \zeta} \tag{3.25}$$

Note that ζ need not correspond to the entire fuel load of the aircraft; it is the fraction of fuel used during a given cruise leg. Assuming that the TSFC remains constant throughout a cruise leg, Eq. (3.23) can be integrated between two points to obtain the range

$$X = -\frac{1}{SFC_T} \int_1^2 \frac{V}{D} dW \tag{3.26}$$

3.4.1 General Range Equations (Jet Aircraft)

Flight conditions throughout a cruise leg must be identified before the integral equation (3.26) can be solved. Three alternatives will be considered here, although a large number of other possibilities exist: 1) cruise at constant velocity and altitude, 2) cruise at constant velocity and lift coefficient, and 3) cruise at constant altitude and lift coefficient.

The cruise at constant velocity and altitude is the easiest to monitor from the point of view of the pilot and the air traffic controllers. Taking the velocity term out of the integral (constant V), Eq. (3.26) becomes

$$X_{V,h} = -\frac{V}{SFC_T} \int_1^2 \frac{E}{W} dW \tag{3.27}$$

After a laborious integration (see Appendix D for details), the preceding equation reduces to

$$X_{V,h} = \frac{2E_m V}{SFC_T} \arctan\left[\frac{\zeta E_1}{2E_m\left(1 - KC_{L_1}E_1\zeta\right)}\right] \tag{3.28}$$

where E_1 and C_{L_1} are the lift-to-drag ratio and lift coefficient at the beginning of the cruise leg. As the fuel is burned during the cruise, the aircraft weight will decrease. Because the altitude and velocity remain constant, the aircraft's induced drag (which is proportional to the weight of the aircraft in steady level flight) will decrease during the cruise. The throttle setting, thus, will have to be reduced to keep a match between drag and thrust to maintain constant altitude and velocity conditions.

The second cruise mode, which yields a somewhat simpler equation, is that of a cruise at constant velocity and lift coefficient such that a constant lift-to-drag ratio is maintained throughout the cruise leg. Equation (3.26) reduces to

$$X_{V,C_L} = -\frac{EV}{SFC_T} \int_1^2 \frac{dW}{W} = \frac{EV}{SFC_T} \ell_n(\mathrm{MR}) = \frac{EV}{SFC_T} \ell_n\left(\frac{1}{1-\zeta}\right) \tag{3.29}$$

This equation is often referred to as the Breguet range equation. As the aircraft weight decreases during cruise, Eq. (3.30) [from the lift Eq. (3.3)] shows that the density ratio must decrease; thus, the altitude must increase as the cruise leg progresses. This cruise mode is, therefore, referred to as the *cruise climb* condition

$$\sigma = \frac{2(W/S)}{\rho_{SL} V^2 C_L} \tag{3.30}$$

At first, this seems to contradict our assumption of a level flight cruise, but an example at the end of the "Range" subsection will show that the climb angle is small enough to be considered negligible. For the last cruise option to be considered, the altitude and the lift coefficient are kept constant so that Eq. (3.26) becomes

$$X_{h,C_L} = -\frac{E}{SFC_T} \int_1^2 \frac{V}{W} \, dW \qquad (3.31)$$

Solving the integral yields (see Appendix D)

$$X_{h,C_L} = \frac{2EV_1}{SFC_T} \left[1 - \sqrt{1 - \zeta} \right] \qquad (3.32)$$

where V_1 is the initial velocity of the cruise leg, which can be obtained from the lift equation and by using the flight conditions at the beginning of the cruise leg

$$V_1 = \sqrt{\frac{2(W_1/S)}{\rho_{\text{SL}}\sigma C_L}} \qquad (3.33)$$

As fuel is burned and the aircraft's weight decreases, the velocity will also decrease if the lift coefficient is to remain constant. The weight at the end of the segment will be

$$W_2 = W_1(1 - \zeta) \qquad (3.34)$$

and, combining Eq. (3.33) with Eq. (3.34), the value of the velocity at the end of the segment is

$$V_2 = V_1\sqrt{1 - \zeta} \qquad (3.35)$$

The main drawback to this cruise option is that the velocity decreases along the cruise segment, which increases flight time. Also, air traffic control regulations for civil aircraft require a constant (± 10 kn) true airspeed during cruise flight.

3.4.2 Best Range Equations (Jet Aircraft)

As alluded to in the introduction of Sec. 3.4, range is an important performance parameter for most aircraft missions. A given aircraft could achieve a longer range by trading payload for additional fuel (larger ζ), but it does not make any economic sense to simply burn more fuel to increase range without considering other flight parameters. At a given altitude and aircraft weight, for a given fuel load, there exists a velocity that will maximize the range by minimizing the fuel burned per distance covered. This will be called the best range conditions of the aircraft for that altitude and aircraft weight. The velocity at which the best range conditions are obtained can be found by differentiating the specific instantaneous range [Eq. (3.24)], with respect to velocity and setting the result equal to zero

$$\frac{\partial}{\partial V}\left[-\frac{\partial X}{\partial W} \right]_{\text{BR}} = \frac{\partial}{\partial V}\left[\frac{V}{SFC_T D} \right]_{\text{BR}} = 0 \qquad (3.36)$$

which is satisfied when (if SFC_T is not a function of velocity)

$$\frac{\partial D}{\partial V} = \frac{D}{V} \tag{3.37}$$

Thus, the best range conditions will be achieved when the slope of the drag curve becomes equal to the ratio of the drag over velocity. This occurs when the slope of the drag curve goes through the origin, as shown in Fig. 3.20. This also corresponds to the minimum ratio of drag over velocity. Using the parabolic drag equation (1.11) and solving for the velocity, the following best range (subscript BR) parameters are obtained (note that the best range velocity is a function of altitude through the density altitude parameter σ):

$$V_{\text{BR}} = \sqrt{\frac{2(W/S)}{\rho_{\text{SL}}\sigma}} \sqrt[4]{\frac{3\,K}{C_{D_0}}} \approx 1.316 V_{E_m} \tag{3.38}$$

$$C_{L_{\text{BR}}} = \sqrt{\frac{C_{D_0}}{3K}} \approx 0.557 C_{L_{E_m}} \tag{3.39}$$

$$E_{\text{BR}} = \frac{\sqrt{3/4}}{2\sqrt{C_{D_0}K}} \approx 0.866 E_m \tag{3.40}$$

Because best range conditions are achieved when the ratio of drag over velocity is minimum, it can be shown [using Eq. (3.3)] that these conditions exist, as shown in Fig. 3.20 when the $(C_L^{1/2}/C_D)$ ratio is maximum. Introducing these values into

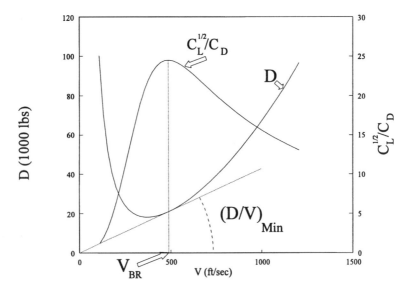

Fig. 3.20   Condition for best range.

the three range equations and setting the initial conditions equal to the best range condition of the aircraft gives

$$X_{V,h}|_{\text{BR}} = \frac{2E_m V_{\text{BR}}}{SFC_T} \arctan\left[\frac{\zeta E_{\text{BR}}}{2E_m\left(1 - KC_{L_{\text{BR}}} E_{\text{BR}}\zeta\right)}\right] \qquad (3.41)$$

$$X_{V,C_L}|_{\text{BR}} = \frac{E_{\text{BR}} V_{\text{BR}}}{SFC_T} \ell_n\left[\frac{1}{1-\zeta}\right] \qquad (3.42)$$

$$X_{h,C_L}|_{\text{BR}} = \frac{2E_{\text{BR}} V_{\text{BR}}}{SFC_T}\left[1 - \sqrt{1-\zeta}\right] \qquad (3.43)$$

Using aircraft A as an example, assuming that it reaches its cruising altitude and velocity at 95% of its maximum gross takeoff weight and that its fuel is consumed in a single cruise leg, a graph of range as a function of fuel-weight fraction is obtained (Fig. 3.21).

The Breguet range equation (constant velocity and lift coefficient) yields the longest range possible and is often the one used for estimating the range of an aircraft during the design stage. During a cruise climb, the altitude gained can be obtained by combining Eqs. (3.30) and (3.34).

$$\sigma_2 = \sigma_1(1 - \zeta) \qquad (3.44)$$

For aircraft A with a maximum fuel-weight fraction of 0.376, the maximum range (cruise climb mode, all fuel used) from an initial altitude of 30,000 ft is 5,330 n mile and the total altitude gain is 12,858 ft, which represents an average climb angle of about 0.023 deg over the entire cruise leg, which is negligible. Thus, the assumption of level flight conditions is appropriate here.

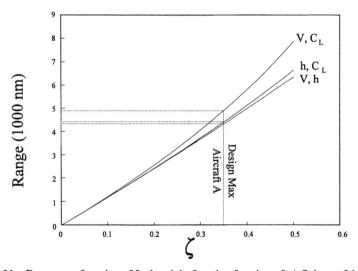

Fig. 3.21   Range as a function of fuel-weight fraction for aircraft A flying at 36,000 ft.

3.4.3   Longest Range (Jet Aircraft)

For a given altitude, the range can be maximized by flying at the best range conditions. Under these conditions, the lift-to-drag ratio is only at approximately 86.6% of its maximum value. An aircraft can maximize its range by maximizing the product of the velocity and aerodynamic efficiency ($V E$) (remembering that we assumed that SFC_T is not a function of altitude). When there are no compressibility effects, this is achieved by flying at best range conditions. Since the best range velocity for a jet aircraft increases with altitude [Eq. (3.38)], the product $V E$ can be maximized by increasing the aircraft altitude. The optimum altitude will be that where the product of velocity and aerodynamic efficiency is maximum. As altitude increases, one must check for compressibility effects, which would change the values of the parasite and the induced drag coefficients as shown in Fig. 1.30. One must also check if there is sufficient thrust to fly at the predicted altitude. This velocity/aerodynamic efficiency/altitude combination will be called the longest range conditions and will probably not be equal to the best range conditions as defined by Eqs. (3.38–3.40), which did not account for compressibility effects and available thrust.

Figure 3.22 illustrates the case of aircraft A at a weight of 270,000 lb and $\zeta = 0.3$; the optimum altitude in this case is approximately 50,000 ft and Mach 0.85 for a longest range of approximately 4,439 n mile. Note again that Eq. (3.38) assumed constant C_{D_0} and K. This means that it does not provide a reliable best range velocity at altitudes beyond which the aircraft's best range velocity equals the Mach drag rise velocity. The drag of aircraft A increases so rapidly at speeds beyond Mach drag rise that the best range conditions, past a critical altitude (determined by $M_{BR} = M_{DR}$), are attained while flying at Mach drag rise and by maximizing the aerodynamic efficiency as illustrated. In this particular case, aircraft A had enough available thrust to fly at the longest range conditions.

The decrease in aircraft weight during cruise will slightly modify the lines of Fig. 3.22 by decreasing the total drag at a given altitude (due to the reduction in induced drag). Furthermore, the best range velocity [Eq. (3.38)] decreases as the aircraft weight decreases. In the case of aircraft A, best range conditions for two different weights (Fig. 3.23) are limited to Mach drag rise beyond a given altitude due to the sharp increase in wave drag. Past that altitude, range can only be maximized by increasing the aerodynamic efficiency to its maximum value, as seen in Fig. 3.22. Other aircraft, such as aircraft C, have a smaller increase in drag past Mach drag rise so that the product of aircraft velocity and aerodynamic efficiency may be maximized in the transonic regime at high altitudes.

Although not discussed at this time, one must recognize the fact that the time required and fuel burned during the climb to reach the optimum cruise altitude may be too long (or uneconomical) for the distance to be traveled. As well, other factors may limit the maximum cruise altitude for a particular aircraft, see Sec. 3.3, "Flight Envelope."

3.4.4   Stepped Climb Cruise

One way to try to match the cruise-climb condition while conforming to air traffic control regulations (constant V and h) is to perform a stepped climb cruise. Upon reaching its initial cruising altitude, the aircraft is set up for a constant velocity (best range) constant altitude cruise. After a given time (and distance),

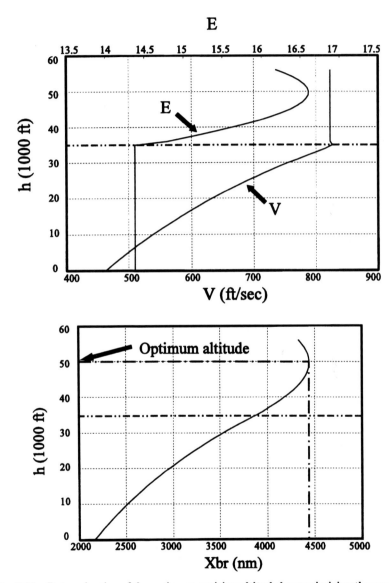

Fig. 3.22 Determination of the optimum cruising altitude by maximizing the product of velocity and aerodynamic efficiency (compressibility effects included).

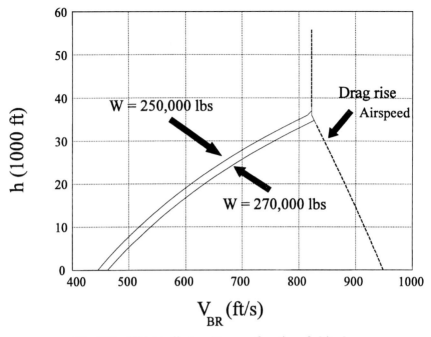

Fig. 3.23   Weight effect on V_{BR} as a function of altitude.

a climb is initiated to a higher altitude to rematch the aircraft's actual velocity to the theoretical best range velocity, which is a function of density ratio and aircraft weight, while avoiding regions of large wave drag increase. The smaller the steps are, the smaller the fuel consumption for each step and, as seen in Fig. 3.24, the smaller the difference between a true cruise climb and a stepped climb cruise at constant velocity constant altitude. At present, air traffic control regulations limit altitudes steps to a minimum of 4,000 ft above 29,000 ft (see Table 3.2) due to the difficulty of controlling actual aircraft position during the climb (this minimizes the number of steps). Airlines would prefer 2,000-ft steps to match as closely as possible the cruise climb conditions and save precious fuel. In any event, smaller steps are impossible because the air traffic control regulations also impose the altitude of travel of aircraft for flight safety and minimum clearance reasons. Table 3.2 is a summary of the permitted cruising altitudes.

3.4.5 Payload–Range Diagram

Another useful graph is the payload–range diagram (Fig. 3.25). It gives the range of an aircraft as a function of its payload weight. This type of graph is usually produced for aircraft that carry payload internally where this increased payload does not affect the aircraft aerodynamic lines (most transport aircraft and military aircraft with internal weapon bays).

Figure 3.25 is obtained as follows (refer to Fig. 3.26). First, the operational empty weight must be determined (empty weight plus weight of crew, residual fluids, etc.); then the maximum payload is loaded onboard the aircraft. The aircraft range is then calculated as the weight of fuel is increased up to the point where

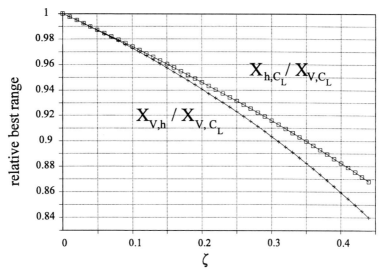

Fig. 3.24   Comparison of ranges as a function of fuel-weight fraction.

the aircraft's maximum takeoff weight is reached. This particular range is also called the harmonic range. Then, if there is still some volume for fuel, the payload weight must be decreased to take on additional fuel and increase the range. When the maximum fuel volume is reached, the only way to still further increase the aircraft range is to further decrease the payload weight, and the maximum range is attained when the payload weight finally reaches zero. This is called the ferry range.

3.4.6   Effect of Deviating from V_{BR} (Jet Aircraft)

It may be impossible to fly at the velocity for best range due to various reasons such as flight schedule, drag rise at assigned altitude, air traffic controller reassignment of arrival time, etc. To determine the impact of flying at some velocity other than V_{BR}, a velocity coefficient (v) is defined as the ratio of an aircraft's velocity to its best range velocity

$$v = \frac{V}{V_{BR}} = \sqrt{\frac{2(W/S)}{\rho C_L}} \Bigg/ \sqrt{\frac{2(W/S)}{\rho C_{L_{BR}}}} = \sqrt{\frac{C_{L_{BR}}}{C_L}} \qquad (3.45)$$

Table 3.2   Permitted cruising altitudes

Altitudes	Going East (000°–179°)	Going West (180°–359°)
Above 29,000 ft (4,000-ft intervals)	29,000 ft; 33,000 ft; etc.	31,000 ft; 35,000 ft; etc.
Below 29,000 ft (2,000-ft intervals)	Odd thousands 19,000 ft; 21,000 ft; etc.	Even thousands 18,000 ft; 20,000 ft; etc.

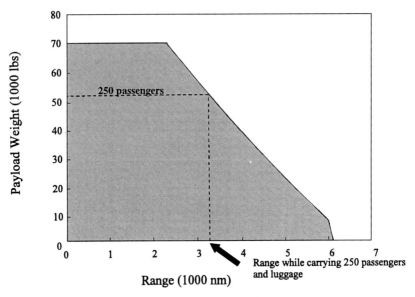

Fig. 3.25   Payload–range diagram.

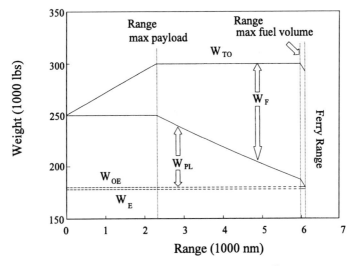

Fig. 3.26   Construction of the payload–range diagram.

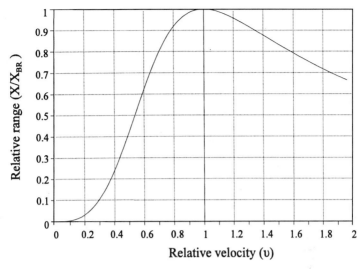

Fig. 3.27   Relative range as a function of relative velocity.

Assuming a parabolic drag polar and using the preceding equation, the aerodynamic efficiency can be written as

$$E = \frac{C_L}{C_D} = \frac{\left(C_{L_{BR}}/v^2\right)}{C_{D_0} + K\left(C_{L_{BR}}/v^2\right)^2} = \frac{2\sqrt{3}E_m v^2}{3v^4 + 1} \tag{3.46}$$

Using the Breguet range equation (constant V, C_L), a relative range equation is obtained

$$\left[\frac{X}{X_{BR}}\right]_{V,C_L} = \frac{EV}{(\sqrt{3}/2)E_m V_{BR}} = \frac{4v^3}{3v^4 + 1} \tag{3.47}$$

which is illustrated in Fig. 3.27.

It can be seen that deviating from the best range velocity ($v = 1$) by $\pm15\%$ would not change the relative range by more than approximately 4%. Any larger deviation will result in an appreciable reduction in range. Again, this is assuming subtransonic flight at all times. Almost all of today's jet airliners have been designed to fly at transonic speeds to reduce flying time and increase range as a result of the larger value of velocity times aerodynamic efficiency. For such high-velocity aircraft, the curve can be expected to be accurate for relative velocities smaller than $v = 1$, but above that point the slope will be steeper than shown because of the large increase in wave drag.

3.4.7 Best Ground Range (Jet Aircraft)

The range, as discussed so far, has been the still air range (i.e., the ground distance traveled through a stationary air mass). Air range is unaffected by the wind, but ground range is! Defining the wind fraction (ω_f) as being the ratio of the wind velocity to the best still air range velocity [Eq. (3.38)] the ground velocity is given as

$$V_{GR} = V_{BR} \pm V_W = V_{BR}(1 \pm \omega_f) \quad \text{where} \quad \omega_f = V_W/V_{BR} \qquad (3.48)$$

The positive sign applies for tail winds. Multiplying both sides of Eq. (3.48) by the duration of flight (for a given amount of fuel burned) gives the ground range

$$X_{GR} = X_{BR}(1 \pm \omega_f) \qquad (3.49)$$

This linear relation shows how ground range can be drastically reduced when flying into a headwind. There is some advantage to flying at a velocity different from the best still air range velocity while in the presence of winds. Clearly, for headwinds, increasing the aircraft airspeed will decrease the time to travel a given distance, thus reducing the aircraft's exposure to the adverse effect of the headwinds. However, deviating from the velocity for best range will decrease the maximum air range. To obtain the best ground range while flying in winds, may they be headwind or tailwind, a compromise must be reached. For best range in still air, as previously seen, a turbojet aircraft must fly at conditions such that its drag over velocity ratio is minimum. In the presence of winds, these conditions must be such that the aircraft's drag over ground velocity ratio is minimum. Thus, the best ground range (X_{BGR}) will occur when D/V_{GR} is minimum and $X_{BGR} = X_{BR}$ only when $V_{GR} = V_{BR}$, that, is when $V_W = 0$.

For an aircraft flying in the presence of wind at a velocity different from V_{BR},

$$V_{GR} = V_{BR}(v \pm \omega_f) \qquad (3.50)$$

and from Eqs. (3.29) and (3.46),

$$X_{GR}|_{V,C_L} = \frac{2\sqrt{3}E_m V_{BR}}{SFC_T}(v \pm \omega_f)\frac{v^2}{3v^4+1}\, \ell_n\,(MR) \qquad (3.51)$$

Then, differentiating the ground range with respect to the relative airspeed will yield the best relative airspeed for maximum range in the presence of winds

$$\left.\frac{\partial X_{GR_{V,C_L}}}{\partial v}\right|_{BGR} = 0 \qquad (3.52)$$

$$\left[3v^5 \pm 6v^4\omega_f - 3v \mp 2\omega_f\right]_{BGR} = 0$$

which gives

$$v_{BGR} = \left(\frac{v_{BGR} \pm 2/3\omega_f}{v_{BGR} \pm 2\omega_f}\right)^{\frac{1}{4}} \qquad (3.53)$$

The solution to this last equation can be found by successive iterations. It defines the optimal value of relative flight velocity that will maximize the ground range for given values of the wind fraction. If the wind fraction is zero, then v_{BGR}

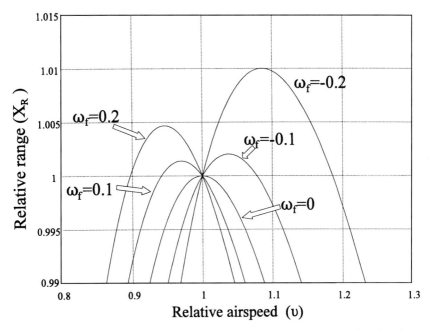

Fig. 3.28   Relative best range as a function of relative airspeed and wind fraction.

becomes unity and the ground range is maximized while flying at V_{BR}. If ω_f is not zero, then υ_{BGR} is the ratio of the best range velocity with winds V_{BGR} to the best range velocity in still air V_{BR}. This ratio (υ_{BGR}) is sometimes called the relative best range velocity. Defining a relative best range (X_R) with winds as the ratio of the best ground range corrected for winds [Eq. (3.51)] to the best ground range uncorrected for winds [Eq. (3.49)], the following equation is obtained:

$$X_R = \left(\frac{4\upsilon^2}{3\upsilon^4 + 1} \right) \left(\frac{\upsilon \pm \omega_f}{1 \pm \omega_f} \right) \tag{3.54}$$

The best ground range can then be obtained by multiplying Eqs. (3.54) and (3.49):

$$X_{BGR} = [X_R]_{\upsilon_{BGR}} [X_{GR}]_{\upsilon=1} \tag{3.55}$$

Note the scale of the relative range in Fig. 3.28. It is seen that the effect of correcting the aircraft's velocity for the presence of winds on the relative range is relatively small. Perhaps a better and clearer way of showing the effects of wind on ground range is to plot the relative range as a function of wind fraction (Fig. 3.29).

This shows that, although the fact of altering the aircraft velocity (deviating from the still air best range airspeed) to compensate for the wind does improve the aircraft's range, its influence on the relative best range is quite small (in this case, 1% increase for a wind fraction equal to −0.2). Again, this is because deviating from the best range (still air) velocity reduces the still air specific instantaneous range of a jet aircraft.

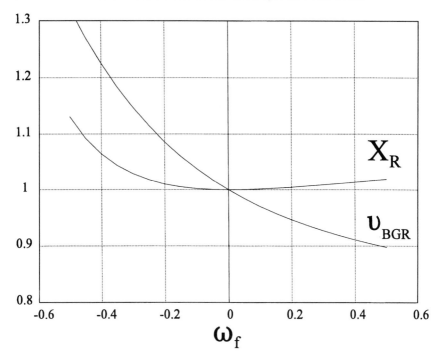

Fig. 3.29   Relative best range X_R and relative best-range velocity υ_{BGR} as a function of wind fraction ω_f.

An alternate way to find the velocity for best ground range as a function of wind fraction is to use the thrust-required curve and the previously stated condition that the best ground range flight conditions are those resulting when the ratio of the drag over the ground velocity is minimum. This is illustrated in Fig. 3.30. By displacing the origin (to the right for a headwind) by an amount equal to the wind velocity, the new horizontal scale will now reflect the aircraft's ground speed. The point of tangency of a line drawn from the new origin to the drag curve now defines the conditions for which D/V_{GR} is minimum, thus the velocity for best ground range and the corresponding thrust required.

From this graph, it is obvious that the lower velocity limit in extremely strong tailwinds is the minimum drag velocity (which, as will be seen in Sec. 3.5.1, provides the longest time in the air, thus using the effect of wind to its maximum). On the other hand, the upper limit, in very strong headwinds, may be restricted by the maximum available thrust or by Mach drag rise.

3.4.8   General Range Equations (Piston-Prop Aircraft)

As mentioned before, the range of an aircraft is the distance it can travel on a given load of fuel and for a given set of flight parameters. Assuming that the decrease in aircraft fuel weight with time is directly proportional to the power P developed by the aircraft's engine(s), the following set of equations for piston-prop

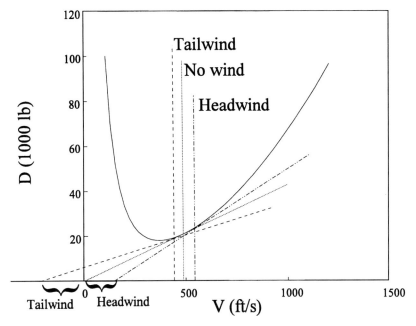

Fig. 3.30   Velocity for best ground range with wind.

aircraft are obtained:

$$\frac{\partial X}{\partial t} = V$$

$$\frac{\partial W}{\partial t} = -SFC_P P \qquad (3.56)$$

$$-\frac{\partial X}{\partial W} = \frac{V}{SFC_P P}$$

where SFC_P is the PSFC. The power required to maintain steady level flight conditions is equal to the thrust required times the velocity of the aircraft, which for a piston-prop aircraft is equal to the power available at the propeller shaft times the efficiency of the propeller

$$P_r = T_r V = \eta_p P \qquad (3.57)$$

Thus, the specific instantaneous range of a piston-prop aircraft takes the following form:

$$-\frac{\partial X}{\partial W} = \frac{V \eta_p}{SFC_P P_r} = \frac{\eta_p}{SFC_P T_r} = \frac{\eta_p E}{SFC_P W} \qquad (3.58)$$

If it is still assumed that the propeller efficiency is independent of aircraft velocity $[\eta_p \neq f(V)]$ and that the PSFC is independent of aircraft velocity and altitude $[SFC_P \neq f(V, h)]$, then it can readily be seen that, unlike the specific instantaneous range of the turbojet aircraft, the specific instantaneous range of the piston-prop aircraft is independent of its velocity. This range equation can be minimized by using a propeller with high efficiency (η_p) and engines with low PSFC, and by flying at minimum drag (at E_m) conditions.

Once again, three alternative flight conditions for the cruise leg will be considered: 1) cruise at constant velocity and altitude, 2) cruise at constant velocity and lift coefficient, and 3) cruise at constant altitude and lift coefficient.

It can readily be seen that because specific instantaneous range is maximized while flying at maximum lift-to-drag conditions (constant lift coefficient for E_m) the three conditions reduce to two, which are 1) cruise at constant lift coefficient and velocity, and 2) cruise at constant lift coefficient and altitude. Furthermore, both flight conditions yield the same equation once integrated:

$$X_{\mathrm{BR}} = \frac{-\eta_p E_m}{SFC_P} \int_{W_1}^{W_2} \frac{\partial W}{W}$$

$$X_{\mathrm{BR}} = \frac{\eta_p E_m}{SFC_P} \ell_n \left(\frac{W_1}{W_2} \right) = \frac{\eta_p E_m}{SFC_P} \ell_n \left(\frac{1}{1 - \zeta} \right)$$

(3.59)

The velocity for best range will be equal to the velocity for minimum drag conditions,

$$V_{\mathrm{BR}} = V_{E_m} \qquad (3.60)$$

Therefore, there is no increase in range, for a piston-prop aircraft, when flying at a higher altitude, but at higher altitudes the velocity for best range will be higher [see Eq. (3.8), V_{E_m}], which translates into a reduced flight time, which in itself is a significant advantage for business. One must ensure that the higher velocity will not lead to compressibility effects that could reduce the propeller efficiency.

3.4.9   Effect of Deviating from V_{BR} (Piston-Prop Aircraft)

Although the velocity is not directly a factor for the range equation of a piston-prop aircraft, it does influence the value of the aerodynamic efficiency. Let

$$v = \frac{V}{V_{\mathrm{BR}}} = \frac{V}{V_{E_m}} \qquad (3.61)$$

Using the lift equation (3.3), it can be shown that the variation in lift coefficient will be

$$\frac{V}{V_{E_m}} = \sqrt{\frac{2W}{\rho_{\mathrm{SL}} \sigma S C_L}} \Big/ \sqrt{\frac{2W}{\rho_{\mathrm{SL}} \sigma S C_{L_{E_m}}}} = \sqrt{\frac{C_{L_{E_m}}}{C_L}} \rightarrow C_L = \frac{C_{L_{E_m}}}{v^2} \qquad (3.62)$$

which leads to the following aerodynamic efficiency variation:

$$E = \frac{C_L}{C_D} = \frac{\left(C_{L_{E_m}} / v^2 \right)}{C_{D_0} + K \left(C_{L_{E_m}} / v^2 \right)^2} = \frac{v^2 \sqrt{C_{D_0}/K}}{v^4 C_{D_0} + K \left(C_{D_0}/K \right)} = \frac{2v^2 E_m}{(v^4 + 1)}$$

(3.63)

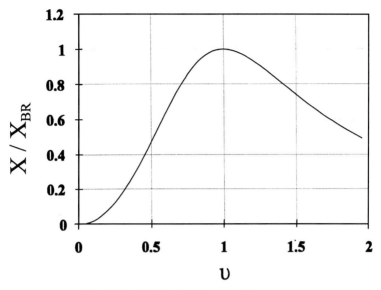

Fig. 3.31   Relative best range piston-prop.

Now, combining this last equation with Eq. (3.59) gives the variation of range as a function of the variation of relative velocity υ

$$\frac{X}{X_{BR}} = \frac{E}{E_m} = \frac{2\upsilon^2}{(\upsilon^4 + 1)} \tag{3.64}$$

The resulting graph (Fig. 3.31) has the same general shape as the one for jet aircraft (Fig. 3.27). Again, there will not be marked effects on the range if the percentage change of velocity, from the best range velocity, is small.

3.4.10   Last Words on Aircraft Range

Range is a very important factor in the design of most aircraft. A long range is required for a bomber to strike deep behind enemy lines. Long range is essential for airliners to cross large stretches of ocean such as the Pacific (with its expanding market). Aircraft range can be increased through the use of in-flight refueling, but other factors such as the number of crew onboard may then limit the aircraft's possible range/time in the air.

In 1962, a Boeing B-52 flew nonstop from Okinawa to Madrid, a distance of 10,872 n mile (12,519 mile). Three B-52s made the first nonstop around the world tour with the help of in-flight refueling in 1957, a total distance of 24,325 n mile in 45 h and 20 min. In 1986, the specially designed scaled composites Voyager performed an around-the-world, nonstop, unrefueled flight in 9 days, 3 min, 44 s covering a distance of 23,168 n mile (still air range of 23,703.7 n mile) at an average speed of 106.04 kn. This particular aircraft had two piston engines in a pusher-tractor configuration to provide the necessary takeoff and climb power (in the record flight configuration it required 14,200 ft to become airborne); then one engine was shut down in flight and its propeller feathered for maximum fuel economy. The Voyager had a wing with an AR of 33.8 and carried 7,011.5 lb of fuel. The Concorde holds the world speed record for an around-the-world flight, without in-flight refueling, which was performed in 32 h and 49 min (including the many stops for refueling) in 1992 to commemorate the 500th anniversary of the discovery of the Americas by Christopher Columbus. More recently (1993), an Airbus A340 set a new world record in its class with its one-stop around the world trip from Paris to Auckland (10,307 n mile) to Paris (10,392 n mile) in 48 h, 22 min, and 6 s including a 5-h stop in Auckland. A Boeing 747-400 also demonstrated a 9,714 n mile nonstop Heathrow to Sydney (Australia) delivery flight in 1989. Finally, two B-1B bombers demonstrated nonstop (with air-to-air refueling) around the world flight in 36 h and 13 min, which is a world record for nonstop flight. This flight took place in 1995.

Before the mid-1980s, it was essential for any commercial airliner crossing the Atlantic or the Pacific to be equipped with more than two engines. This largely came from the belief that if one engine stopped or malfunctioned in-flight and the pilot inadvertently shut down the wrong one, there would still be at least one good operating engine. The concept of extended-range twin-engine operations (ETOPS) appeared with the introduction of the long-range twin-engine Boeing 767. To obtain its clearance, the aircraft had to demonstrate a high reliability of operation of its engines and equipment through in-service monitoring. Afterwards, flight routes were restricted to those where, at any one point, the aircraft was never more than 120-min single-engine flying time away from a suitable airport so that the airline pilots could gain experience. Finally, the 180-min single-engine flying time was allowed, which, in practice, covers all routes over the entire globe. Since then, other twin-engined aircraft have joined the 767 on such routes, namely, the Boeing 757 (up to 120 min), the Airbus A310 (up to 120 min), the Boeing 737 (up to 120 min) and the Airbus A330. The Pratt & Whitney equipped Boeing 777 was cleared for ETOPS on its introduction to service.

3.5 Endurance

The endurance of an aircraft is defined as the length of time an aircraft can fly on a given quantity of fuel under a given set of flight conditions. Loiter is a mission leg where endurance is most important and range is either secondary or not important. Loiter is important for such missions as combat air patrol (CAP) where a fighter may have to stay aloft for a long period of time to patrol a specific area, or for antisubmarine and surveillance patrols (airborne warning and control systems, E-8C Joint STARS, etc.).

3.5.1 Maximum Endurance (Jet Aircraft)

Maximum endurance will be achieved by flying at a set of conditions where the instantaneous rate of fuel consumption is minimum or the inverse of the fuel consumption is maximum

$$-\frac{\partial t}{\partial W} = \frac{1}{SFC_T D} = \frac{E}{SFC_T W} \qquad (3.65)$$

This equation indicates that flying at conditions for minimum drag will maximize endurance [for $SFC_T \neq f(V, h)$]. The minimum drag conditions occur at maximum lift-to-drag ratio; thus, at a constant lift coefficient

$$E_{t_{\max}} = E_m \qquad V_{t_{\max}} = V_{E_m} \qquad (3.66)$$

Maximum endurance can easily be calculated by integrating Eq. (3.65) while maintaining E at its maximum value throughout the flight

$$t_{\max} = \frac{E_m}{SFC_T} \ell_n (\text{MR}) = \frac{E_m}{SFC_T} \ell_n\left(\frac{1}{1 - \zeta}\right) \qquad (3.67)$$

This equation shows that for a given fuel weight fraction the higher the aerodynamic efficiency and the lower the SFC, the longer the endurance will be. From Fig. 3.5 it can be seen that as the weight decreases so must the velocity (for constant altitude flights) if the conditions for maximum endurance are to be maintained. The endurance will not be affected by the wind speed because distance is not important here, only the time in the air. Nor will it be affected by the altitude, although as the altitude increases so will the velocity for minimum drag; consequently, the distance traveled will be greater at higher altitudes.

Taking aircraft A as an example, the velocity for maximum endurance will vary from 219KCAS at its maximum takeoff weight to 170 KCAS at its operational empty weight, two extremes that would not normally be encountered during the leg of the mission where endurance counts most.

If the maximum endurance of the aircraft is compared to the time to fly its best range for a given quantity of fuel,

$$\frac{t_{\max}}{t_{X_{BR}}} = \frac{E_m}{E_{BR}} = \frac{1}{\sqrt{3/4}} \approx 1.155 \qquad (3.68)$$

The aircraft, therefore, will stay in the air approximately 15.5% longer if it flies at its maximum endurance velocity rather than at its best range velocity. The distance traveled in both cases can also be compared,

$$\frac{X_{t_{\max}}}{X_{BR}} = \frac{E_m V_{E_m}}{E_{BR} V_{BR}} = \frac{1}{\sqrt{3/4}\sqrt[4]{3}} \approx 0.877 \qquad (3.69)$$

Thus, the aircraft will fly a distance equal to approximately 87.7% of its best range while flying at its maximum endurance velocity even though it is remaining in the air 15.5% longer because it is flying at a velocity some 26% slower [Eq. (3.38)].

Equation (3.67) was derived using a flight at constant lift coefficient and constant velocity. For a flight at constant altitude and constant velocity, the maximum endurance equation takes the following form:

$$t_{\max} = \frac{2E_m}{SFC_T} \arctan\left(\frac{0.5\zeta}{1 - 0.5\zeta}\right) \qquad (3.70)$$

3.5.2   Effect of Deviating from $V_{t,\,max}$ on Endurance (Jet Aircraft)

Just as it was done in the section on aircraft range, it can be useful to determine how fast the aircraft endurance performance is degraded while moving away from the velocity for maximum endurance. Because the maximum endurance velocity for a jet aircraft corresponds to the velocity of minimum drag (V_{E_m}), the analysis of Sec. 3.4.1 can be used to obtain the following equation:

$$\frac{t}{t_{\max}} = \frac{2v^2}{v^4 + 1} \tag{3.71}$$

where v is the ratio of the velocity flown to the maximum endurance velocity. This equation is plotted in Fig. 3.31. Thus, small changes in velocity will not affect the endurance by much (a 5% change in velocity will correspond to less than 1% change in endurance).

3.5.3   Maximum Endurance (Piston-Prop Aircraft)

The endurance for a piston-prop aircraft is maximized while flying at the conditions for minimum power required as indicated,

$$-\frac{\partial t}{\partial W} = \frac{1}{SFC_P P} = \frac{\eta_p}{SFC_P P_r}$$

$$-\frac{\partial t}{\partial W} = \frac{\eta_p E}{SFC_P V W}$$

If η_p and the SFC are assumed constant, then

$$t = -\frac{\eta_p}{SFC_P} \int_{W_1}^{W_2} \frac{E}{V W} \, \partial W = -\frac{\eta_p}{SFC_P} \int_{W_1}^{W_2} \frac{\partial W}{P_r} \tag{3.72}$$

Remember that the minimum power required occurs when the ratio ($C_L^{3/2}/C_D$) is maximum [see Eqs. (3.13–3.15)]. This maximum occurs when the induced drag coefficient ($K C_L^2$) is equal to three times the minimum drag (C_{D_0}). Thus, the maximum endurance for a piston-prop aircraft is

$$t_{\max} = \frac{\eta_p E_{\mathrm{MP}}}{SFC_P V_{\mathrm{MP}}} \, \ell_n \left(\frac{1}{1-\zeta} \right) \tag{3.73}$$

where

$$V_{t_{\max}} = V_{\mathrm{MP}} \qquad E_{t_{\max}} = E_{\mathrm{MP}}$$
$$C_{L_{t_{\max}}} = C_{L_{\mathrm{MP}}} \qquad C_{D_{t_{\max}}} = C_{D_{\mathrm{MP}}} \tag{3.74}$$

Because V_{MP} increases with altitude the endurance of a piston-prop aircraft will be maximized while flying at low altitudes. But minimum power-required conditions correspond to a velocity that is below the velocity for minimum drag. Thus, a piston-prop aircraft must fly on the back side of the drag curve to achieve maximum endurance. This is labor intensive for the pilot, especially if the pilot has to maintain that flight condition for extended periods of time (long endurance). Also, while flying at minimum power-required conditions, the stall margin is low, and turns are even more critical. Any disturbances could stall the aircraft and, if it is flying low to increase its endurance, could cause the aircraft to crash.

3.5.4 Effect of Deviating from $V_{t,\,max}$ on Endurance (Piston-Prop)

To improve the stall margin of the aircraft, it may be desirable to increase its velocity from minimum power-required velocity to minimum drag velocity. But this increase in velocity will affect the endurance. Let

$$v = \frac{V}{V_{t_{max}}} = \frac{V}{V_{MP}} = \sqrt{\frac{C_{L_{MP}}}{C_L}} \rightarrow C_L = \frac{C_{L_{MP}}}{v^2} \tag{3.75}$$

where

$$C_{L_{MP}} = \sqrt{3}C_{L_{Em}} \qquad C_{D_{MP}} = 4C_{D_0}$$

$$E_{MP} = \sqrt{3/4}E_m \qquad V_{MP} = \frac{V_{E_m}}{\sqrt[4]{3}} \tag{3.76}$$

then

$$E = \frac{C_{L_{MP}}/v^2}{C_{D_0} + K\left(C_{L_{MP}}^2/v^4\right)} = \frac{v^2 C_{L_{MP}}}{v^4 C_{D_0} + K\left(3C_{D_0}/K\right)} = \frac{v^2 C_{L_{MP}}}{C_{D_0}(v^4 + 3)}$$

$$E = \frac{v^2\sqrt{3}}{\sqrt{C_{D_0}K}(v^4 + 3)} = \frac{2\sqrt{3}v^2 E_m}{(v^4 + 3)} \tag{3.77}$$

$$\frac{E}{E_{MP}} = \frac{2E}{\sqrt{3}E_m} = \frac{4v^2}{(v^4 + 3)}$$

which finally leads to

$$\frac{t}{t_{max}} = \frac{E/V}{E_{MP}/V_{MP}} = \frac{E}{E_{MP}}\frac{1}{v} = \frac{4v}{(v^4 + 3)} \tag{3.78}$$

Thus, as the velocity is increased to the minimum drag velocity (V_{E_m}, $v = 1.316$) for safety, the endurance will be reduced to 87.7% of its maximum value. Reducing the velocity, from V_{E_m}, to a value somewhere between V_{MP} and V_{E_m} will be a compromise between stall margin and long endurance.

Problems

3.1. Aircraft A is flying at Mach 0.8 at 36,000 ft. In preparation for descent, the pilot retards the throttle to a point where the engines provide 35% of maximum available thrust at altitude; at what velocity will the aircraft settle ($W = 220,000$ lb)?

3.2. What is the thrust required for aircraft C at combat weight (16,000 lb) to fly at its stall velocity and 15,000 ft (Assume $\alpha_T = 0$ deg)?

3.3. Aircraft A, at 80% of maximum takeoff weight, is flying at Mach 0.8 at 39,000-ft altitude. The pilot wishes to slow down to a velocity where the thrust required will be minimum; what is that velocity and what is the thrust required at that velocity?

3.4. Aircraft A, at 90% of maximum takeoff weight, is flying at 29,000 ft. It must travel a given distance while respecting air traffic control regulations. What will be the maximum distance traveled (at the best range conditions) while using 50,000 lb of fuel? What will be that distance if the aircraft is flying at Mach 0.8 at the same altitude?

3.5. If aircraft A is flying at the same initial conditions as in problem 4 (including V_{BR}) and initiates a cruise climb, what distance will it travel using the same amount of fuel? What will be the final altitude and average climb angle? Will the aircraft fly at velocities above M_{DR} if V_{BR} conditions are maintained?

3.6. With the same initial conditions as in problem 3.4, what will be the ground range if the aircraft encounters headwinds of 50 kn (use V_{BR} uncorrected for winds and the cruise-climb range equation)? What will be the relative best range velocity (v_{BR}) and the relative best range (X_R) with winds? If the pilot were to adjust the flight speed to compensate for the presence of wind to achieve the best ground range, then what should be the flight speed and the resulting ground range?

3.7. Aircraft C must do a CAP mission over a given region. The CAP altitude is 25,000 ft (constant). What is the maximum time on station (in the CAP region) if the aircraft weight at the beginning of the patrol is 17,000 lb and the aircraft can use 3,000 lb of fuel before returning to its base? What will be the patrol velocity? If the altitude is reduced to 20,000 ft, what is the maximum time on station and the patrol velocity?

3.8. A turbojet equipped aircraft will cruise at a constant Mach number of 0.7 and a constant lift coefficient. Determine the altitude for longest range. The aircraft weighs 10,000 lb and has a wing area of 100 ft² and a drag polar defined by $C_D = 0.018 + 0.1C_L^2$.

3.9. A search and rescue aircraft loiters at 5,000 ft. After using 1,000 lb of fuel, the pilot spots something in the forest below and decides to go down to 1,000 ft to check it out. Once at 1,000 ft, another 1,000 lb of fuel is used before the aircraft has to return to its base. What is the ratio of the time spent at 5,000 ft to that spent at 1,000 ft? Use aircraft B and the following initial conditions: flight at the minimum drag velocity at 5,000 ft, initial weight equal to 90% of maximum, and fuel weight equal to 95% of maximum. Assume zero fuel burn for the descent and use the same velocity for the flight at 1,000 ft ($V_{1000ft} = V_{5000ft}$).

3.10. Aircraft A, $W = 250,000$ lb, is flying at 36,000 ft and Mach 0.8 when it loses the thrust from one engine (due to failure). The equivalent increase in minimum drag is $\Delta C_{D_0} = 0.01$ due to flight controls deflection and engine windmilling. Can the aircraft maintain its altitude and speed with only one engine operating? If not, at what altitude must the aircraft descend to maintain a flight speed of Mach 0.8? Assuming there is still 5,000 lb of fuel available for cruise what distance can the aircraft cover (use Breguet range equation)?

3.11. The absolute ceiling of a turbojet equipped aircraft is 38,000 ft while its ceiling airspeed is 600 ft/s. Determine its best still air range at an altitude of 22,000 ft. Determine the ratio of the fuel consumptions at those two conditions, that is, $\dot{m}_{f_{\text{ceiling}}}/\dot{m}_{f_{V_{\text{BR,22,000ft}}}}$.

3.12. The model of an aircraft has been tested in a wind tunnel and the following graph shows the value of the lift-to-drag ratio produced by the model at various airspeeds. The conditions in the wind-tunnel test section during the tests were identical to those at 2,000 ft in a standard atmosphere. The model's wing loading was 20 lb/ft².

a) Determine the value of the drag polar coefficient C_{D_0} and K for the model.

b) Assuming a value of 0.8 for the Oswald's coefficient (e), determine the AR of the model.

c) If a prototype aircraft has the same aerodynamic characteristics as the model and its engine's specific fuel consumption is 0.6 lb$_{\text{fuel}}$/h/lbf, determine the best range of the prototype in still air for a cruise beginning at 15,000 ft for a MR value of 0.2.

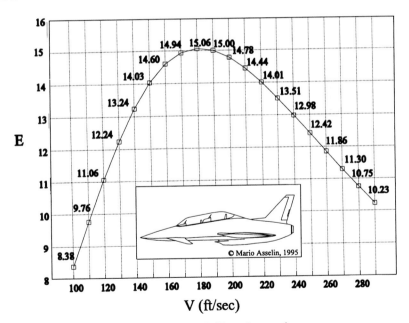

Fig. P3.12   Model lift-to-drag ratio.

4
Flight in the Vertical Plane

4.1 Introduction

TO alter its altitude, an aircraft has to climb to gain altitude or dive to lose altitude. This new flight path will now be at an angle γ with the horizontal ($\gamma > 0$ for a climb). The lift and drag vectors will still be perpendicular and parallel, respectively, to the flight path, but the weight vector will remain vertical, thus not directly opposed to the lift force. Part of the weight will either act in the same direction as the drag (during a climb) or the thrust depending on the angle γ. From the force diagram (Fig. 4.1) of an aircraft in steady-state climb, the following equations apply:

$$T - D - W \sin(\gamma) = 0 \qquad (4.1)$$

$$L - W \cos(\gamma) = 0 \qquad (4.2)$$

$$\frac{\mathrm{d}X}{\mathrm{d}t} = V \cos(\gamma) \qquad (4.3)$$

$$\frac{\mathrm{d}h}{\mathrm{d}t} = V \sin(\gamma) \qquad (4.4)$$

where $\mathrm{d}h/\mathrm{d}t$ is the vertical velocity component (the rate of climb RC or rate of descent RD) of the aircraft. Therefore, using Eq. (4.1),

$$\sin(\gamma) = (T - D)/W \qquad (4.5)$$

or

$$\gamma = \arcsin[(T - D)/W] \qquad (4.6)$$

Equations (4.5) and (4.6) show that the climb angle is a function of the specific excess thrust of the aircraft, which is the thrust above that required to counteract the drag. A qualitative feel for the magnitude of the climb angle can be obtained by setting the drag to zero. Evidently, the drag will never be equal to zero, but it shows that the steady-state climb angle cannot exceed that which could be produced with the maximum thrust-to-weight ratio and zero drag.

On the other hand, the descent angle is a function of specific thrust deficit. Any steady-state descent angle could be maintained as long as the maximum design speed is not exceeded. Aircraft A, which is typical of the transport aircraft category, has a sea-level maximum thrust-to-weight ratio of 0.33 on takeoff (for $W_{TO} = 300,000$ lb) resulting in a steady-state climb angle not exceeding 19.5 deg. On the other hand, aircraft C, a typical fighter aircraft, has a maximum thrust-to-weight ratio of 1.125 at typical combat weight (16,000 lb). Although a value of $\sin(\gamma)$ greater than unity is mathematically impossible, a T/W ratio of 1.125

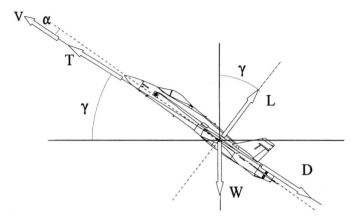

Fig. 4.1 Aircraft in climbing flight.

indicates that the aircraft would be able to enter an accelerating climb at any angle up to the vertical, if the drag was negligible. By combining Eq. (4.5) with Eq. (4.4), the rate of change of altitude with time is

$$\frac{dh}{dt} = \frac{TV - DV}{W} = \frac{P_a - P_r}{W} \tag{4.7}$$

It is evident here that the rate of climb of an aircraft is a function of its specific excess power, whereas the rate of descent is a function of the aircraft specific power deficit. For the sake of standardizing the discussion on climb and descent, let an increase in altitude with time (dh/dt positive) correspond to a positive rate of climb RC and a negative rate of descent RD, i.e., $RD = -RC$.

From Eq. (4.2), the drag equation during climb, for an aircraft with a parabolic drag polar, is

$$D = qSC_{D_0} + \frac{KW^2\cos^2(\gamma)}{qS} \tag{4.8}$$

The larger the climb angle, the larger the component of thrust acting opposite the aircraft's weight will be and the smaller the lift force needs to be. Therefore, this will reduce the lift-dependant drag component. Ultimately, at a climb (or descent) angle of 90 deg, the induced drag will equal zero and only the zero-lift drag will remain. Obviously, an aircraft would require a thrust-to-weight ratio greater than unity to sustain a steady 90-deg climb angle.

4.2 Maximum Climb Angle

The steepest climb is defined as a climb at maximum climb angle or maximum altitude gain for a given horizontal distance traveled. By studying Eq. (4.5), it can be seen that to maximize the climb angle, one must maximize the thrust-to-weight ratio and minimize drag. Rewriting Eq. (4.5):

$$\sin(\gamma) = \frac{T}{W} - \frac{\cos(\gamma)}{E} \tag{4.9}$$

Therefore, flying at the maximum lift-to-drag ratio, that is, at the minimum drag condition, will result in the maximum climb angle. For climb angles below approximately 15 deg where $\cos(\gamma) \approx 1$, the equations for the maximum climb angle and corresponding rate of climb, for a given throttle setting, become

$$\sin(\gamma_{max}) \simeq \frac{T}{W} - \frac{1}{E_m}$$

$$RC_{\gamma_{max}} \simeq V_{E_m} \sin(\gamma_{max})$$

(4.10)

The absolute maximum climb angle at a given altitude will then result when the maximum thrust is used.

4.3 Maximum Rate of Climb

Another climb mode, which is particularly important for fighter and interceptor aircraft, is the fastest climb, that is, the climb that would produce the maximum rate of climb. It results in the greatest altitude gain over a given period of time, which occurs when the aircraft's vertical velocity component is largest. The velocity for fastest climb is determined by differentiating the rate-of-climb equation (4.7) with respect to velocity and equating the result to zero. It should be noted from Eq. (4.7) that the rate of climb will be maximum when the excess power $(TV - DV)$ is maximum

$$\left[\frac{\partial RC}{\partial V}\right]_{FC} = \frac{\partial}{\partial V}\left[\frac{TV - DV}{W}\right]_{FC} = \left[T - D - V\frac{\partial D}{\partial V}\right]_{FC} = 0 \qquad (4.11)$$

From this equation the dynamic pressure for maximum rate of climb (subscript FC for fastest climb) can be determined:

$$T - \frac{1}{2}\rho_{SL}\sigma V^2 SC_D - V\frac{1}{2}\rho_{SL}\sigma S\frac{\partial}{\partial V}(VC_D) = 0$$

$$T - qSC_{D_0} - \frac{KW^2}{qS} - V\left[\frac{1}{2}\rho_{SL}\sigma 2V SC_{D_0} - \frac{2KW^2}{\frac{1}{2}\rho_{SL}\sigma V^3 S}\right] = 0$$

$$T - qSC_{D_0} - \frac{KW^2}{qS} - 2qSC_{D_0} + \frac{2KW^2}{qS} = 0$$

Therefore,

$$3C_{D_0}q_{FC}^2 - (T/S)q_{FC} - K(W/S)^2 = 0 \qquad (4.12)$$

$$q_{FC} = \frac{(T/S)}{6C_{D_0}}\left[1 \pm \sqrt{1 + \frac{12C_{D_0}K}{(T/W)^2}}\right] \qquad (4.13)$$

The minus sign does not need to be considered because the dynamic pressure cannot be negative. Let us define a climb coefficient Γ as follows:

$$\Gamma = 1 + \sqrt{1 + \frac{12C_{D_0}K}{(T/W)^2}} = 1 + \sqrt{1 + \frac{3}{[E_m(T/W)]^2}} \qquad (4.14)$$

where the values of Γ will be between 2 (for high values of $[E_m(T/W)]$, i.e., maximum throttle setting at sea level) and 3 (at the aircraft's absolute ceiling). With this definition, the dynamic pressure for the maximum rate of climb becomes

$$q_{\mathrm{FC}} = \frac{(T/S)\Gamma}{6C_{D_0}} \tag{4.15}$$

and the velocity for fastest climb is

$$V_{\mathrm{FC}} = \sqrt{\frac{(T/S)\Gamma}{3\rho_{\mathrm{SL}}\sigma C_{D_0}}} \tag{4.16}$$

The angle of climb for fastest climb is then given by

$$\sin(\gamma_{\mathrm{FC}}) = \frac{T}{W}\left(1 - \frac{\Gamma}{6}\right) - \frac{3}{2\Gamma E_m^2(T/W)} \tag{4.17}$$

and

$$RC_{\max} = V_{\mathrm{FC}}\sin(\gamma_{\mathrm{FC}}) \tag{4.18}$$

The conditions for maximum rate of climb and maximum climb angle can be illustrated by using typical aircraft A at 80% maximum takeoff weight (240,000 lb) and maximum throttle setting (Table 4.1).

4.4   Velocity Hodograph

Of course, the steady-state equations for fastest climb and steepest climb assume that the climb angle will be relatively small (approximately <15 deg), which is true for most aircraft with the exception of fighter aircraft. These high-performance aircraft can perform very steep steady-state climbs (sometimes as high as 90 deg). In the analysis of climb performance, it can be useful to produce a plot of the vertical velocity against the horizontal velocity for a given altitude. This plot is called a velocity hodograph, and an example of such a graph is given in Fig. 4.2 for aircraft A.

Table 4.1   Illustration for conditions for maximum rate of climb and maximum climb angle

	V, ft/s	γ, deg	RC, ft/s
	Maximum climb angle		
Sea level	331.2	20.9	118.0
25,000 ft	495.0	10.2	87.7
	Maximum rate of climb		
Sea level	716.7	15.6	192.9
25,000 ft	820.7	8.3	117.9

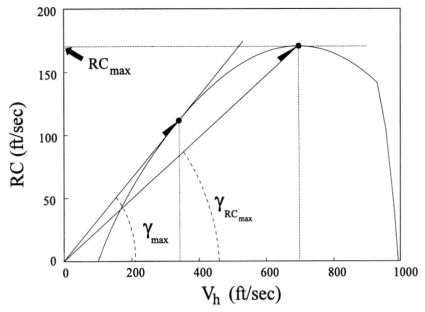

Fig. 4.2   Hodograph of aircraft A at sea level, maximum thrust condition.

The steepest climb parameters can be found by drawing a line from the origin tangent to the hodograph line. The point of tangency identifies the vertical and horizontal velocities for that condition. The angle of climb is equal to inverse tangent of the vertical velocity over the horizontal velocity, and the airspeed is the vectorial sum of the two. The maximum rate-of-climb conditions are obtained in a similar way at the point along the hodograph line where the rate of climb is maximum.

Comparing a velocity hodograph and a graph of rate of climb (Fig. 4.3) as a function of aircraft velocity for a given set of parameters, the differences in the curves and the information that each type of graph can provide is readily observed. The first one provides the horizontal and vertical velocity components (here providing the maximum horizontal velocity possible) and the corresponding angle of climb/dive. As well, the extension of the curve all the way to a vertical dive, when the horizontal velocity component is zero, provides the maximum vertical velocity, which is often referred to as the terminal dive velocity.

The second graph, on the other hand, provides the actual velocity the aircraft is flying and the associated rate of climb. Note that the velocity for zero climb on both graphs is the same because both now refer to the level flight velocity.

The time to climb from one altitude to another is found by integrating the rate of climb from the initial to the final altitudes. The case of a climb at the maximum rate of climb is illustrated in Fig. 4.4. For cases where the climb is performed at some other velocity, a similar graph would need to be produced for the actual flight conditions. Note that the time to climb from one altitude to another is represented

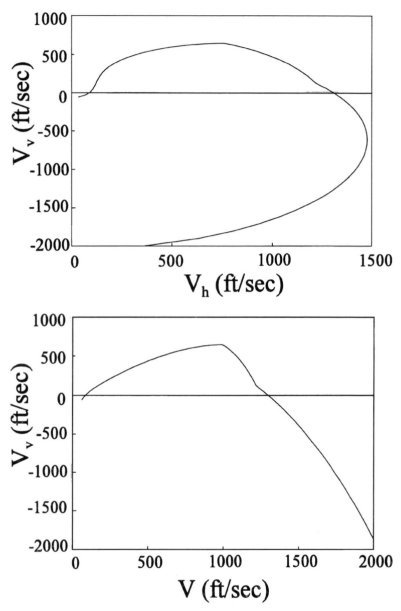

Fig. 4.3   Comparison of velocity hodograph and graph of rate of climb as a function of aircraft velocity.

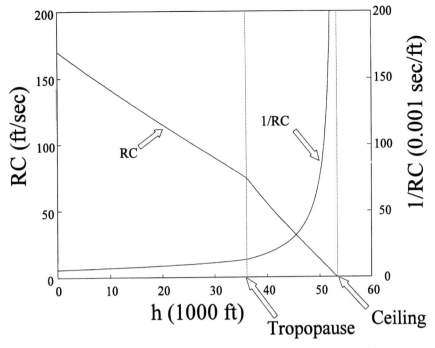

Fig. 4.4   Maximum rate of climb for aircraft A at a weight of 270,000 lb.

by the area under the curve $1/RC$ in accordance with Eq. (4.19),

$$t = \int_{h1}^{h2} \frac{dh}{RC} \tag{4.19}$$

For a given aircraft weight and throttle setting, the available thrust-to-weight ratio will decrease as the altitude is increased. This translates into reduced climb angle and rate of climb (as shown in Fig. 4.4). When both values reach zero, the aircraft is said to be flying at its absolute ceiling altitude for the given weight and throttle setting. In theory, it would take an infinite time for an aircraft to reach its absolute ceiling in quasisteady-state climb because the rate of climb approaches zero as the ceiling altitude is neared and $1/RC$ will tend to infinity (Fig. 4.4). An aircraft operational ceiling corresponds to the altitude where the maximum rate of climb (for a given weight, maximum throttle setting) is 100 ft/min. The combat ceiling is the altitude where the maximum rate of climb is 500 ft/min.

In many instances, the rate of climb vs altitude curve can be approximated by a linear function or broken into segments over which linear functions may be used so that the integral of Eq. (4.19) can be completed quite easily. From Fig. 4.4, two straight lines could be used as an approximation for the time to climb: one line from sea level to 36,089 ft (tropopause), and one from 36,089 ft to the absolute ceiling.

4.5 Quasisteady-State Climb Analysis

One of the assumptions used to obtained the equations of this chapter is that the aircraft is in steady-state climb (i.e., no acceleration, constant velocity flight, constant thrust, etc.). However, a change in altitude will alter the thrust and drag of the aircraft, even if the velocity is maintained to a constant value, due to the reduction in air density as altitude increases. If the velocity remains constant during the climb, our steady-state assumption is correct, but the steepest climb velocity (Sec. 4.2) and maximum rate-of-climb velocity (Sec. 4.3) do change with altitude, which contradicts our steady-state assumption. To investigate the effects of this change in velocity on the steepest climb and maximum rate-of-climb conditions, one must account for this change in velocity. Rewriting Eq. (4.1),

$$T - D - W\sin(\gamma) = \frac{W}{g}a = \frac{W}{g}\frac{dV}{dt} \tag{4.20}$$

The velocity is, from Eq. (4.4), the rate of climb divided by $\sin(\gamma)$. Thus, the rate of change of velocity will be

$$\frac{dV}{dt} = \frac{dV}{dh}\frac{dh}{dt} = \frac{dV}{dt}RC = \left[\frac{1}{\sin(\gamma)}\frac{dRC}{dh} - RC\frac{\cos(\gamma)}{\sin^2(\gamma)}\frac{d\gamma}{dh}\right]RC$$

Therefore,

$$\frac{dV}{dt} = \frac{RC}{\sin(\gamma)}\frac{dRC}{dh} - \frac{RC^2}{\sin(\gamma)\tan(\gamma)}\frac{d\gamma}{dh} \tag{4.21}$$

A feel for the effects of this change in velocity can be obtained by using the data from Table 4.1 (aircraft A at 240,000 lb). For the maximum climb angle, using a linear approximation between the sea level values and the values at 25,000 ft, the following information is found:

$$\frac{dRC}{dh} = \frac{87.7 - 118.0}{25,000} = -1.212 \times 10^{-3}\ \text{s}^{-1}$$

$$\frac{d\gamma}{dh} = \frac{10.2 - 20.9}{25,000} = -486 \times 10^{-6}\frac{\text{deg}}{\text{ft}} = -7.47 \times 10^{-6}\frac{\text{rad}}{\text{ft}}$$

Using these results and Eq. (4.20), new values for the rate of climb and climb angles are found:

$$\left[\frac{dV}{dt}\right]_{\text{SL}} = 0.363\frac{\text{ft}}{\text{s}^2}$$

$$\gamma = \arcsin\left[\frac{T - D - (W/g)(dV/dt)}{W}\right] = 19.78\ \text{deg}$$

$$RC = V\sin(\gamma) = 331.2\sin(\gamma) = 114.3\ \text{ft/s}$$

After a few iterations, the stable values are

$$\left[\frac{dV}{dt}\right]_{SL} = 0.393\frac{\text{ft}}{\text{s}^2}$$

$$\gamma = \arcsin\left[\frac{T - D - (W/g)(dV/dt)}{W}\right] = 19.72 \text{ deg}$$

$$RC = V\sin(\gamma) = 331.2\sin(\gamma) = 114.0 \text{ ft/s}$$

Doing the same exercise for the maximum rate-of-climb conditions, the following values are obtained: $\gamma = 14.69$ deg and $RC = 183.7$ ft/s. Thus, the steady-state assumption slightly overestimates some climb performance because it is assumed that the aircraft has zero acceleration along its flight path. Actually, to maintain steepest climb and maximum rate-of-climb conditions, the aircraft will accelerate slightly as it climbs to maintain the conditions for steepest climb or maximum rate-of-climb velocities, Eqs. (4.10) and (4.16), respectively. Thus, the aircraft is using part of its excess thrust/power to accelerate, in the preceding case at about 0.4 ft/s^2 (≈ 0.0124 g).

4.6 Factors Influencing the Rate of Climb

The rate-of-climb equation (4.7) will now be rewritten using the lift and drag equations to identify the factors influencing it. As a first approximation, the cosine value will be set equal to one (low angle of climb assumption),

$$RC = \left[\frac{T}{W} - \frac{qC_{D_0}}{(W/S)} - \frac{K(W/S)}{q}\right]V \tag{4.22}$$

Aircraft C (fighter) at full takeoff weight (clean) will be used to produce graphs of its rate of climb under different conditions. Decreasing the thrust of the aircraft, while keeping the other parameters constant (i.e., decreasing the throttle setting), will decrease the rate of climb of the aircraft (Fig. 4.5). Decreasing the wing loading of the aircraft by considering a larger wing surface during design tradeoff studies while maintaining the other parameters constant will increase the rate of climb of the aircraft in the low-velocity region due to the decrease in induced drag. However, at speeds above the minimum drag velocity a decrease in wing loading will reduce the rate of climb due to the increase in parasite drag (Fig. 4.6).

During the course of a mission, the weight of the aircraft will decrease as fuel is burned, in this instance decreasing the wing loading and increasing the thrust-to-weight ratio. This will have a positive effect on the rate-of-climb curve over the entire velocity range. This is shown in Fig. 4.7 with aircraft C at sea level, comparing maximum takeoff weight and combat weight climb performance.

The influence of C_{D_0} and K on climb performance is clearly seen in Figs. 4.8 and 4.9. Changes in C_{D_0} will alter the rate-of-climb curve in the high-velocity region, whereas the induced drag coefficient will affect it in the low-velocity region as was noted during the level flight analysis.

Changes in altitude while flying at a given velocity and aircraft weight will result in changes in the dynamic pressure and the thrust-to-weight ratio. These effects are clearly visible in Fig. 4.10. A decrease in the dynamic pressure is

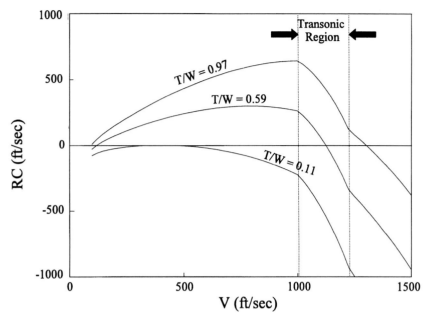

Fig. 4.5   Rate-of-climb as a function of thrust, given in terms of T/W; sea level conditions.

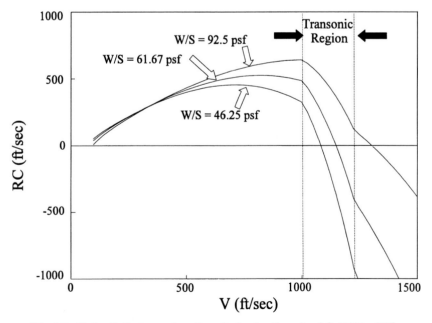

Fig. 4.6   Rate-of-climb as a function of wing loading: $h = 0$ ft, $T/W = 0.97$.

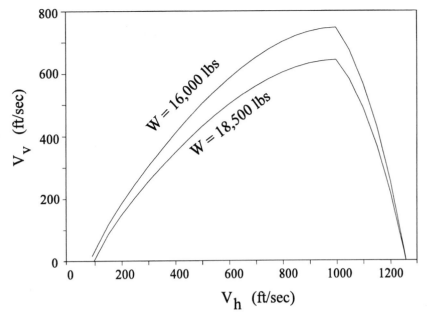

Fig. 4.7   Effect of weight on climb performance, sea level conditions.

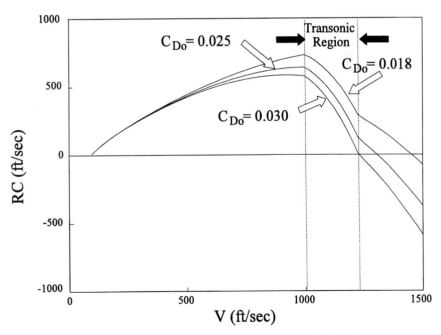

Fig. 4.8   Influence of the minimum drag coefficient on climb performance.

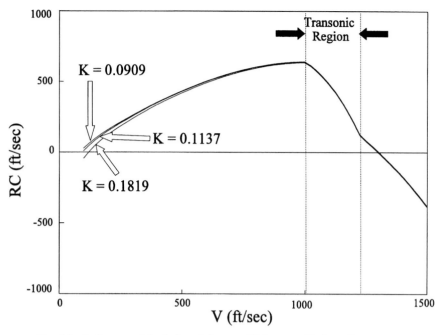

Fig. 4.9   Influence of the induced drag coefficient on climb performance.

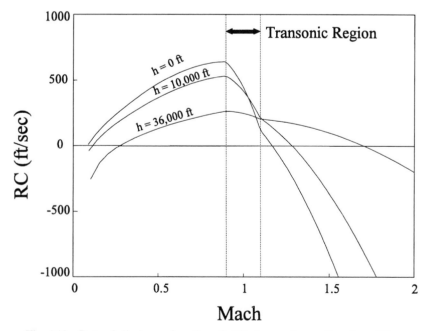

Fig. 4.10   Rate of climb as a function of altitude, maximum thrust condition.

beneficial in decreasing the magnitude of the parasite drag but the adverse effects are that the induced drag will increase, due to an increased C_L to maintain lift, and the thrust-to-weight ratio will decrease as altitude increases. This means that the rate of climb should decrease in the low-velocity region where the induced drag has the most influence, and the rate of climb should increase in the high-velocity region where the parasite drag has the most influence.

4.7   Piston-Prop Climb

The same approach as for the turbojet aircraft can be used for analyzing the climb performance of a piston-prop aircraft (Fig. 4.11). The maximum angle of climb is achieved while flying at maximum excess thrust conditions, and the maximum rate of climb is achieved while flying at maximum excess power conditions:

$$\sin(\gamma) = \frac{T - D}{W} = \frac{P_a - P_r}{VW} = \frac{\eta_p P - P_r}{VW} \tag{4.23}$$

$$RC = \frac{\mathrm{d}h}{\mathrm{d}t} = \frac{TV - DV}{W} = \frac{P_a - P_r}{W} = \frac{\eta_p P - P_r}{W} \tag{4.24}$$

For a piston-prop aircraft, the two cases occur close to or at the stall speed of the aircraft as shown in Figs. 4.12 and 4.13. Generally, the maximum thrust available occurs below the stall velocity; thus, this flight condition is not possible. The maximum climb angle for a piston-prop aircraft, therefore, occurs while the aircraft is flying at its stalling speed.

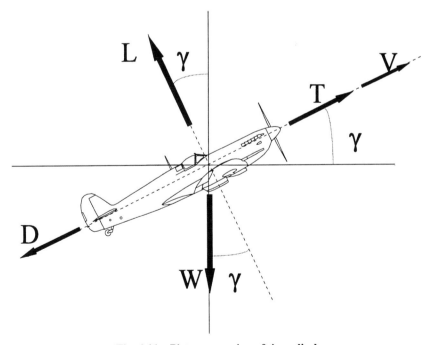

Fig. 4.11   Piston-prop aircraft in a climb.

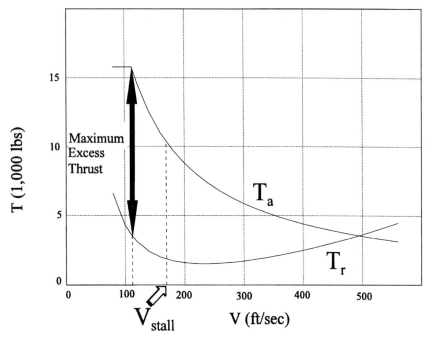

Fig. 4.12   Thrust curves for aircraft B.

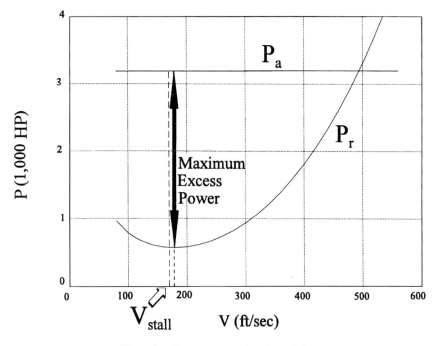

Fig. 4.13   Power curves for aircraft B.

It can readily be seen in Figs. 4.12 and 4.13, that the steady-state climb performance of piston-prop aircraft rapidly decreases as the aircraft velocity is increased. Again this is assuming constant propeller efficiency and shaft power for a given altitude. The actual shape of the power-available curve will dictate best climb conditions, but the steepest climb angle will still be at the conditions of maximum excess thrust and the maximum rate of climb at the conditions of maximum excess power.

For a turbocharged engine aircraft, for the altitudes below the critical altitude, the climb performance at a given airspeed (above the velocity for minimum drag) will increase with increasing altitude due to a reduction in parasite drag with altitude.

4.8 Gliding Flight

A special case of flight in the vertical plane is that of steady-state gliding flight or power-off flight. A gliding flight is one where no thrust or power is developed by the aircraft engines. There are then only three forces acting on the aircraft as shown in Fig. 4.14.

$$D = -W \sin(\gamma) \tag{4.25}$$

$$L = W \cos(\gamma) \tag{4.26}$$

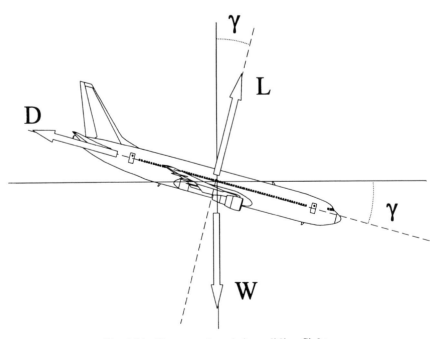

Fig. 4.14   Forces and angle in a gliding flight.

Considering the forces along the flight path, the equilibrium glide angle during an unaccelerated glide is

$$\frac{-\sin(\gamma)}{\cos(\gamma)} = \frac{D}{L} \quad \text{or} \quad \gamma = -\arctan\left(\frac{1}{E}\right) \tag{4.27}$$

Equation (4.27) indicates that the glide angle is solely a function of the lift-to-drag ratio E. The greater the value of the aerodynamic efficiency of the aircraft (E), the shallower the glide angle is. The minimum glide angle will thus occur while flying at E_m conditions. For an aircraft that loses all power in flight, the maximum gliding distance is often desired to improve the chances of reaching an airfield or the best possible alternative. Note that the aircraft's minimum drag coefficient may increase when an engine is inoperative in-flight because the engine may act as a windmill.

From Eqs. (4.27), (4.3), and (4.4)

$$dX = E \, dh \tag{4.28}$$

The longest gliding distance for a given altitude loss will be achieved when the lift-to-drag ratio is maximum. The longest gliding range (subscript LG) is, thus,

$$X_{LG} = \frac{\Delta h}{-\tan(\gamma_{LG})} = E_m \Delta h \tag{4.29}$$

where Δh is the altitude lost during the glide. The longest glide range is, thus, independent of aircraft weight and of the initial and final altitudes as long as the maximum aerodynamic efficiency is maintained during the glide. Other parameters are

$$\begin{array}{ll} V_{LG} = V_{E_m} & E_{LG} = E_m \\ C_{L_{LG}} = C_{L_{E_m}} & C_{D_{LG}} = C_{D_{E_m}} \end{array} \tag{4.30}$$

The vertical velocity of the aircraft as it descends, commonly called the sink rate SR or rate of descent RD, is

$$RD = -V \sin(\gamma) = \sqrt{\frac{2W \cos(\gamma)}{\rho_{SL} \sigma S C_L}} \left(\frac{C_D}{C_L}\right) \cos(\gamma) \tag{4.31}$$

Remember that a positive rate of descent corresponds to a negative rate of climb ($RD = -RC$); thus, a positive rate of descent has a negative γ value. For shallow angles of glide (less than approximately 15 deg) it can be assumed that the cosines in the preceding equation equal one so that Eq. 4.31 reduces to

$$RD = \sqrt{\frac{2W}{\rho_{SL} \sigma S}} \left(\frac{C_D}{C_L^{3/2}}\right) \tag{4.32}$$

The rate of descent, unlike the glide angle, depends on the weight and the altitude of the aircraft. If the rate of descent is to be minimum (thus maximum duration glide) for a given weight and altitude, the ratio $(C_D/C_L^{3/2})$ must be minimum, which is the same condition as previously determined for level flight minimum power required (see Fig. 4.15). In fact, from an energy conservation point of view, the power required to overcome the drag during a glide must be equal to the rate of loss of potential energy of the aircraft

$$V_{RD_{min}} = V_{MP}$$

$$-\tan(\gamma_{RD_{min}}) = \frac{1}{E_{MP}} = \frac{1}{\sqrt{3/4}E_m} \tag{4.33}$$

The rate of descent, as can be seen in Eq. (4.32), is equal to the sea level value divided by the square root of the density ratio. Integrating the rate of descent with respect to time will yield the time to go from one altitude to a lower one,

$$RD = \frac{RD_{SL}}{\sqrt{\sigma}} = -\frac{dh}{dt}$$

$$\int_{t_1}^{t_2} dt = -\frac{1}{RD_{SL}} \int_{h_1}^{h_2} \sqrt{\sigma}\,dh \tag{4.34}$$

For a gliding flight that takes place entirely within the troposphere (see Appendix B for values of σ), the time required to descend from an altitude

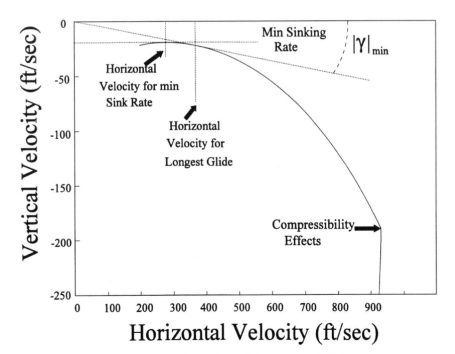

Fig. 4.15 Velocity hodograph for power-off flight.

h_1 to an altitude h_2 is (with h in feet)

$$t = 1/RD_{SL}(4.644 \times 10^4[1 - (6.875 \times 10^{-6})h]^{3.13205})_{h_1}^{h_2} \qquad (4.35)$$

Of course, this assumes that the velocity during the descent would be adjusted as altitude decreases, which in turn would affect the rate of descent. For a rate of descent of 10 ft/s (600 ft/min) at sea level, the time to go from an altitude of 36,000 ft to sea level will be around 2,739 s (approximately 46 min) and the rate of descent at 36,000 ft will be 18.3 ft/s (1,098 ft/min).

For a flight within the stratosphere, the time to go from h_1 to h_2 is (with h in feet)

$$t = 1/RD_{SL}[-5.3975 \times 10^4 \exp(-0.000024055h)]_{h_1}^{h_2} \qquad (4.36)$$

Remember that the rate of descent is a negative value. For the case where h_1 is in the stratosphere and h_2 is in the troposphere, one would use Eq. (4.36) to determine to time required to descend from h_1 to the tropopause and Eq. (4.35) to determine the time to descend from the tropopause to h_2.

The longest glide condition occurs while flying at the minimum drag condition (V_{E_m}) and the minimum rate of descent occurs at the minimum power-required condition ($V = V_{MP} \approx 0.76V_{E_m}$). If a compromise between the two is desired, what will the actual gliding distance and time to descend be? Let

$$\upsilon = \frac{V}{V_{LG}} = \frac{V}{V_{E_m}} \qquad (4.37)$$

Using an analysis similar to that developed in Chapter 3, Sec. 3.4.1, the relative gliding distance and time in to descent can be expressed as

$$\frac{X}{X_{LG}} = \frac{E\,dh}{E_m\,dh} = \frac{2\upsilon^2}{\upsilon^4 + 1} \qquad (4.38)$$

$$\frac{t}{t_{LG}} = \frac{X/V}{X_{LG}/V_{LG}} = \frac{X}{X_{LG}}\frac{1}{\upsilon} = \frac{2\upsilon}{\upsilon^4 + 1} \qquad (4.39)$$

It can be seen in Fig. 4.16 that, as the aircraft slows down from longest glide conditions, its relative time in the air will increase but its glide distance will decrease. Decreasing the relative velocity (υ) to 0.76 (minimum rate-of-descent conditions) will maximize the time in the air while providing approximately 87% of the longest glide range. If υ is only lowered to 0.87, this situation would provide a 10% increase in time in the air while decreasing the range by only 4%. On the other hand, increasing the velocity from longest glide conditions will decrease both the glide distance and time in the air.

Performing a similar analysis, but this time with respect to the minimum rate-of-descent conditions, will give (see similar analysis in Chapter 3, Sec. 3.5.4):

$$\upsilon = V/V_{MP} = V/V_{RD_{min}} \qquad (4.40)$$

Conditions LG

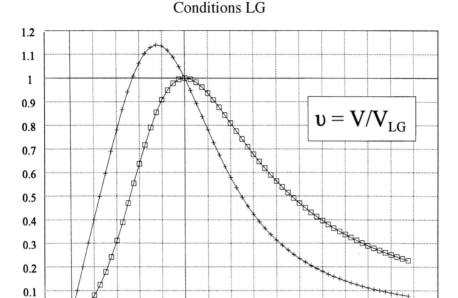

$$\upsilon = V/V_{LG}$$

$$\square \, X / X_{LG} \quad + t / t_{LG}$$

Fig. 4.16 Effects of velocity change on longest glide performance.

$$X/X_{RD_{min}} = 4\upsilon^2/(\upsilon^4 + 3) \tag{4.41}$$

$$\frac{t}{t_{RD_{min}}} = \frac{X}{X_{RD_{min}}}\frac{1}{\upsilon} = \frac{4\upsilon}{\upsilon^4 + 3} \tag{4.42}$$

Note that the relative time, Eq. (4.42), is identical to Eq. (3.78). If the velocity is increased from that for minimum rate of descent, the relative range will increase up to the point where minimum drag conditions are reached, after which the relative range will decrease. If the relative velocity υ is decreased from the minimum RD conditions, both relative range and relative time in the air will decrease. But, the conditions for minimum RD being equal to those for minimum power required, thus below the minimum drag conditions (slow flight region), the aircraft's stall margin may be too small to reduce the velocity (Fig. 4.17).

4.9 Maneuvering in the Vertical Plane

As an aircraft changes attitude, to initiate a climb or terminate a dive (Figs. 4.18 and 4.19), it will encounter an additional force, which will alter the steady climb or dive angle. The following simplified analysis will assume zero acceleration along the flight path.

Conditions RD_{min} (t_{max})

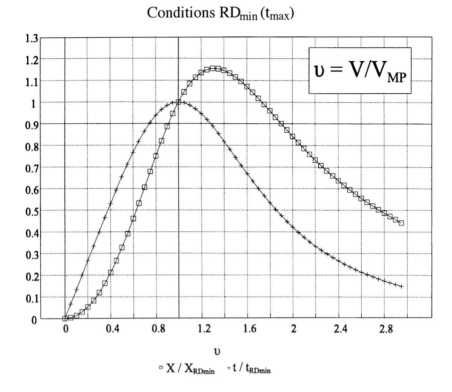

Fig. 4.17   Effects of velocity change on the minimum rate-of-descent performance.

Starting from a level unaccelerated flight path, the lift force must be suddenly increased to pull the aircraft upward, thus curving the flight path. Considering the forces acting on the aircraft and the acceleration to which it is subjected,

$$L - W = \frac{W}{g}\frac{V^2}{r} \qquad (4.43)$$

The load factor n, or g loading as it is commonly referred to, is defined as the lift-to-weight ratio. A load factor of 1.0 means that the lift force is equal to the

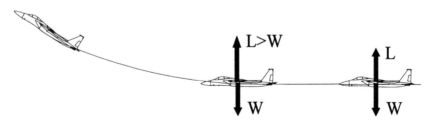

Fig. 4.18   Pullup maneuver.

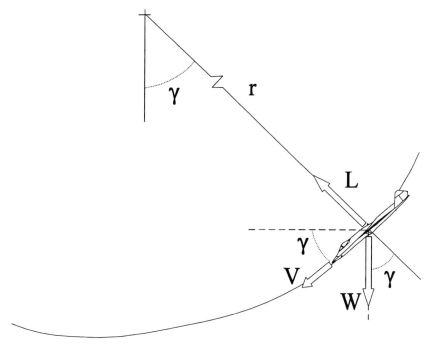

Fig. 4.19   Pullout of dive.

weight of the aircraft. Equation (4.43) becomes

$$n = \frac{L}{W} = \left(\frac{V^2}{gr} + 1 \right) \tag{4.44}$$

where V is the aircraft velocity and r is the instantaneous radius of turn in the vertical plane. Once the aircraft starts going up, this equation is no longer valid because the weight vector is not directly opposed to the lift vector. Combining Eqs. (4.2) and (4.43) gives a more general form of the load factor equation

$$n = (V^2/gr) + \cos(\gamma) \tag{4.45}$$

Although the analysis of this dynamic condition is not within the scope of this book, a graph illustrating the approximate size of loops performed at constant radius and constant load factor is provided for general interest (Fig. 4.20).

In many cases the aircraft attempting to perform such loops will not have sufficient thrust to complete a perfect loop (due to the increase in induced drag and a weight component acting in the same direction as drag) so that the actual loop will have a different shape (as shown in Fig. 4.21).

When an aircraft terminates a climb (or initiates a dive), the pilot will push on the control stick to bring the nose down. In such a case, the load factor will take

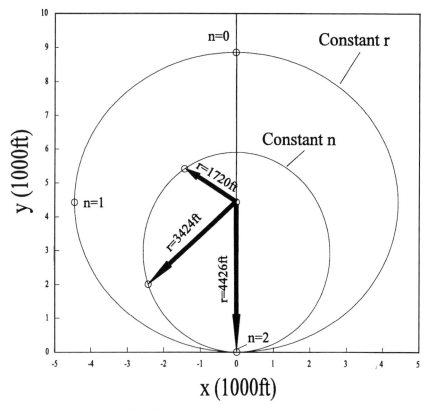

Fig. 4.20   Loop at constant *r* and constant *n*.

the following general form (see Fig. 4.22):

$$n = \cos(\gamma) - (V^2/rg) \qquad (4.46)$$

During the training of astronauts, initiation to microgravity can be performed by a specially modified transport aircraft (padded walls and floor in training section) performing a near zero-*g*, constant velocity flight from an initial climbing angle to a final diving angle. Because these flights are never exactly at zero load factor, there might be slight acceleration in any direction, thus the requirement for padded walls and floor. To maintain zero acceleration, the pilot must fly at constant airspeed and at conditions where [from Eq. (4.46)]:

$$r = V^2/g \cos(\gamma) \qquad (4.47)$$

As γ decreases from and initial positive climbing value, the radius of curvature of the flight path must decrease to maintain constant speed, zero load factor conditions. Past the horizontal, as the dive angle becomes more and more negative, the radius of curvature of the flight path must increase to maintain the desired flight conditions. Thus, the resulting flight path will resemble a parabola. From a pilot

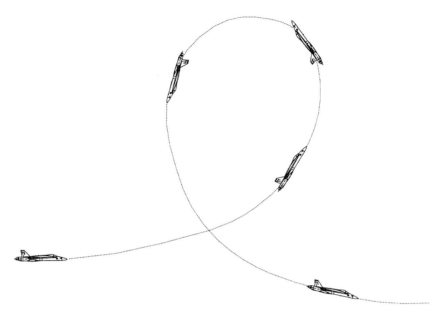

Fig. 4.21   Hypothetical shape of a loop performed by an aircraft at low-thrust setting.

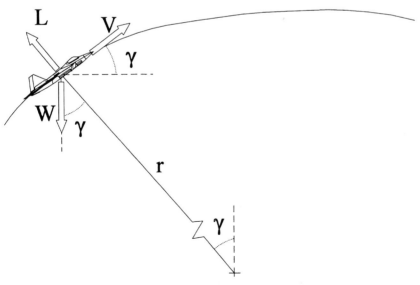

Fig. 4.22   Push down.

point of view, because there is no way to directly measure the radius of curvature of the flight path, it will be easier to follow a pitch rate $(d\gamma/dt = V/r)$

$$\frac{d\gamma}{dt} = \dot{\gamma} = \frac{V}{r} = \frac{g\cos(\gamma)}{V} \qquad (4.48)$$

Thus, the pitch rate must first increase as the climb angle is decreased from a positive climb value, and then decrease as the horizontal is passed.

4.10   Influence of the Wind on the Climb Angle

The climb angle calculated in Sec. 4.2, maximum climb angle, was determined for a stationary air mass. The actual climb angle with respect to an observer on the ground will be affected by the velocity of the wind. Figure 4.23 illustrates the case of a headwind. A headwind will shorten the ground distance traveled for a given altitude gain, thus an effective increase in climb angle with respect to a ground observer. Note also that the aircraft attitude will be the same with or without winds.

The angle of climb with respect to a ground observer (γ_{ground}) is

$$\gamma_{\text{ground}} = \arctan\left[\frac{V\sin(\gamma)}{V\cos(\gamma)\pm V_w}\right] = \arctan\left[\frac{\sin(\gamma)}{\cos(\gamma)\pm \omega_f}\right] \qquad (4.49)$$

where

$$\omega_f = V_{\text{wind}}/V \qquad (4.50)$$

and V is the aircraft true airspeed, γ is its climb angle with respect to the air mass, and V_w is the wind velocity. A headwind has a negative value in the preceding equation, thus increasing the climb angle with respect to a ground observer. For the case where there is no wind, $\gamma_{\text{ground}} = \gamma$.

Another way of determining the climb angle with respect to a ground observer when there is wind present is to use the velocity hodograph (Sec. 4.4) as illustrated in Fig. 4.24. One will notice that in the case of a headwind, the maximum climb angle with respect to a ground observer will be somewhat larger than in the case with no wind. One will also notice that a tailwind is detrimental to the

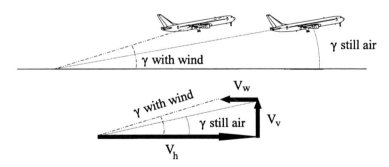

Fig. 4.23   Effect of a headwind on the climb angle with respect to the ground.

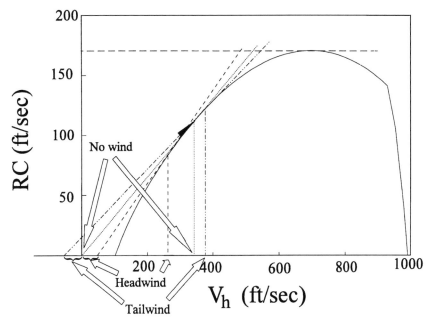

Fig. 4.24   Use of velocity hodograph to determine the climb angle with respect to a ground observer; case illustrated corresponds to maximum climb angle.

climb performance of an aircraft with respect to a ground observer or ground obstacle.

Problems

4.1.   A CF-18A (see Appendix E) is flying at Mach 0.8 at sea level. What will be the maximum climb angle (sustainable) at that velocity if the pilot advances the throttle to the maximum thrust available before starting the climb? What would be that angle if the aircraft initial altitude is 10,000 ft?

4.2.   Repeating problem 1 with an F-15C (see Appendix E), find the rate of climb instead of the climb angle. Could an F-16C reach a steady rate of climb at sea level conditions and Mach 0.8 using full thrust?

4.3.   Assuming a linear variation of the maximum rate of climb from sea level to the maximum altitude (absolute ceiling) of an F-15C (see Appendix E), what would be the minimum time required to climb from sea level to 30,000 ft?

4.4.   The engine of aircraft C flames out (shuts down) while it is flying at 30,000 ft and 35 miles from shore (i.e., over water). The aircraft must reach an altitude of 10,000 ft before the pilot can eject. The aircraft weight at flame out is 15,500 lb. What is the minimum sinking rate of the aircraft at 30,000 ft? While flying at minimum sinking rate conditions, what will be the time required to go from

30,000 to 10,000 ft? If the pilot adopts the minimum sinking rate velocity, will the pilot be able to reach the shore before ejecting at 10,000 ft? If the pilot dumps 2,000 lb of fuel at 30,000 ft will the pilot be able to reach the shore while flying at the minimum sinking rate conditions? If the pilot decides on ejecting at 5,000 ft instead of 10,000 ft, will the pilot make it to shore (still minimum sinking rate conditions)? If the pilot elects to adopt the longest glide conditions, can the pilot make it to shore and eject at 10,000 ft (initial altitude 30,000 ft and weight 15,500 lb)?

4.5.   Aircraft C is in level flight at 35,000 ft and Mach 0.6. The pilot initiates a pullup maneuver resulting in an instantaneous 2-g load factor. What is the instantaneous radius of turn in the vertical plane? Will the radius decrease or increase if the velocity is increased to Mach 0.7 before the pullup is initiated? If the velocity is Mach 0.6 but the altitude is 11,000 ft, what will be the instantaneous radius of turn in the vertical plane?

4.6.   A CF-18A pilot is doing a high-speed, low-altitude pass during an airshow. While at full thrust, a pullup all of the way to the vertical is initiated. When 90-deg climb angle is reached, the speed and altitude are 300 kn and 600 ft. If the pilot maintains a 90-deg climb at full throttle, at what altitude will the airspeed be 200 kn?

Fig. P4.6.

5
Turning Flight

5.1 Introduction

T HERE are many ways to change heading in flight. For example, an aircraft can go to the vertical, roll around its longitudinal axis, and come back wing level in a different direction. This chapter concentrates on the constant velocity coordinated turn, at a constant altitude, which is a pure rotation about a fixed vertical axis. In the analysis that follows there is no tangential acceleration by the aircraft, the aircraft does not sideslip, and the thrust AoA is neglected. The following equations are obtained:

$$T - D = 0 \tag{5.1}$$

$$L \sin(\phi) - (W/g)V\dot{\chi} = 0 \tag{5.2}$$

$$L \cos(\phi) - W = 0 \tag{5.3}$$

where ϕ is the aircraft angle of bank (as shown in Fig. 5.1) and $\dot{\chi}$ its angular velocity. The load factor, which is defined as the lift-to-weight ratio, will be a function of the bank angle (ϕ) during a turning flight. Combining Eqs. (5.3), (3.5) and (5.1) gives a sustained (steady state), constant altitude, turning flight load factor:

$$n = \frac{1}{\cos(\phi)} = \frac{T}{W}E \tag{5.4}$$

Equation (5.4), which combines the load factor with the bank angle, thrust-to-weight ratio, and lift-to-drag ratio, indicates that there can be no constant altitude coordinated turn without a bank angle. It also shows that, for a bank angle of 90 deg, the load factor becomes infinite. Therefore, an aircraft cannot maintain a steady-state 90-deg constant altitude turn. A turn at a 90-deg bank is possible only by sideslipping or losing/gaining altitude and/or airspeed because the engine thrust remains finite.

The aircraft's angular velocity, called turn rate or rate of turn, can be calculated by combining Eqs. (5.2–5.4)

$$\dot{\chi} = \frac{g \tan(\phi)}{V} = \frac{g\sqrt{n^2 - 1}}{V} \tag{5.5}$$

where the correlation between the bank angle and the load factor is obtained by simple trigonometry. The tangential velocity is equal to the angular velocity $\dot{\chi}$ times the radius of turn r. Therefore, the radius of turn is given by

$$r = \frac{V}{\dot{\chi}} = \frac{V^2}{g\sqrt{n^2 - 1}} \tag{5.6}$$

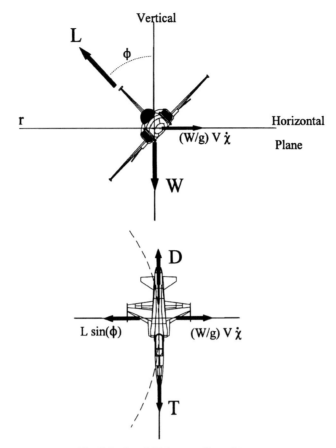

Fig. 5.1   Level flight coordinated turn.

It is important to note that these last two equations are not a function of the aircraft characteristics but that they depend on its instantaneous flight parameters (namely, the aircraft's velocity and its load factor). A general grid incorporating the radius of turn, the turning rate, the load factor, and the velocity can thus be created, which is valid for all aircraft. This grid (Fig. 5.2) shows the interdependence between all four parameters.

5.2   Instantaneous vs Sustained Turn Performance

While maneuvering, it is often desired for an aircraft to have the smallest radius of turn as well as the highest rate of turn possible. From the preceding two equations and Fig. 5.2, it is seen that these conditions will be achieved when the load factor is highest and the velocity lowest. The maximum safely usable load factor is the structural limit set during the design and the construction of the aircraft. The minimum velocity is the stall speed, which is attained at the

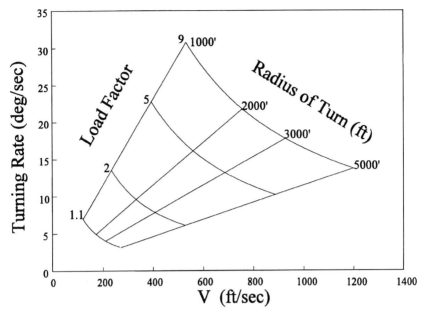

Fig. 5.2 General grid for aircraft turning performance.

maximum lift coefficient, noting that the required lift is now

$$L = nW$$

$$V_{\text{stall},n} = \sqrt{\frac{2nW}{\rho_{\text{SL}} \sigma S C_{L_{\max}}}} = \sqrt{n} V_{\text{stall}} \tag{5.7}$$

The maneuvering envelope of the aircraft can be illustrated with a V–n diagram where the limits are imposed by the design load factors and velocities (V_{NE}, $q_{\max}$, or engine operating velocity limits). Note that the maximum negative lift coefficient may not be the same absolute value as the maximum positive lift coefficient.

The minimum velocity ($V_{\text{stall},n}$) at which the actual applied load factor is equal to the structural limit load factor ($n_{\max,\text{struc}}$) is called the corner speed or the maneuvering speed and is identified as point A in Fig. 5.3. Below V_A, the aircraft will stall before the load factor imposed on the structure exceeds the structural design limit. At V_A the radius of turn [Eq. (5.6)] will be minimum, and the turn rate [Eq. (5.5)] maximum. The following equations are obtained:

$$V_{\text{corner}} = \sqrt{\frac{2n_{\max_{\text{struc}}}(W/S)}{\rho_{\text{SL}} \sigma C_{L_{\max}}}} \tag{5.8}$$

$$\dot{\chi}_{\text{corner}} = \frac{g\sqrt{n_{\max_{\text{struc}}}^2 - 1}}{V_{\text{corner}}} \tag{5.9}$$

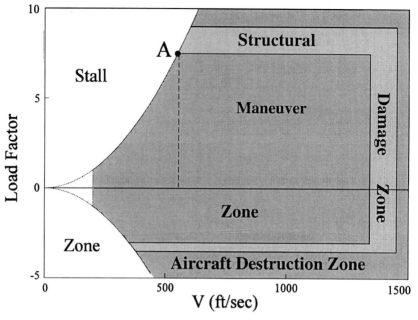

Fig. 5.3   *V–n* diagram.

$$r_{\text{corner}} = \frac{V_{\text{corner}}^2}{g\sqrt{n_{\text{max}_{\text{struc}}}^2 - 1}} \tag{5.10}$$

These values are limited both by the maximum lift coefficient and the maximum structural design load factor. They can, however, be attained during sustained (steady) level turn only if the thrust available is greater or equal to the drag. Substituting the drag equation into Eq. (5.1), knowing that the lift is equal to nW, gives the following equation:

$$S^2 C_{D_0} q^2 - T Sq + K n^2 W^2 = 0 \tag{5.11}$$

which is solved for the dynamic pressure

$$q = \frac{(T/S)}{2C_{D_0}} \left[1 \pm \sqrt{1 - \frac{4K C_{D_0} n^2}{(T/W)^2}} \right] \tag{5.12}$$

to yield the velocity in a turn

$$V = \sqrt{\frac{(T/S)}{\rho_{\text{SL}} \sigma C_{D_0}} \left[1 \pm \sqrt{1 - \frac{n^2}{[E_m (T/W)]^2}} \right]} \tag{5.13}$$

This last equation indicates that for a given thrust, the flight speed will decrease as the load factor increases. Conversely, as the load factor is increased, the thrust must be increased to maintain a constant speed. As calculated with this equation, the aircraft flight envelope will be reduced as shown in Fig. 5.4.

In the event that the thrust is insufficient to compensate for the drag, the aircraft's deceleration can be calculated by summing up the horizontal forces,

$$(T - D) = \frac{W}{g} \frac{dV}{dt} \qquad (5.14)$$

$$\therefore \quad \frac{dV}{dt} = g \frac{(T_a - T_r)}{W} \qquad (5.15)$$

It should be noted here that T_r is the thrust required to compensate for the aircraft's drag during the turn, that is, with the aircraft subjected to the load factor imposed on the aircraft during the turn. T_a is the thrust available from the engine(s). A thrust deficiency will result in a loss of velocity.

On the other hand, if a constant velocity is maintained, the aircraft will have to lose altitude. The sink rate can be determined from energy considerations by trading potential energy to counter the excess drag. The rate of altitude loss is given by Eq. (4.7) from the preceding chapter. There can also be a mix of deceleration/acceleration and sink/climb rate. A more general way to account for the rates of change in altitude and velocity is to express them in terms of the

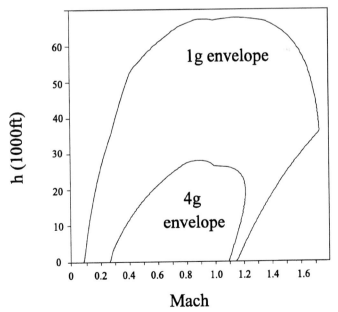

Fig. 5.4   Decrease in size of the flight envelope of aircraft C due to the load factor.

instantaneous specific power by combining Eqs. (4.7) and (5.15) multiplied by the velocity

$$V \frac{(T_{\text{avail}} - T_{\text{req}})}{W} - \frac{dh}{dt} - \frac{V}{g} \frac{dV}{dt} = 0 \tag{5.16}$$

The first term represents the specific excess power of the aircraft, the second term the specific rate of change of the potential energy (rate of change of altitude), and the third term the specific rate of change of kinetic energy (acceleration or deceleration). This approach will be discussed in more detail in Chapter 8.

A point to remember is that the flight envelope represents the boundaries of level unaccelerated flight. Transient excursions outside the boundaries of the envelope can be performed, but not sustained, such as a dive to go faster than the maximum velocity in level flight (for a given load factor) or a zoom to an altitude outside the defined envelope. These transient flights always involve a certain degree of velocity and altitude changes. Using Eq. (5.11) and solving for the load factor gives

$$n = \frac{q}{(W/S)} \sqrt{\frac{1}{K} \left[\frac{(T/S)}{q} - C_{D_0} \right]} \tag{5.17}$$

which is the load factor, thus the bank angle, required for a sustained turn at constant altitude, velocity, and thrust-to-weight ratio. From Eq. (5.4) it can be seen that the maximum sustainable load factor (no stall) occurs when both the thrust-to-weight and lift-to-drag ratios are at their maximum:

$$n_{\text{max}_{\text{aero}}} = \frac{T_{\text{max}}}{W} E_m \tag{5.18}$$

For high-performance aircraft in certain flight regimes, this maximum sustainable load factor may exceed the structural design load factor; in which case the pilot, or a g limiter within the aircraft's avionics, will limit the load factor to the structural design load factor.

5.3 Maximum Sustained Turning Rate

The maximum constant altitude turning rate that can be sustained by an aircraft is found by setting the derivative of Eq. (5.5) with respect to the velocity equal to zero:

$$\left[\frac{\partial \dot{\chi}}{\partial V} \right]_{\text{FT}} = 0$$

$$\frac{g}{\sqrt{n^2 - 1}} 2n \frac{\partial n}{\partial V} V - g\sqrt{n^2 - 1} = 0$$

$$\therefore \quad n^2 - 1 - 2nV \frac{\partial n}{\partial V} = 0$$

Or, in terms of the dynamic pressure,

$$n^2 - 1 - n\rho V^2 \frac{\partial n}{\partial q} = 0$$

Solving the preceding equation (see Appendix D) for the dynamic pressure gives the value for the fastest sustained level turn (subscript FT):

$$q_{FT} = \left[\frac{W}{S}\right] \sqrt{\frac{K}{C_{D_0}}} \tag{5.19}$$

$$V_{FT} = \sqrt{\frac{2(W/S)}{\rho_{SL}\sigma}} \left(\frac{K}{C_{D_0}}\right)^{\frac{1}{4}} \tag{5.20}$$

The fastest turn velocity is thus equal to the level flight minimum drag velocity. This would be expected because that is the velocity where the maximum excess thrust is available to counter the increase in induced drag in a turn. Using Eqs. (5.17) and (5.19) gives

$$n_{FT} = \sqrt{2\left(\frac{T_{max}}{W}\right) E_m - 1} = \sqrt{2n_{max_{aero}} - 1} \tag{5.21}$$

which then gives the fastest sustained level turn rate

$$\dot{\chi}_{FT} = g\sqrt{n_{FT}^2 - 1} / V_{FT} \tag{5.22}$$

Thus, for a high-turn rate, an aircraft requires a large thrust-to-weight ratio (T/W), a low-wing loading (W/S), a small K (large aspect ratio and Oswald number), and a low-parasite drag coefficient (C_{D_0}). It should be observed in Eq. (5.21) that even if $n_{max,aero}$ exceeds the structural design load factor, the maneuver at n_{FT} is still permissible as long at it does not exceed $n_{max,struc}$.

Unfortunately, the preceding equations for fastest turn do not take into account the value of maximum lift coefficient. For an aircraft with a large thrust-to-weight ratio (such as aircraft C in Fig. 5.5), the value for fastest turn may fall outside the envelope of maximum lift ($C_{L,FT}$ exceeds $C_{L,max}$). Then the velocity for fastest unaccelerated turn would fall on the lift limit curve, which represents a stalled turn (discussed later in this chapter). Also, the maximum allowable structural load factor is not taken into account. This means that there may exist a range of velocities where the aircraft could be subjected to load factors that exceed the maximum structural design load factor of the aircraft if the thrust is sufficient. The region where both the lift and the structural limits are sufficiently high but the T/W is inadequate is a region of decelerating and/or descending flight. It is called the region of instantaneous turn performance. In this region high-turning rates can be achieved for a brief period of time as the aircraft decelerates (or descends), whereas the region below the maximum T/W curve is one of sustained turn performance. This means that, in that region, the aircraft can maintain a turn rate without decelerating or losing altitude.

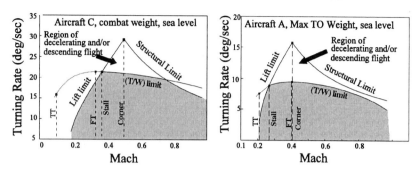

Fig. 5.5   Turn rate at constant altitude.

Another limit not considered in the preceding discussion is the fact that the maximum load factor is usually weight dependent. This is because the lift force that must be developed by the wing is proportional to the load factor times the weight of the aircraft ($L = nW$), but the wing root can only sustain a given bending moment without damage. The aircraft may structurally be able to attain 9 g safely at combat weight (aircraft C) but may be limited to 7.5 g at full up weight. Also, the pilot may not be able to sustain such high-g loading. This will be further discussed in Chapter 9.

The effect of an increase in altitude on the turn performance of an aircraft is to increase the stall velocity for a given load factor and to decrease the available thrust-to-weight ratio. An altitude gain has no effect on the maximum structural design load factor as such because it is only structural strength dependent, but it will increase the corner velocity, minimum radius of turn, and decrease the turning rate as seen in Fig. 5.6.

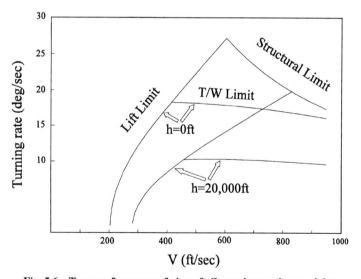

Fig. 5.6   Turn performance of aircraft C, maximum clean weight.

5.4 Minimum Sustained Radius of Turn

The minimum constant altitude radius of turn that can be sustained by an aircraft can be found by differentiating Eq. (5.6) with respect to velocity and setting it to zero

$$\left[\frac{\partial r}{\partial V}\right]_{TT} = \left(n^2 - 1 - qn\frac{\partial n}{\partial q}\right)_{TT} = 0 \tag{5.23}$$

Solving for the dynamic pressure (see Appendix D) yields the tightest turn (subscript TT) possible

$$q_{TT} = \frac{2(W/S)K}{(T/W)} \tag{5.24}$$

and

$$V_{TT} = 2\sqrt{\frac{(W/S)K}{\rho_{SL}\sigma(T/W)}} \tag{5.25}$$

Combining Eqs. (5.17) and (5.24) gives the load factor for the tightest turn

$$n_{TT} = \sqrt{2 - \left(1/n^2_{max_{aero}}\right)} \tag{5.26}$$

$$r_{TT} = \frac{V^2_{TT}}{g\sqrt{n^2_{TT} - 1}} = \frac{4K(W/S)}{\rho_{SL}\sigma g(T/W)\sqrt{1 - \left(1/n^2_{max_{aero}}\right)}} \tag{5.27}$$

A small radius of turn calls for a large thrust-to-weight ratio, a low-wing loading, a low K (large AR e), and low altitude. Unfortunately, the theoretical velocity for minimum radius is usually lower than the level flight stalling speed of most aircraft. This is due to the strong dependence of the radius of turn on the velocity [the radius of turn being proportional to the velocity squared as seen in Eq. (5.6)] and to the fact that the maximum lift coefficient of the aircraft is not taken into account in the preceding equations. The tightest sustained level turn will, thus, usually be performed while the aircraft is bordering on the stall and while at maximum thrust. Knowing that the thrust will equal the drag at stall, the following equations result:

$$V_{ST} = \sqrt{\frac{2(T_{max}/S)}{\rho_{SL}\sigma\left(C_{D_0} + KC^2_{L_{max}}\right)}} \tag{5.28}$$

$$n_{ST} = \frac{(T_{max}/W)C_{L_{max}}}{C_{D_0} + KC^2_{L_{max}}} \tag{5.29}$$

The corresponding radius of turn and turning rate can be obtained from Eqs. (5.6) and (5.5) respectively. Figure 5.7 illustrates the results.

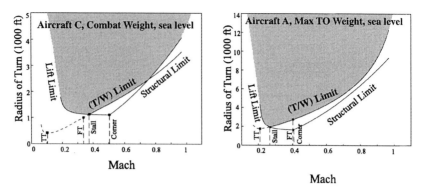

Fig. 5.7   Radius of turn at constant altitude.

Problems

5.1. Aircraft A at 75% of maximum takeoff weight is slowing down before initiating a descent, the pilot intends to slow to Mach 0.65 at 30,000 ft. Once the aircraft is at Mach 0.65, the pilot initiates a 2-g, constant altitude turn. Will the pilot stall the aircraft? If not, what is the stall margin (expressed as the difference between the actual velocity and the stall velocity)? What is the turn rate and radius of turn?

5.2. Two aircraft (a CF-18A and a MiG-29, see Appendix E) are both flying at 25,000 ft and Mach 0.75. Which one has a turning rate advantage over the other at that velocity (maximum sustainable turn and maximum instantaneous turn at that velocity)? What is the minimum radius of turn, at constant altitude, that both aircraft can achieve at that velocity (minimum sustainable and instantaneous)?

5.3. An F-15C encounters an Su-27 (see Appendix E) at 31,000 ft and both engage into tight turns, which reduces their velocity. If both aircraft reach their respective corner velocity at the same time, which aircraft will have the turning rate advantage over the other? Which one will have the radius of turn advantage over the other? What will be the deceleration rate of each aircraft if they both perform a maximum instantaneous rate of turn at the corner velocity?

5.4. Aircraft C initiates a 5-g turn while flying at a constant altitude of 10,000 ft and Mach 0.8 (combat weight of 16,000 lb). What is the rate of deceleration of the aircraft at that velocity? At what velocity will the aircraft settle if the 5-g turn is maintained long enough? Note that the throttle is not moved during the turn and is kept at the position for flying at Mach 0.8 and 10,000 ft.

5.5. Aircraft C, at combat weight, is turning at 15 deg/s at 20,000 ft of altitude and at a velocity of 700 ft/s. Is the aircraft stalled? If a velocity of 700 ft/s is maintained, can the aircraft maintain its altitude if the engine provides full thrust at altitude? If not, what is the rate of descent?

5.6. What is the difference between the fastest turn load factor for aircraft C, at combat weight and 25,000 ft, and the maximum sustainable turn load factor if the aircraft is flying at 1,000 ft? If the aircraft adopts the fastest turn conditions at 25,000 ft (velocity and turn rate), will it stall?

5.7. During an air-to-air encounter between an F-16C and a MiG-29 (see Appendix E), both aircraft try to turn inside the radius of turn of the opponent. What is the minimum sustainable radius of turn of each aircraft if the combat is done at a constant altitude of 22,000 ft? Which aircraft has the turn advantage in terms of minimum instantaneous radius of turn? Which one has the advantage in terms of maximum sustainable rate of turn?

6
Aircraft In Ground Effect

6.1 Introduction

AN aircraft is said to be in ground effect (IGE) when it is flying very close to the ground, usually within a distance of one to two times the aircraft's wing span (relative height $h/b \approx$ 1–2). The close proximity of the ground affects the flow about the aircraft sufficiently to change the streamlines and the pressure distribution around the aircraft.

As the aircraft flies closer and closer to the ground, the lift coefficient increases for a given AoA and there is also an increase in nose-down pitching moment. At first this seems to be a good thing because you have more lift plus a stabilizing moment, but there are also penalties to pay. The maximum lift coefficient is reduced and so is the stall AoA. This means that the stalling speed will be higher IGE than out-of-ground effect (OGE), as seen in Fig. 6.1.

Another important effect for an aircraft flying IGE is a large reduction in induced drag (see Fig. 6.2). The increase in the lift coefficient and the decrease in drag coefficient, for a given AoA, translate into an increase in lift-to-drag ratio. This drag reduction will reduce the aircraft's rate of deceleration (and may even increase its velocity if the drag becomes smaller than the thrust available). Combined with the increased lift coefficient, the aircraft will tend to float and refuse to touchdown unless the airspeed is further reduced or until it is forced onto the ground with forward stick inputs from the pilot.

The magnitude of these effects increase hyperbolically as the relative height is reduced. For aircraft with large span, ground effects will occur at higher heights. Furthermore, all of these effects depend on the aircraft layout and its configuration, as well as its dimensions.

Results from a computer simulation using a two-dimensional panel method[5] indicate that the effects of ground proximity are to increase the pressure under the wing and increase the suction at the leading edge. The increased suction at the leading edge is the major drag reduction factor in that it tilts the resultant force vector forward, closer to the two-dimensional value, thus reducing the effects of the three-dimensional flow around the wing, and consequently, the induced drag.

A few words will be added here on recent (late 1980s, early 1990s) research activities on ground effects, mainly on the use of thrust reversers while flying IGE and on dynamic ground effects. Modern fighter and attack aircraft may have to use thrust reversers in flight to reduce their landing distance and ground roll. This may be necessary because today's air forces have better weapons to attack and disable airstrips, leaving only small lengths of usable runway to land on. The use of thrust reversing in flight introduces adverse aerodynamic effects, and at low speed while on the runway there is a possibility of reingestion of the hot exhaust gases. Fighter aircraft that will be equipped with thrust reversers are likely to be configured like the F-15 short takeoff and landing (STOL) and maneuver technology demonstrator (S/MTD) (U.S. Air Force sponsored flight-test program,

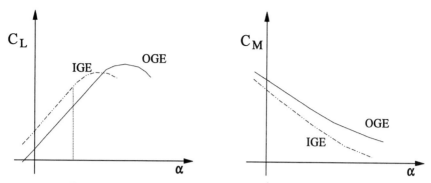

Fig. 6.1   Change in lift and moment coefficients while IGE.

1988–1991, to investigate the in-flight thrust vectoring and reversing and other systems) with upper and lower jets.

Upper jets, although they can decrease fin and rudder efficiency, will be marginally affected by the ground proximity, as shown in Fig. 6.3. Lower jets, on the other hand, will generate many problems while IGE, especially close to touchdown. During flight OGE the lower jets are bent aft due to the airflow dynamic pressure. This effectively keeps the jets away from the aircraft fuselage and wings and as such creates no specific interference problems. Below a certain relative height, which again is a function of the aircraft's configuration as well as its engine power setting, as the lower jets turn forward, they will affect the flowfield under the aircraft. Some identified problems are extremely stabilized aerodynamic rolling moment for small sideslip angles (which is as bad as destabilizing the aircraft because the flight control system may not be able to provide enough aerodynamic rolling power), destabilizing yawing moment, aircraft pitch-up tendency, and decreased lift. This is a dangerous situation to be in.

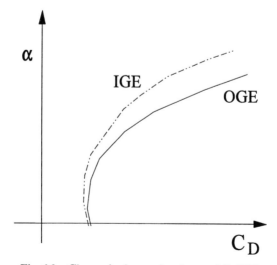

Fig. 6.2   Change in drag polar shape while IGE.

Possible re-ingestion
of exhaust gases

Fig. 6.3   Using thrust reversers during the approach.

Directing the jets outward will decrease or eliminate these problems, but it increases the thrust reversers' weight and complexity while reducing their effectiveness. As well, upon failure of one engine, the thrust reversing must be terminated due to the large yawing moment that would be created by the operative engine. Moving the thrust reversers away (farther aft) from the wing will also decrease the magnitude of the problems. Another way to eliminate in-flight problems is to keep the lower jets facing aft until touchdown and then turning them forward.

The F-15 S/MTD used rotary vanes to permit rapid transition from forward thrust to reverse thrust (Fig. 6.4). The vanes controlled the direction of the exhaust jet from each engine while the throttle was maintained at a high setting. This configuration avoided the engine spool up time of conventional installations, which require the throttle setting to transition from flight condition setting, to idle, and then to full reverse thrust.

Dynamic ground effects are present when the aircraft possesses a rate of descent. This rate of descent will reduce the influence of ground effects for a given relative height compared to static ground effects (flying at a constant altitude) at the same relative height. It will also reduce the negative effects of thrust reversing in-flight near the ground as the aircraft constantly moves away from the engines reversed

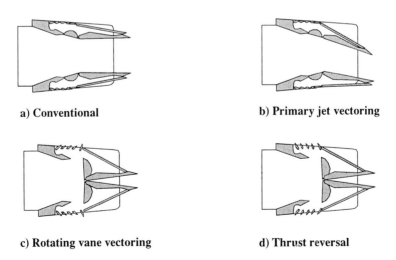

a) Conventional   b) Primary jet vectoring

c) Rotating vane vectoring   d) Thrust reversal

Fig. 6.4   F-15 S/MTD vectoring nozzle and rotary vanes.

jet plumes.[6] This is mainly because although the AoA in both cases (static and dynamic) may be the same, the aircraft attitude angle will differ, i.e., a smaller angle for the dynamic case means that the lower jets will be angled more toward the ground than for the static case.

6.2 Development of an IGE Equation for Performance Analysis

Let us quantify ground effects to see how exactly the aircraft's performance is affected. Only the wing will be considered here, but all parts of the aircraft are affected to some degree. An elliptic loading (Γ_{ell}) will be used to represent the lift spanwise distribution on a straight wing of moderate AR ($\approx$5) plus a virtual image to simulate the presence of the ground (Fig. 6.5). A nonviscous flow approach is used because the interaction between the aircraft and the ground does not interfere with the BL (i.e., the BL of the aircraft does not come into contact with the ground)

$$\Gamma_{\text{ell}}(y) = \Gamma_0\sqrt{1 - (y/s)^2} \qquad \text{where } s = b/2 \qquad (6.1)$$

The vortex sheet behind the wing produces, for an elliptic loading only, a constant downwash (ω) along the span b. The downwash is reduced by the presence of the ground and will ultimately become 0 for $h = 0$

$$\omega(y) = \omega_{\text{ell}} = \Gamma_0/4s \qquad \text{for } h = \infty \qquad (6.2)$$

The ground (or virtual image) produces an upwash ($\Delta\omega_{\text{IGE}}$) that reduces the downwash to various degrees along the span. It can be demonstrated that the contribution of the image vortex sheet to the downwash at a point P located at y_1 (Fig. 6.5) is

$$\omega(y_1) = \omega_{\text{ell}} + \Delta\omega_{\text{IGE}} \qquad (6.3)$$

$$\Delta\omega_{\text{IGE}}(y_1) = \frac{-1}{4\pi} \int_{-s}^{s} \frac{\dot{\Gamma}}{r} \cos(\theta)\mathrm{d}y \qquad (6.4)$$

$$\dot{\Gamma} = \frac{\mathrm{d}\Gamma}{\mathrm{d}y} = \frac{-(-\Gamma_0)}{s^2}\frac{y}{\sqrt{1 - (y/s)^2}}$$

$$r = \sqrt{4h^2 + (y_1 - y)^2} \qquad (6.5)$$

$$\cos(\theta) = (y_1 - y)/r$$

We use $\cos(\theta)$ in Eq. (6.4) because only the vertical component of $(\mathrm{d}\Gamma/\mathrm{d}y)$ at y_1 is of interest here. Substituting Eq. (6.5) into Eq. (6.4) gives

$$\Delta\omega_{\text{IGE}}(y_1) = \frac{-\Gamma_0}{4\pi s^2} \int_{-s}^{s} \frac{y(y_1 - y)}{\sqrt{1 - (y/s)^2}(4h^2 + (y_1 - y)^2)}\mathrm{d}y \qquad (6.6)$$

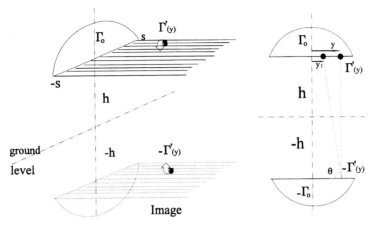

Fig. 6.5   Image of wing vortex.

or

$$\Delta\omega_{\text{IGE}}\left(\frac{y_1}{s}\right) = \frac{-\Gamma_0}{4\pi s} \int_{-1}^{1} \left\{ \left[\frac{y}{s}\left(\frac{y_1}{s} - \frac{y}{s}\right) d\left(\frac{y}{s}\right) \right] \bigg/ \sqrt{1 - \left(\frac{y}{s}\right)^2} \right.$$

$$\left. \times \left[4\left(\frac{h}{s}\right)^2 + \left(\frac{y_1}{s} - \frac{y}{s}\right)^2 \right] \right\} \tag{6.7}$$

Solving this integral is at best laborious; by doing it numerically, the curves of $\Delta\omega_{\text{IGE}}(y_1/s)$ as a function of relative height (h/s) shown in Fig. 6.6 are obtained.

The graph of Fig. 6.6 shows that the strength of the upwash increases as the relative height (h/s) decreases. The effects are most visible in the middle section $(y/s = 0)$ where the influence of the vortex sheet is the greatest. The effect of this upwash on the aircraft's normal downwash is plotted in Fig. 6.7. This approach, however, does not provide a simple way to see the influence of ground effects on aircraft performance.

An easier way to estimate the influence of the ground effects on aircraft performance is to use a simplified horseshoe vortex. The upwash due to the ground proximity is then

$$\Delta\omega_{\text{IGE}}(y_1) = \frac{\Gamma_0}{4\pi}\left(\frac{s' + y_1}{r_1^2} + \frac{s' - y_1}{r_2^2}\right)$$

where

$$r_1^2 = 4h^2 + (s' + y_1)^2$$

$$r_2^2 = 4h^2 + (s' - y_1)^2) \tag{6.8}$$

$$s' = \pi/4s$$

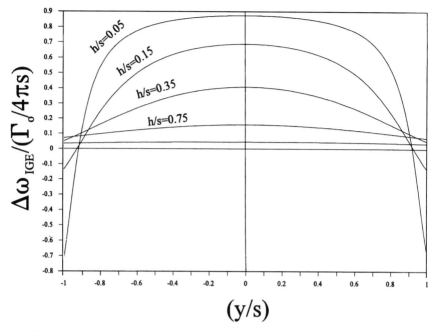

Fig. 6.6   Upwash distribution along the span as a function of relative height.

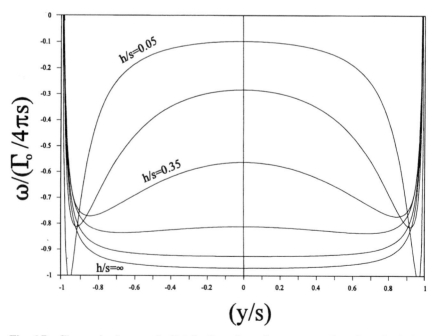

Fig. 6.7   Change in downwash distribution along the span as a function of relative height.

The induced drag will be reduced by a certain value, which equals

$$\Delta D_i = \frac{-\rho \Gamma_0^2}{4\pi} \int_{-s'}^{s'} \left(\frac{s'+y_1}{r_1^2} + \frac{s'-y_1}{r_2^2} \right) dy_1 \tag{6.9}$$

$$\Delta D_i = \frac{-\rho \Gamma_0^2}{4\pi} \ell_n \left[1 + \left(\frac{s'}{h} \right)^2 \right] \tag{6.10}$$

Reduced to its nondimensional coefficient form, the induced drag reduction is

$$\Delta C_{D_i} = \frac{-C_L^2}{\pi \, \text{AR}} \left\{ \frac{2}{\pi^2} \ell_n \left[1 + \left(\frac{s'}{h} \right)^2 \right] \right\} \tag{6.11}$$

The induced drag coefficient (for an elliptical loading) thus takes the following form:

$$C_{D_i} = \frac{C_L^2}{\pi \, \text{AR}} \left[1 - \frac{2}{\pi^2} \ell_n \left(1 + \left(\frac{s'}{h} \right)^2 \right) \right] = \Phi C_{D_{i,\text{OGE}}} \tag{6.12}$$

where

$$\Phi = \left\{ 1 - \frac{2}{\pi^2} \ell_n \left[1 + \left(\frac{s'}{h} \right)^2 \right] \right\} = \left\{ 1 - \frac{2}{\pi^2} \ell_n \left[1 + \left(\frac{\pi b}{8h} \right)^2 \right] \right\} \tag{6.13}$$

Although this approach has a theoretical lower limit of $h/s' \approx 0.0851$ ($h/b \approx 0.0334$ or a height above ground of 3% of the span) where the induced drag would be zero, it serves to illustrate the point that ground effects will decrease rapidly as the relative height increases. Typically, the wing of an aircraft, as it sits on the ground with landing gear extended, will usually be at a minimum h/b varying from 0.057 (high AR low wing) to 0.20 (low AR high wing), which means that the preceding equation would still be useful. Figure 6.8 compares Eq. (6.13) with another commonly used IGE equation. Actual flight test of a specific aircraft in a given landing configuration will most likely produce results that fall somewhere in between these curves.

The effects of ground proximity can be seen on the lift curve as well. The downwash angle ε is reduced when flying IGE. As $h/s' \Rightarrow \infty$,

$$\varepsilon = C_L / \pi \, \text{AR} \tag{6.14}$$

Once modified for IGE,

$$\varepsilon = \Phi (C_L / \pi \, \text{AR}) \tag{6.15}$$

This effectively increases the slope of the lift curve, which means that more lift will be generated by the wing for a given AoA (Fig. 6.9).

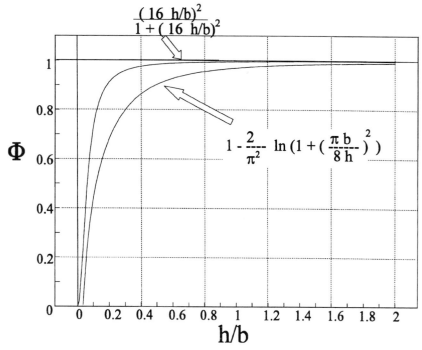

Fig. 6.8   Two ground effects equations.

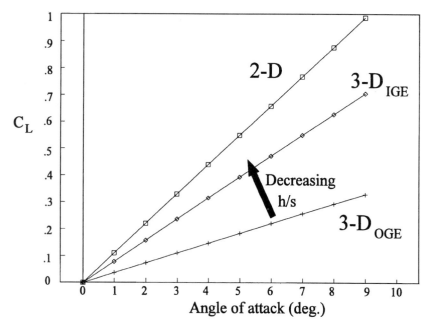

Fig. 6.9   Increasing slope of lift curve IGE.

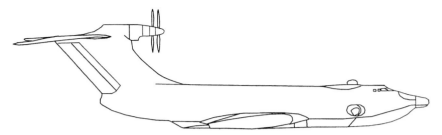

Fig. 6.10   Ekranoplan.

The effects of AR are evident by inspecting the preceding equations. As the AR increases, the magnitude of the ground effects will decrease for a given relative height because the downwash angle is smaller to start with. The proximity of the ground effectively increases the AR of a wing. If AR_{IGE} is the effective AR of the wing IGE, then

$$C_{D_i} = C_{D_{i,IGE}} \qquad (6.16)$$

when

$$AR_{IGE} = AR/\Phi \qquad (6.17)$$

The Russians have experimented extensively with IGE aircraft and have developed many. One such aircraft is the A.90.150 Ekranoplan (wing-IGE craft) (Fig. 6.10). Its propulsive system consists of one Kuznetsov NK-12 turboprop engine powering two counter-rotating propellers and two NK-8 takeoff turbofans mounted in the nose of the aircraft. The turbofans are fitted with vectoring exhaust nozzles to direct the airstream under the wing during takeoff and rearward during transition to cruise flight. The longitudinal stability, which depends on AoA as well as height above ground when flying IGE, is maintained by the large swept tailplane, which is in the propellers slipstream (this further helps by increasing the local dynamic pressure). The Ekranoplan dimensions are length 58 m, wingspan 31.5 m, and height 16 m. It can cruise at 400 km/h and has a range of about 2,000 km.

From the preceding basic analysis, it can be concluded that ground effects will definitely affect aircraft performance when the aircraft is flying in very close proximity to the ground. The influence on the aircraft drag was summed up into one simple equation [Eq. (6.13)], which will be used in the next chapter.

Problems

6.1. Aircraft A is doing some IGE testing. For this, the pilot does a series of low-level flight passes at the following altitudes: 250, 200, 150, 100, and 50 ft. The landing gear is extended ($C_{D,\text{gear}} = 0.015$) and the thrust setting is such that the aircraft flies at a steady $1.4V_{\text{stall}}$ while OGE. What is the effective AR IGE of the aircraft for the different altitudes? What is the steady level flight airspeed while IGE if the thrust setting is not changed from its OGE position? Assume an aircraft weight of 220,000 lb.

6.2. In preparation for landing, the pilot of aircraft A lowers the landing gear ($C_{D,\text{gear}} = 0.015$) and the landing flaps ($C_{D,\text{flaps}} = 0.02$, $C_{L,\text{max}} = 2.2$). The pilot flares a little too high and starts flying parallel to the runway at a height of 30 ft and at $1.1 V_{\text{stall}}$. The thrust setting was enough to maintain $1.1 V_{\text{stall}}$ while OGE. What is the instantaneous acceleration of the aircraft? Assume a landing weight of 200,000 lb. What corrective actions could be done by the pilot to avoid the increase in velocity?

7
Takeoff and Landing

7.1 Introduction

T HE analysis of takeoff and landing performance is very important because airfield performance, the ability to safely takeoff or land within a given runway length, is often a driving factor for aircraft design. As it will be seen, the design requirements for acceptable takeoff and landing distances are different from those for good cruise performance.

7.2 Takeoff

The takeoff is the first phase of flight even though a large portion of it occurs on the ground. The length of this phase will vary according to how its completion is defined. It may be considered completed as soon as the wheels leave the ground, or when the aircraft clears 50 ft above ground altitude, or even when gears and flaps are fully retracted. Applicable regulations for the design of an aircraft takeoff performance include: FAR 23.45, 23.51 and FAR 25.101, 25.107 for civilian aircraft, and MIL-C-005011B (paragraphs 3.4.2.4 and 3.4.5) and AS-5263 (paragraphs 3.5.2.4, 3.5.2.5, and 3.5.5) for military aircraft.

Each of these regulations uses different values for liftoff speed, minimum height clearance, and climb performance. We will define the takeoff phase as starting from brake release, with zero initial velocity, up to the point where the aircraft reaches an altitude of 50 ft above the ground. The aircraft will rotate to increase the wing's AoA at a velocity V_R where it is assumed that the elevator will provide sufficient force to lift the nose gear. The liftoff (main wheels leaving the ground) is then followed by a transition stage where the aircraft goes from horizontal flight to steady climb angle flight (Fig. 7.1).

7.2.1 Jet Aircraft Takeoff

Having set the initial and final conditions, let us break down the takeoff phase into subphases and examine them separately to determine the takeoff distance. A jet aircraft with tricycle gear (aircraft A) will be illustrated. The first subphase is the ground acceleration where the aircraft starts with zero initial velocity and all wheels on the ground. The forces acting on the aircraft at this point are as shown in Fig. 7.2.

The ground surface will be assumed level so that the weight vector is perpendicular to the ground and the thrust vector is parallel to the ground. Summing up the vertical forces, it can be seen that as long as there is some weight on wheels, there will exist an equal and opposing force, called the normal force, acting on the landing gear. This opposing force is equal to

$$N = W - L \qquad (7.1)$$

151

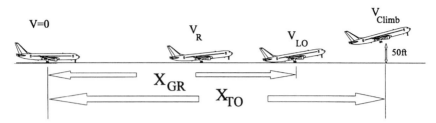

Fig. 7.1   Takeoff phase.

Mechanical losses because of the contact between the wheels and the ground, as well as in the bearings, are accounted for in a horizontal force, which opposes the motion of the aircraft on the ground. This force is called the rolling friction force f, which is proportional to the normal force through the use of a rolling friction force coefficient μ_r. Typical values of μ_r are given in Table 7.1. As the aircraft gains speed, the lift force will increase, therefore decreasing the normal force, which translates into reduced friction force

$$f = \mu_r(W - L) \tag{7.2}$$

Summing up the horizontal forces, shown in Fig. 7.2, gives the following equation:

$$T - D - f = \frac{W}{g}\frac{dV}{dt} \tag{7.3}$$

Substituting the lift and drag equations,

$$T - \tfrac{1}{2}\rho_{SL}\sigma V^2 SC_{D_T} - \mu_r\!\left(W - \tfrac{1}{2}\rho_{SL}\sigma V^2 SC_{L_T}\right) = \frac{W}{g}\frac{dX}{dt}\frac{dV}{dX} \tag{7.4}$$

The takeoff lift and drag coefficients (C_{L_T} and C_{D_T}) depend on the aircraft configuration, that is, the flap setting, the aircraft incidence on the ground, external stores, etc. The AoA during the ground roll up to the point of rotation when the

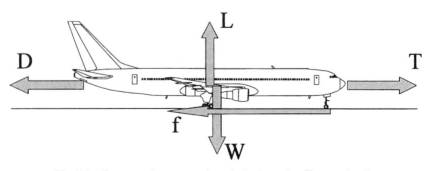

Fig. 7.2   Forces acting on an aircraft during takeoff ground roll.

Table 7.1   Typical values of rolling friction force coefficient μ_r

Runway type	μ_r, brake off	μ_r, brake on
Concrete, asphalt: dry	0.02–0.05	0.3–0.5
wet	0.05	0.15–0.3
iced	0.02	0.06–0.10
Hard turf, short grass	0.05	0.4
Long grass	0.10	0.4
Soft turf	0.07–0.30	0.2–0.4

nose wheel lifts off will remain fairly constant, and so will both the lift and drag coefficients.

To avoid inadvertent liftoff at a velocity that is too low for safe flight, the aircraft attitude is usually such that the AoA is very small up to the point of rotation. During the last part of the ground run, the aircraft is rotated at a rate usually around 2–3 deg/s so as to reach the takeoff AoA at liftoff speed.

To provide an adequate safety margin, the liftoff speed is usually set at 1.2 times the power-off stalling speed (in the takeoff configuration, i.e., takeoff flaps/slats) even though it would be possible for the aircraft to actually lift off at a lower velocity:

$$V_{LO} = 1.2\sqrt{\frac{2(W/S)}{\rho_{SL}\sigma C_{L_{max,TO}}}} = \sqrt{\frac{2(W/S)}{\rho_{SL}\sigma C_{L_{LO}}}} \tag{7.5}$$

where

$$C_{L_{LO}} = \left[1/(1.2)^2\right]C_{L_{max,TO}} \approx 0.694C_{L_{max,TO}} \tag{7.6}$$

The 1.2 factor is used to avoid the possibility of flying in the reversed-command region of the aircraft's thrust-required curve. It also provides a stall margin so that the aircraft does not stall by inadvertently overrotating on takeoff or as a result of a large gust of wind. Finally, aircraft with high-stall AoA [swept wings, with slats/leading-edge (LE) flaps] may be restricted to a maximum AoA during rotation due to ground clearance to avoid a tail strike (Fig. 7.3). Note that, as the aircraft AoA increases, some of the thrust will be directed upward, thus providing a lift contribution. This means that the aircraft may become airborne before reaching the takeoff AoA.

The design of retractable high-lift devices for takeoff is, thus, very important to provide the temporary increase in maximum lift coefficient required for takeoff

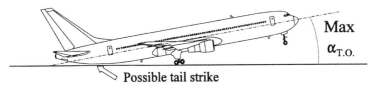

Possible tail strike

Fig. 7.3   Possible tail strike at large takeoff angles.

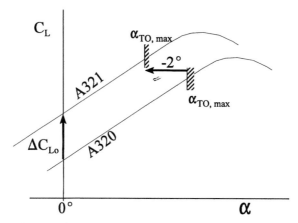

Fig. 7.4   Requirement for a larger coefficient of lift at a reduced angle for the A321.

and landing while minimizing the impact of cruise drag once retracted. As an example,[7] when Airbus industry decided to design the A321, a stretched version of the 150-seat A320, it was desired that the new aircraft should have the same airfield performance as the A320. The new aircraft, the A321, had a greater takeoff weight (approximately 13% larger) and a reduced ground rotation angle (by about 2 deg). The increase in lift coefficient (Fig. 7.4) was achieved by replacing the single-slotted trailing edge flaps with double-slotted ones. Careful design of these new flaps also minimized the reduction in L/D in takeoff conditions, which is important during the climb portion.

Figure 7.5 illustrates the variation of drag and rolling friction force during the takeoff ground roll with constant thrust. Before rotation, the aircraft AoA, thus its lift coefficient, is maintained at a low value, and the total dissipative forces (drag and rolling friction force) remain relatively small. This provides a large excess thrust, which is translated into a relatively large acceleration. As rotation is initiated, the lift coefficient is increased so as to reach its takeoff value at liftoff speed. The rolling friction force will decrease rapidly due to the increase in lift (proportional to C_L) but, at the same time, the induced drag will increase more rapidly (proportional to C_L^2). The net result is, therefore, an increase in the total dissipative force and a decrease in excess thrust and aircraft acceleration in the last seconds before liftoff.

Let us quantify the ground roll subphase. The takeoff ground roll distance depends mainly on the thrust of the aircraft. Assuming that the thrust is constant and neglecting the drag and friction forces (dissipative forces are usually less than 10–20% of the aircraft maximum takeoff thrust), gives the following equation:

$$T = \frac{W}{g}V\frac{dV}{dX} \rightarrow X_{GR} = \frac{1}{g(T/W)}\int_0^{V_{LO}} V\,dV$$

$$X_{GR} = \frac{V_{LO}^2}{2g(T/W)} = \frac{(1.2)^2(W/S)}{\rho_{SL}\sigma C_{L_{max,TO}}g(T/W)}$$

(7.7)

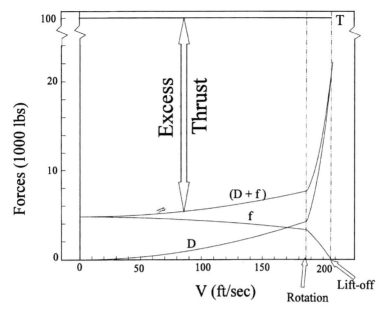

Fig. 7.5 Horizontal forces variation with velocity during ground roll.

which will be the shortest ground roll distance possible. This simple equation is very useful as a first estimate of the takeoff ground run required because it requires only that the liftoff velocity and thrust-to-weight ratio of the aircraft be known. It will also be very useful in determining the effects of altitude, weight, wind conditions, and runway slope on the takeoff ground run distance. Here, the aircraft is subjected to a constant acceleration; thus, the time required to cover the ground roll distance to liftoff is

$$t_{\mathrm{LO}} = \frac{2X_{\mathrm{GR}}}{V_{\mathrm{LO}}} = \frac{V_{\mathrm{LO}}}{g(T/W)} \tag{7.8}$$

If the runway conditions are known (i.e., the rolling friction force coefficient of Table 7.1), one may assume that the dissipative forces are approximately constant throughout the ground run and equal to the rolling friction force (see Fig. 7.5). This approximation yields the following equation (see Appendix D for the development):

$$X_{\mathrm{GR}} = \frac{V_{\mathrm{LO}}^2}{2g[(T/W) - \mu_r)]} \tag{7.9}$$

This last equation accounts for the reduction in excess thrust, due to constant dissipative forces, that could be converted into acceleration. Thus, this will be a useful equation to determine what the effects of runway friction are on the takeoff ground roll distance.

If further information is known about the aircraft, such as the values of the drag and lift coefficients during the ground roll, Eq. (7.4) can be fully integrated to account for all of the dissipative forces during the ground roll. The only assumption that will be made here is that the lift and drag coefficients remain constant during the ground roll (neglect the rotation part, which represents only a small part of the total ground roll distance). Rewriting Eq. (7.4) gives

$$dX = \frac{W V \, dV}{g\left(T - \frac{1}{2}\rho_{SL}\sigma S V^2 C_{D_T} - \mu_r\left(W - \frac{1}{2}\rho_{SL}\sigma S V^2 C_{L_T}\right)\right)} \qquad (7.10)$$

where

$$C_{D_T} = C_{D_0} + \Delta C_{D_{gear}} + \Delta C_{D_{flaps}} + C_{D_i}$$

$$C_{D_i} = \Phi K C_{L_T}^2$$

$$\Phi = 1 - \frac{2}{\pi^2}\ell_n\left(1 + \left[\frac{\pi}{8}\frac{b}{h}\right]^2\right) \qquad (7.11)$$

$$\frac{\partial W}{\partial t} = -SFC_T T$$

Once integrated (Appendix D), the equation for the ground roll distance is

$$X_{GR} = \frac{\Omega^2}{2g}\ell_n\left[\frac{(T/W) - \mu_r}{(T/W) - \mu_r - (V_{LO}/\Omega)^2}\right] \qquad (7.12)$$

where

$$\Omega = \sqrt{\frac{W}{\frac{1}{2}\rho_{SL}\sigma S\left(C_{D_T} - \mu_r C_{L_T}\right)}} \qquad (7.13)$$

After liftoff, there exists a transition region where the aircraft goes from horizontal flight to steady climbing flight. During this transition, the aircraft will accelerate to reach $1.3V_{stall}$ at 50-ft altitude above ground level (AGL). The higher velocity will allow the pilot to trade AoA for climb angle while maintaining lift. During transition, the aircraft will be in a pullup where it can be assumed that the radius r is constant and the load factor will generally vary from 1.01 to 1.25 depending on the aircraft type and the pilot. During transition, the aircraft may or may not reach 50-ft AGL. In the first case, the takeoff can be considered complete; in the latter case, the aircraft achieves steady climb angle (while still accelerating to reach $1.3V_{Stall}$) up to 50-ft AGL. Rewriting Eq. (5.16) in terms of horizontal distance covered yields

$$\frac{(T - D)}{W}V = \frac{dh}{dt} + \frac{V}{g}\frac{dV}{dt} \rightarrow \frac{T - D}{W} = \frac{dh}{dX} + \frac{V}{g}\frac{dV}{dX}$$

$$\therefore dX = \frac{W}{(T - D)}\left[dh + \frac{V}{g}dV\right] \qquad (7.14)$$

If we assume that the drag component will be approximately constant during this subphase and equal to the liftoff value, we get the following equation for the horizontal distance of the climb–acceleration subphase:

$$X_{climb/acc} \approx \frac{W}{(T - D_{LO})} \left[\int_0^{50ft} dh + \frac{1}{g} \int_{1.2V_{stall}}^{1.3V_{stall}} V \, dV \right]$$

$$X_{climb/acc} \approx \frac{W}{(T - D_{LO})} \left[50\,ft + \frac{V_{stall}^2}{8g} \right]$$

(7.15)

The total distance for the takeoff phase is then the summation of the ground roll and initial climb to 50-ft phases. Note that the aircraft could reach the 50-ft AGL in a smaller distance if it maintained its velocity to its liftoff value (no excess power used to accelerate). For aircraft A at maximum takeoff weight and on a sea level runway, this represents an 82% reduction in the horizontal distance required to climb to 50 ft (226 ft instead of 1,260 ft) after liftoff. This can be useful to clear obstacles on takeoff, but this does not provide the additional safety of the increased stall margin of flying at $1.3V_{stall}$.

Consider aircraft A at one end of the runway with brakes on and zero forward velocity. The flaps are set so as to get $C_L = \mu_r/(2\Phi K)$ and the aircraft is at 80% of its maximum weight. The runway is at sea level. Because of the extended flap, which changes the lift distribution over the wing span, the value of e will decrease by about 5% ($e = 0.76$). $C_{L_{max}}$ with flaps is 2.0. The wing is at a relative height $h/b = 0.079$ when the aircraft is on the ground so that ground effects can be accounted for. The minimum drag coefficient increase due to the extended landing gear is 0.02 and due to the flaps is 0.015. The runway is made of asphalt with a rolling friction force coefficient of 0.02. The thrust will be assumed constant throughout the takeoff run and set at its maximum value.

In general, aircraft have enough speed to rotate when they reach 90% of the liftoff speed. When the aircraft reaches its liftoff speed it will leave the ground and climb at an incidence such that it reaches $1.3V_{stall}$ at 50 ft above the ground. The landing gear will not be retracted before the 50-ft above-ground altitude.

Using Eqs. (7.7), (7.9), and (7.12), the estimates for takeoff ground roll are 1,759, 1,848, and 1,927 ft, respectively (Fig. 7.6). Thus, the simple Eq. (7.7) provided an estimate that is only 8.7% lower than the more complicated Eq. (7.12) even though it required only the aircraft liftoff speed and thrust-to-weight ratio. But the gap between Eq. (7.7) and Eq. (7.12) will increase as the ratio of the thrust to the dissipative forces (rolling friction force and aerodynamic drag) decreases.

For the climb portion, the distance required to reach an altitude of 50-ft AGL from the end of the ground roll may vary according to the aircraft's climb angle. FAR 23 requires a minimum rate of climb of 300 ft/min gear up at sea level; FAR 25 recommends a climb gradient of 1/2% at liftoff with the landing gear extended and of 3% at climb speed with the landing gear up, whereas MIL specifications require 500 ft/min for all engines operating and 100 ft/min for one inoperative engine.

The reduction in weight during the ground roll results from the fuel consumed. As the weight decreases, so does the rolling friction force, which helps to reduce the ground roll distance. The weight at liftoff is, in this case, 99.8% of the weight at the brake release point. Thus, the assumption of constant weight during takeoff has very little effect on the estimated takeoff distance.

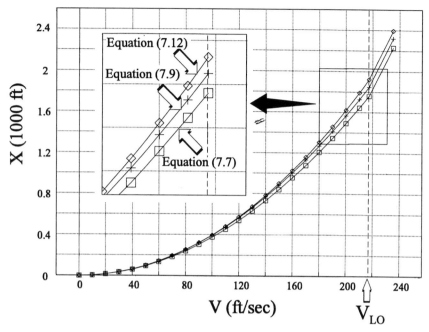

Fig. 7.6   Takeoff distance.

The elevation of the runway will have a pronounced effect on the takeoff distance as well as on the liftoff speed [Eq. (7.5)] and on the available thrust [Eq. (2.58)] compared to a sea level elevation. This is due to the change in air density. By using Eq. (7.7), the increase in takeoff distance can be estimated,

$$X_{LO_{alt}}/X_{LO_{SL}} \simeq 1/\sigma^{1.7} \tag{7.16}$$

A change in ambient temperature and pressure, away from the standard conditions, and an increase in humidity (from zero for the standard atmosphere) will also affect the air density. Generally, an increase in temperature and/or humidity will increase the takeoff distance because of the decrease in air density. The takeoff performance is better on a dry winter day than on a humid summer day. If an equivalent density ratio can be found, the change in takeoff distance can be approximated with Eq. (7.16). An increase in weight results in an increase in the takeoff distance. Again using Eq. (7.7), the effect of weight on takeoff distance is seen to be

$$X_{LO,W_2}/X_{LO,W_1} = W_2^2/W_1^2 \tag{7.17}$$

Because the aircraft aerodynamic forces are a function of the airspeed and not of the ground speed, a headwind during the takeoff run will reduce the takeoff distance. This, in effect, means that the aircraft will takeoff at a lower ground speed than during a no-wind takeoff. The effect on the takeoff distance can be approximated from Eq. (7.7). The result is the ratio of the takeoff ground speed with wind to the takeoff ground speed without wind (V_W for a headwind being negative):

$$\frac{X_{LO_{wind}}}{X_{LO}} \approx \frac{(V_{LO} \pm V_W)^2}{V_{LO}^2} \tag{7.18}$$

Another item that will affect the takeoff distance is the slope of the runway. Its effect is equivalent to having an additional force in the direction of the aircraft motion due to the component of aircraft weight parallel to the runway:

$$\frac{X_{LO_\theta}}{X_{LO_{\theta=0}}} = \frac{T_{\theta=0}}{T + W \sin(\theta)} \tag{7.19}$$

Using the information from Table 7.1 and Eq. (7.9), the effects of runway surface condition on takeoff ground roll distance can be evaluated:

$$\frac{X_{\mu_2}}{X_{\mu_1}} = \frac{(T/W) - \mu_1}{(T/W) - \mu_2} \tag{7.20}$$

There exists a lift coefficient that will produce the lowest resistance to acceleration during the takeoff ground roll, thus resulting in a shorter liftoff distance. Differentiating Eq. (7.3) with respect to the lift coefficient and equating the result to zero will give

$$\frac{\partial T}{\partial C_L} - \frac{\partial D}{\partial C_L} - \frac{\partial f}{\partial C_L} = 0$$

$$0 - qS(0 + \Phi K 2C_L) - \mu_r(0 - qs) = 0 \tag{7.21}$$

Thus, the lift coefficient required for the shortest takeoff distance is

$$C_L = \frac{\mu_r}{2\Phi K} \tag{7.22}$$

The possibility of engine failure is always in the mind of every pilot and is most critical during the takeoff run or during the initial climb. For a single-engine aircraft there is no other alternative than to brake while on the ground or to land if in the air. For a multiengine aircraft, the runway must be long enough for the aircraft to safely takeoff after an engine failure or to bring the aircraft to a complete stop. The minimum runway length that satisfies this requirement is known as the balanced field length (BFL), FAR 25.113. Under FAR 25.113 rules, that distance must also be equal or greater than 115% of the horizontal distance along the takeoff path, with all engines operating, from the start of the takeoff to the point at which the aircraft is 35 ft (50 ft for FAR 23 and military) above the takeoff surface, as determined by a procedure consistent with FAR 25.111.

The BFL is determined by the point of intersection of the two curves shown in Fig. 7.7. One curve gives the distance of the ground roll with all engines operating up to the engine failure speed plus the distance required to continue the acceleration up to liftoff and climb to 50 ft (35 ft for FAR 25) AGL with the remaining operating

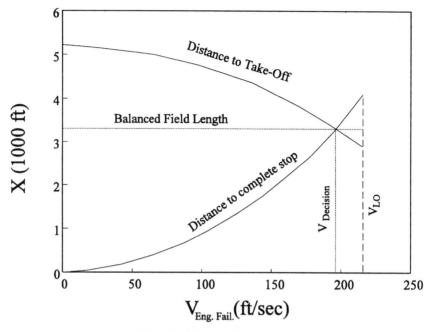

Fig. 7.7   Balanced field length.

engine(s) (upper curve). The lower curve represents the distance required for the aircraft to accelerate up to the engine failure speed, plus the distance covered with the remaining engine(s) developing maximum thrust while the pilot reacts to the situation (approximately 1 s), plus the distance required to bring the aircraft to a complete stop with brakes ($\mu_{\text{brake}} \approx 0.5$). During this deceleration, the remaining operating engine(s) would be at idle ($T_{\text{idle}} \approx 5\%T_{\text{max}}$).

It is observed, from Fig. 7.7, that if an engine failure occurs before V_{decision} (also called V_1), the aircraft can be safely brought to a stop if the runway length is equal to the BFL, whereas if an engine failure occurs at a velocity greater than V_{decision}, the aircraft can safely continue its takeoff and get airborne before reaching the end of the runway.

If the runway used for takeoff is longer than the BFL (see Fig. 7.8) three specific zones can be identified. In zone 1, the aircraft must be brought to a stop because the distance required to liftoff on the remaining engine(s) is greater than the runway length available, whereas in zone 3 the aircraft must complete its takeoff. Within zone 2 the pilot has the option to abort or to complete the takeoff. For a runway shorter than the BFL, zone 4 is identical to zone 1 and zone 6 is identical to zone 3, but zone 5 is a danger zone where the remaining runway length at engine failure will be insufficient to allow the aircraft to lift off or to be brought to a complete stop. Thrust reversers cannot be used in the calculation because of the inherent danger of an asymmetric force created by one engine being inoperative. The pilot must always use a runway that is longer than the aircraft's BFL and use the decision velocity (V_1) as a reference for go/no go decisions. It is clear, from the equations just seen, that the only control the pilot has over the runway

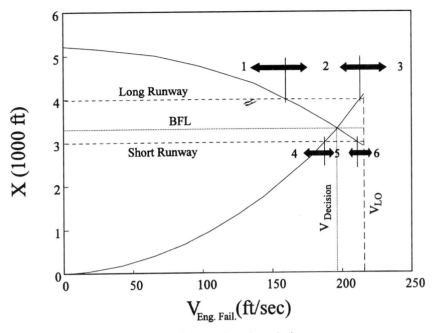

Fig. 7.8   Runway length analysis.

length required for a safe takeoff attempt is through the control of the weight of the aircraft prior to initiating the takeoff.

Note that a twin-engine aircraft will have a larger BFL than a four-engine aircraft with the same thrust-to-weight class and liftoff speed because of the larger percentage of lost thrust when one engine goes down (50% reduction compared to 25%).

7.2.2 Piston-Prop Aircraft Takeoff

The analysis of takeoff for a piston-prop aircraft is basically the same as the one for a jet aircraft (Fig. 7.9). The main difference here is with our assumption that the thrust is constant up to Mach 0.1, then the thrust decreases as the inverse of the velocity ($T = P_a / V$, see Chapter 2, propulsion assumptions). The equations of the takeoff analysis for a piston-prop are identical to Eqs. (7.1–7.3), reproduced here for convenience

$$N = W - L \tag{7.23}$$

$$f = \mu_r(W - L) \tag{7.24}$$

$$T - D - f = \frac{W}{g}\frac{dV}{dt} \tag{7.25}$$

For the first part of the takeoff ground run, any one of Eqs. (7.7), (7.9), or (7.12) can be used to estimate the distance required to accelerate to Mach 0.1

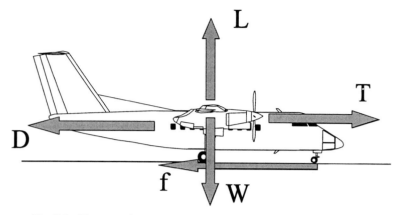

Fig. 7.9   Forces acting on a piston-prop aircraft during ground roll.

(constant thrust assumption as per Chapter 2). For the ground run from Mach 0.1 to the liftoff velocity, the power available will be constant but the thrust will decrease with increasing velocity. Rewriting Eq. (7.25), and neglecting the effects of rotation,

$$\frac{\eta_P P}{V} - D - f = \frac{W}{g} V \frac{dV}{dX}$$

$$dX = \frac{W}{g} \int_{V_{\text{Mach0.1}}}^{V_{\text{LO}}} \frac{V^2 dV}{\eta_P P - \frac{1}{2}\rho_{\text{SL}}\sigma V^3 S\left(C_{D_T} - \mu_r C_{L_T}\right) - \mu_r W V}$$

To simplify the integration, it can be assumed that the dissipative forces will remain approximately constant during this part of the ground run and they will be equal to the value of the sum of the dissipative forces at Mach 0.1 (see Fig. 7.10). The resulting equation is

$$X_{\text{Mach0.1} \to V_{\text{LO}}} = \frac{W}{g[\eta_P P - (D+f)_{\text{Mach0.1}}]} \int_{V_{\text{Mach0.1}}}^{V_{\text{LO}}} V^2 \, dV$$

$$\therefore X_{\text{Mach0.1} \to V_{\text{LO}}} = \frac{W\left(V_{\text{LO}}^3 - V_{\text{Mach0.1}}^3\right)}{g[\eta_P P - (D+f)_{\text{Mach0.1}}]}$$

(7.26)

Note that the assumption of constant thrust during part of the takeoff will introduce some errors in the calculation of the takeoff distance. Because thrust is a combination of engine power and propeller efficiency, it is very hard to estimate without flight testing. In fact, an error as small as 2% in the prediction of thrust may have the same impact on the calculation of the takeoff distance as an error of 50% in the prediction of the aircraft drag and ground friction coefficient due to the relative importance of these last forces compared to the thrust during takeoff. At low-forward speed, the propeller blades are partially stalled because of the low-advance ratio (J) value (see Chapter 2).

The liftoff velocity of a piston-prop aircraft is still defined as being 20% over the power-off stall velocity in the takeoff configuration for the aircraft [Eq. (7.5)].

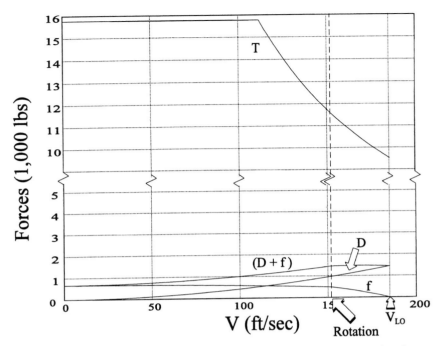

Fig. 7.10   Force variations with velocity during the takeoff ground run of a piston-prop aircraft.

However, the lift generated by the wing, in most cases for piston-prop aircraft, will be affected by the propeller slipstream. Anything located in the propeller slipstream will be subjected to a higher local dynamic pressure. Thus, the wing may be generating more lift than the expected lift due to the forward speed. Note also that the drag may be slightly higher for the same reason.

Tail draggers, aircraft with a tail wheel, have a slightly different takeoff run compared to those with a nose wheel. They start their takeoff run with a high AoA (see Fig. 7.11), with the tail wheel on the ground. When the velocity is high enough to give aerodynamic pitch control to the horizontal tail, the tail wheel is raised. This first phase is relatively short. Raising the tail lowers the AoA and the aircraft drag due to lift. Although the ground roll friction force may increase, the overall effect is generally to reduce the retarding forces (combined drag and ground roll friction forces), thus allowing for a faster acceleration. Provided the runway is long enough, the aircraft can be kept in this low AoA attitude and takeoff once

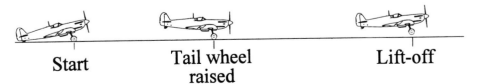

Fig. 7.11   Tail dragger takeoff sequence.

it has reached a velocity that provides enough lift, or the aircraft can be rotated at a velocity near liftoff speed ($1.2V_{\text{stall,TO}}$) just as in the case of a nose wheel aircraft.

7.3 Landing

The landing phase extends from 50 ft above the ground to a complete stop on the runway. Applicable regulations (FAR 23.75; FAR 25.125; MIL-C-005011B paragraphs 3.4.2.11, 3.4.2.12, and 3.4.7; and AS-5263 paragraphs 3.5.2.12, 3.5.2.13, and 3.5.7) use different definitions of approach speed varying from 1.2 to 1.3 times the approach configuration stall speed.

For the analysis of the landing performance, this phase will be divided into an approach segment, where the aircraft comes in at a constant descent angle. Then there is a roundout segment, where the rate of descent is reduced to zero. This is followed by a deceleration segment just above the runway where the airspeed is reduced prior to touchdown. Finally, there is the ground roll segment, which extends from the point of touchdown to that where the aircraft has been brought to a complete stop (Fig. 7.12).

To determine the total landing distance, the distance traveled in each of the four segments listed will be evaluated. The first segment, the approach, is fairly simple. It is performed at a constant angle of descent and constant airspeed ($1.2V_{\text{stall}}$) from an initial altitude of 50 ft. The final altitude is h_2. The distance covered along the ground (X_a) and the rate of descent are then

$$X_a = \frac{-(50\,\text{ft} - h_2)}{\tan(\gamma_a)}$$

$$SR = V_a \sin(\gamma_a)$$

(7.27)

The angle of descent during the approach is usually around 3 deg but could be more if dictated by a specific airport rule [such as for London City Airport, a 3380-ft runway in the heart of the city with a 5.5-deg instrument landing system approach] or less if the residual thrust of the aircraft is such that the descent angle cannot be increased while maintaining a constant approach speed (Fig. 7.13). This approach angle can be determined by using Eq. (4.6), repeated here for convenience

$$\gamma_a = \arcsin\left(\frac{T_{\text{residual}} - D}{W}\right)$$

(7.28)

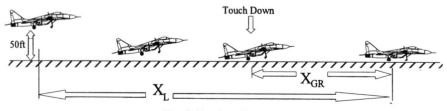

Fig. 7.12   Landing phase.

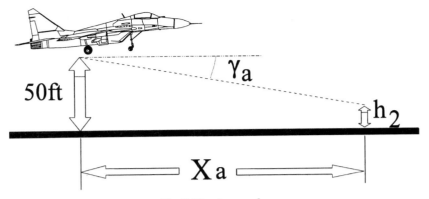

Fig. 7.13   Approach.

To find the value of h_2, an analysis of the roundout segment of the landing, also known as the flare, is required (Fig. 7.14). This segment consists in decreasing the rate of descent to zero. This is done in the same way as pulling out of a dive (Chapter 4, Sec. 4.9). Here, the airspeed and the load factor, as well as the radius, will be considered constant.

At h_2, the load factor is

$$n = \left(V_a^2/rg\right) + \cos(\gamma_a) \tag{7.29}$$

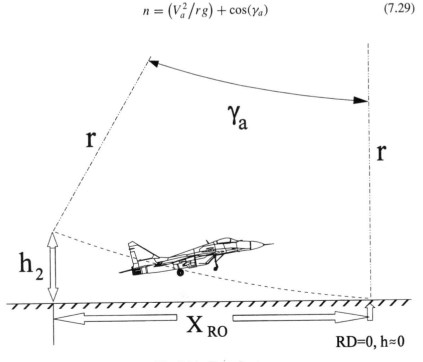

Fig. 7.14   Roundout.

The height h_2 and the roundout horizontal distance are, respectively,

$$h_2 = r[1 - \cos(\gamma_a)]$$
$$X_{RO} = r \sin(\gamma_a)$$

(7.30)

For the last in-flight segment, the aircraft is barely above the ground, and it decelerates from $V_a(= 1.2V_{\text{stall}})$ to $V_{TD}(= 1.1V_{\text{stall}})$. The horizontal distance required to slow the aircraft will depend on the excess drag it has. Usually, at this point the pilot retards the throttle completely (if this was not done during the approach) leaving only the residual thrust. The horizontal distance is (assuming the drag term remains approximately constant and equal to the approach value)

$$\frac{T_{\text{residual}} - D}{W} = \frac{V}{g}\frac{dV}{dX}$$

(7.31)

$$X_{\text{decel}} \approx \frac{W\left(V_{TD}^2 - V_a^2\right)}{g(T_{\text{residual}} - D_{\text{approach}})}$$

(7.32)

It should be noted that this is only one way to calculate the variables during the flare and that the actual trajectory will depend on how the pilot performs this part of the landing phase. The load factor is expected to increase gradually from a value slightly inferior to 1.0 (an approach at a steady 3-deg descent angle gives a load factor of 0.9986) to a maximum value (usually between 1.01 and 1.25) and to go back down to 1.0 after touchdown. If the pilot flares too high, the aircraft may also start floating above the runway because of ground effects.

Aircraft designed for aircraft carrier operations, such as the F/A-18, are not required to flare before landing because the landing gear was designed to absorb the shock of landing, provided the aircraft weight is below a specified value. In this case the entire deceleration occurs on the ground.

During the last part of the landing phase, the ground roll segment (Fig. 7.15), the forces acting on the aircraft are essentially the same as during the takeoff ground roll (see Fig. 7.2). Therefore, the equation of motion during the ground roll is

$$T - \tfrac{1}{2}\rho V^2 SC_{D_L} - \mu\left(W - \tfrac{1}{2}\rho V^2 SC_{L_L}\right) = \frac{W}{g}\frac{dX}{dt}\frac{dV}{dX}$$

(7.33)

To shorten the ground roll as much as possible, the pilot will retard the throttle to its minimum setting (idle thrust $\approx$ 5–10% of maximum thrust). In some cases, the pilot has the option of using thrust reversers to deflect the flow from the engines forward, effectively creating a force component acting in the same direction as the drag (negative thrust). This will be discussed a little more at the end of this section but, for this analysis, the residual thrust will be simply the idle thrust from the engines.

The rolling friction force coefficient during braking is usually between 0.4 and 0.5 on a dry concrete/asphalt runway (see Table 7.1). It can be seen that, again to shorten the ground roll distance, the pilot may want to increase the rolling friction force. One way to accomplish this is to decrease lift as much as possible. This

Touch Down

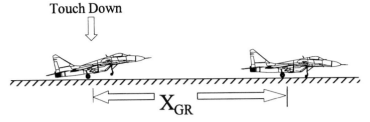

Fig. 7.15   Ground run.

can be done by bringing the nose of the aircraft down as quickly as possible after landing and by using spoilers if they are available because they have a combined effect of reducing the lift while increasing the minimum drag of the aircraft.

Some aircraft are able to keep the landing drag coefficient high by using aerodynamic braking (Fig. 7.16). This is accomplished by keeping the aircraft's AoA high during the ground roll, thus keeping the induced drag coefficient high. This may be effective in the initial part of the ground roll where the velocity is still high (and where a pilot may not want to use excessive braking to avoid brake overheating). However, in the latter part, the ground roll friction is more effective and lift must be minimized. It should be noted that many aircraft cannot use this technique because they do not possess sufficient controllability from wheel braking power only while aerobraking. Other aircraft use parachutes for braking at high speeds.

From Fig. 7.17, it is seen that the aerodynamic drag force component is relatively small after the nose gear touchdown and that the ground roll forces can be approximated using ground roll friction force f only. Assuming zero lift upon touchdown, thus maximum ground roll friction, Eq. (7.25) then reduces to

$$T_{\text{res}} - \mu_r W = \frac{W}{g} V \frac{\mathrm{d}V}{\mathrm{d}X}$$

$$\mathrm{d}X = \frac{V}{[(T_{\text{res}}/W) - \mu_r]g} \, \mathrm{d}V$$

$$\int_0^{X_{\text{GR}}} \mathrm{d}X = \frac{1}{g[(T_{\text{res}}/W) - \mu_r]} \int_{V_{\text{TD}}}^0 V \, \mathrm{d}V$$

$$\therefore X_{\text{GR}} = \frac{-V_{\text{TD}}^2}{2g[(T_{\text{res}}/W) - \mu_r]} = \frac{-1.21(W/S)}{\rho_{\text{SL}}\sigma C_{L_L} g[(T_{\text{res}}/W) - \mu_r]} \tag{7.34}$$

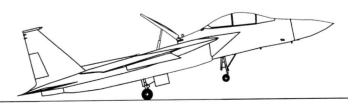

Fig. 7.16   Aerodynamic braking.

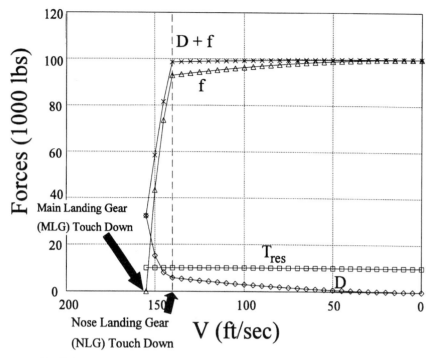

Fig. 7.17 Landing ground roll, aircraft A at 200,000-lb landing weight.

Consider an example. Aircraft A at 200,000-lb landing weight has the following parameters: idle thrust of 10% of maximum thrust, landing maximum lift coefficient of 2.5, minimum drag coefficient increase of 0.02 due to the landing gear and of 0.04 due to the landing flaps, Oswald coefficient value of 0.72, rolling friction force coefficient of 0.5, and level runway at sea level. The graph of Fig. 7.18 results.

As can be seen from the graph in Fig. 7.18, the aircraft will cover a lot less ground at low-forward speed for the same deceleration ($dV/dt = V\, dV/dX$). The ground roll distance obtained from Eq. (7.34) was 912 ft. Doing the same exercise but with a residual thrust of 5%, the ground run is 864 ft (5.3% difference), whereas if the residual thrust is zero, the result is 821 ft (10% difference). Equation (7.34) assumes that there is a constant braking force from the point of touchdown, which is not quite true (see Fig. 7.17). During the first portion of the ground roll where the velocity is still relatively high, the aircraft will cover a large distance before the total braking force reaches its maximum value.

The use of thrust reversers during the initial part of the ground roll segment is commonly used on most airliners and on some fighters (such as the Tornado). It consists of deflecting the thrust vector forward (or at an angle) so as to contribute a force vector opposed to the movement of the aircraft. In the previous example, if 25% of the maximum thrust is now deflected forward ($T_{res} = -0.25T_{max}$) from touchdown to a velocity equal to 50% of the touchdown velocity, after which there is a residual forward thrust equal to 10% of maximum, the ground roll distance would only be 721 ft (20.9% reduction from the first case).

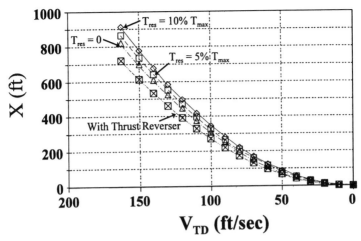

Fig. 7.18 Ground roll distance to travel before complete stop as a function of velocity.

An aircraft that uses thrust reversing during the ground roll may be faced with reingestion problems. The risk of ingesting hot exhaust gases, or debris from the runway, depends on the aircraft configuration, such as how close the exhaust nozzles are from the engine inlets, and on the dynamic pressure ratio (q ratio, that is, the ratio of the jets dynamic pressure to the freestream dynamic pressure). The q ratio is

$$q \text{ ratio} \equiv q_{\text{jet}}/q_{\infty} \tag{7.35}$$

A typical q ratio of 27 can be expected during the landing approach.[8] This ratio will increase sharply as the aircraft slows down (decreasing freestream q) with the engine at a constant power setting. Today's transport aircraft using thrust reversing usually limit the use of such device to the high-speed portion of the ground roll to prevent reingestion and because the higher reverse thrust power will be generated at the higher aircraft velocity. Future fighters may have to use thrust reversing all the way from approach to a complete stop on the runway.

Equation (7.34) is useful in determining the approximate effects of the various landing parameters during the landing ground roll in the same manner that Eq. (7.7) and (7.9) did for the takeoff analysis.

Problems

For this section, use Eq. (7.7) for the ground roll during the takeoff and Eq. (7.34) for the ground roll during landing. Use Eq. (7.22) for the lift coefficient value during the ground run.

7.1. Aircraft C is at the beginning of a sea level runway at maximum takeoff weight (clean). The pilot advances the throttle to maximum thrust (afterburner). What is the takeoff distance (ground roll plus climb to 50 ft and $1.3V_{\text{stall}}$)? What will be the takeoff distance if the pilot does not use an afterburner? What would those distances be if the runway was at 5,000 ft?

7.2. The first 3,000 ft of a 10,000-ft runway are sloped down 1 deg. Aircraft A at full takeoff weight is at the beginning of the sea level runway with takeoff flaps extended ($C_{D,\text{flaps}} = 0.01$, $C_{L,\text{max}} = 2.0$, $C_{D,\text{gear}} = 0.015$). If the pilot advances the throttle to maximum thrust, will the aircraft lift off in less than 3,000 ft? What is the effect of the slope on the takeoff ground run? If the runway is at an altitude of 5,000 ft, can the aircraft lift off in less than 3,000 ft?

7.3. Aircraft A is in approach for landing on a sea level runway. The landing gear is extended ($C_{D,\text{gear}} = 0.015$), as are the landing flaps ($C_{D,\text{flaps}} = 0.02$, $C_{L,\text{max}} = 2.2$), and the aircraft weight is 200,000 lb. What is the normal approach velocity (V_a) and touchdown velocity (V_{TD} or V_L)? What is the value of the lift coefficient during the approach (C_{L_a}) and upon touchdown ($C_{L,L}$)? Assuming that the rolling friction force coefficient during the ground roll is 0.35 and 5% residual thrust, what is the ground roll distance?

8
Energy Method

8.1 Introduction

T HE methods used up to now are quite effective for most subsonic aircraft, but for high-performance aircraft in accelerated flights, especially during climbs and turns, these methods may fall short of determining the optimal performance. This chapter introduces the concepts of specific energy and energy maneuverability as a different approach to aircraft performance.

8.2 Aircraft Specific Energy

Instead of the forces acting on an aircraft, consider its energy state. While flying at a given altitude the aircraft will possess a certain amount of potential energy (function of mass and altitude) and kinetic energy (function of mass and velocity). An aircraft's specific energy (energy per unit weight) is its total energy divided by its weight; thus,

$$H_e = \frac{mgh + \frac{1}{2}mV^2}{W} = h + \frac{V^2}{2g} \tag{8.1}$$

Specific energy H_e has units of altitude (it is sometimes referred to as energy height) and is independent of any aircraft's characteristics. Figure 8.1 shows lines of constant energy height as a function of velocity and of Mach number.

An aircraft flying at 400 ft/s and at an altitude of 7,500 ft has an energy height of 9,984.5 ft. This means that it could theoretically (if the aircraft was only subjected to the force of gravity) reach a maximum altitude of 9,984.5 ft by trading all of its kinetic energy for potential energy, or it could achieve a velocity of 802 ft/s at zero altitude by trading its potential energy for kinetic energy. However, this neither takes into account any energy dissipated by the aircraft's drag nor any energy input from the aircraft's thrust. The rate of change per unit time of energy height is called specific power P_s, which has units of velocity,

$$P_s = \frac{dH_e}{dt} = \frac{dh}{dt} + \frac{V}{g}\frac{dV}{dt} \tag{8.2}$$

Thus, specific power is the sum of the rate of change of altitude of an aircraft and its acceleration multiplied by its velocity along a given flight path. It is sometimes referred to as the aircraft specific excess power, which was discussed in previous sections (aircraft specific power available minus the specific power required). Let us rewrite Eq. (5.16) here

$$\frac{P_a - P_r}{W} = \frac{dh}{dt} + \frac{V}{g}\frac{dV}{dt} = P_s \tag{8.3}$$

171

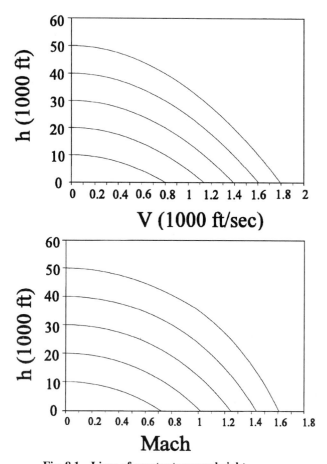

Fig. 8.1   Lines of constant energy height.

The amount of specific excess power (or the specific power) available during flight can be used to determine the maneuvering capability of an aircraft. A plot of constant specific power as a function of altitude and airspeed can be calculated as long as the thrust-to-weight ratio (T/W), the wing loading (W/S), and the load factor of the aircraft are specified. Figure 8.2 illustrates the lines of constant specific power (solid lines) for aircraft C at a throttle setting that provides 8,000 lb of thrust at sea level. The aircraft is at maximum weight (clean) and has a load factor of 1. The lines of constant energy height are superimposed.

The curves shown can be interpreted as follows: at any point along the curve where P_s is zero, the aircraft is in steady level flight and this represents the aircraft's flight envelope, for the given throttle setting, previously analyzed and shown in Figs. 3.14–3.18 (note, however, that the aircraft stall velocity may restrict flight at the lower flight speeds). In the region where P_s is greater than zero, the aircraft may accelerate, climb, or do both simultaneously, or may trade off one against the other. A point outside of the curve $P_s = 0$, where P_s is in fact negative, may be

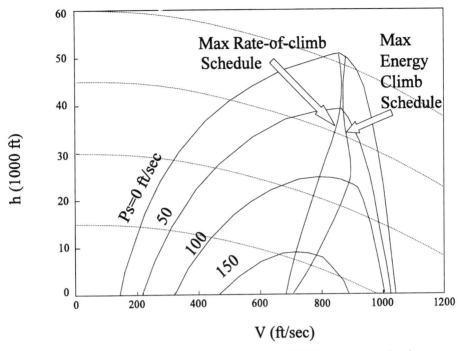

V (ft/sec)

Fig. 8.2   Specific excess power for aircraft C at 8,000-lb thrust at sea level.

reached only by zooming the aircraft, that is, trading kinetic energy for potential energy or vice versa. It must be remembered that, in that region, the aircraft can only go from an energy height state to one of lower level because P_s is negative; it is an unsteady-state condition.

The rate of climb of an aircraft, for a given climb schedule, can be determined with the help of the energy plot (Fig. 8.2) and by rewriting Eq. (8.2) in the following form:

$$RC = \frac{dh}{dt} = P_s \bigg/ \left(1 + \frac{V}{g}\frac{dV}{dh}\right) \tag{8.4}$$

where P_s is the specific excess power at a given velocity and altitude and dV/dh is the slope of the climb schedule trajectory on the energy plot. Two climb schedules are of particular interest: 1) the schedule to reach an altitude in minimum time (the maximum rate-of-climb schedule) and 2) the schedule to gain the maximum amount of energy height in minimum time.

The maximum rate of climb at a given altitude is determined by maximizing the specific excess power at that altitude (as per Chapter 4), that is, flying through the points on the P_s curves that are tangent to the lines of constant altitude. Mathematically, the maximum rate of climb will occur when

$$\frac{\partial}{\partial V}(P_s)_{h=\text{const}} = 0 \tag{8.5}$$

On the other hand, the maximum energy climb is found by maximizing the specific excess power with respect to the energy height, that is, flying through the points on the P_s curves that are tangent to the lines of constant energy height. Mathematically, it is expressed as follows:

$$\frac{\partial}{\partial V}(P_s)_{H_e=\text{const}} = 0 \qquad (8.6)$$

For a pure subsonic aircraft these two curves are usually close to one another. In Fig. 8.2, compressibility effects are readily apparent on the schedule for both climbs as observed from the discontinuities in the slope of each curve.

In the case of high-performance supersonic aircraft (such as fighters and interceptors), it is often important to reach a given altitude and velocity in minimum time to establish a tactical advantage. One of the problems faced by these aircraft, especially 1950s and 1960s designs, is the large drag increase in the transonic region such that the excess thrust in that speed regime is marginal. One way to alleviate this problem is to operate the aircraft, in the transonic regime, in such a way that the highest value of specific power is maintained without decreasing the energy height. When an energy-height contour is reached that is tangent to two equal-valued P_s contours (as shown in Fig. 8.3), the aircraft is put into a constant

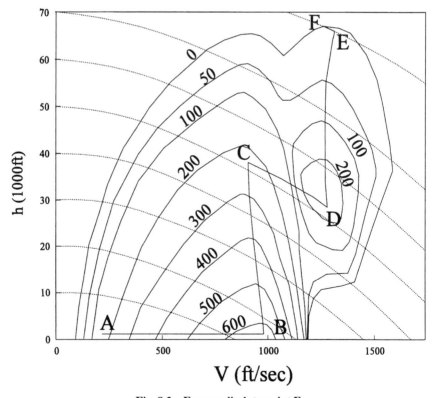

Fig. 8.3   Energy climb to point F.

energy-height dive until normal climb can be resumed. A zero-g dive yields the most rapid increase in specific energy because drag due to lift is zero and the total drag coefficient of the aircraft is at a minimum. Furthermore, in a dive, a component of the aircraft's weight acts in the direction of motion, thus helping to combat the large increase in drag encountered in the transonic region and, also, the available thrust increases as the altitude decreases. It should be noted that in an actual climb, pullouts and nose overs will decrease the specific power because of the rotational energy required to perform the maneuvers. This will also modify the flight path somewhat. The dark lines in Fig. 8.3 illustrate a maximum energy climb following a sea level takeoff (point A) to the aircraft's ceiling (point F).

As an example, an F-4 Phantom requires approximately 640 s to reach Mach 2.2 at 35,000 ft from the beginning of takeoff (0 ft/s and sea level) using the maximum rate-of-climb schedule followed by a level flight acceleration to Mach 2.2, whereas it requires only around 275 s using the maximum energy climb schedule. On Feb. 29, 1992, the B-1B established 11 time-to-climb records for its weight class. For the 335,000-lb climb to 40,000 ft, the climb schedule included a nose over to accelerate through the transonic zone up to Mach 1.2 before pulling up again and zooming to 40,000 ft.

8.3 Energy Maneuverability

Conventional turn performance equations can be used to indicate whether one aircraft has a turn advantage over another. However, such equations provide only a relative measure of sustained (fastest turn, stall turn) or instantaneous [corner speed equation 5.8] maneuverability. This is because the changes in altitude and/or velocity resulting from the increase in induced drag from increased load factor are not accounted for. A feel for an aircraft's maneuverability (both sustained and instantaneous) can be investigated with the use of specific energy curves.

Points to keep in mind during air combat maneuvering (ACM) is that a fighter pilot will have an energy maneuverability advantage over the opponent under the following conditions.

1) The pilot enters the engagement at a higher energy level and maintains more energy than the opponent during combat.

2) The pilot enters the engagement at a lower energy level but can gain energy faster than the opponent.

To determine whether one fighter aircraft has the advantage, the aircraft specific excess power curves can be superimposed on a similar plot over that of the opponent. The specific excess power equation (8.3) can be rewritten as

$$P_s = V\left[\frac{T}{W} - \frac{qC_{D_0}}{W/S} - \frac{Kn^2(W/S)}{q}\right] \tag{8.7}$$

The energy maneuverability concept may be better illustrated using a specific example. Aircraft C will be compared to another aircraft (aircraft D defined subsequently). The conventional turn performance of both aircraft at a 15,000-ft altitude will be compared. The following equations give the performance of aircraft C, and they were also used to obtain the performance of aircraft D (the

Table 8.1 Example illustrating energy maneuverability

Parameters	Aircraft C	Aircraft D
$T_{SL,max}$, lb	18,000	29,000
W, lb	16,000	24,000
S, ft^2	200	300
C_{D_0}	0.025	0.018
K	0.11368	0.13263
$C_{L_{max}}$	1.9	1.65
n_{max}	9	9
$T/W|_{max}$	1.125	1.208
W/S	80	80
E_m	9.38	10.23

results for both aircraft are shown in Table 8.2):

$$\sigma = 0.7382 \qquad \sigma^{0.7} = 0.8086 \qquad T_{avail_{max}} = 14,555 \text{ lb}$$

$$V_{FT} = \sqrt{\frac{2(W/S)}{\rho_{SL}\sigma}} \sqrt[4]{\frac{K}{C_{D_0}}} = 441 \text{ ft/s}$$

$$n_{FT} = \sqrt{2\left(\frac{T_{max}}{W}\right)E_m - 1} = 4.01$$

$$\dot{\chi}_{FT} = g\frac{\sqrt{n_{FT}^2 - 1}}{V_{FT}} = 0.2926 \text{ rad/s} = 16.76 \text{ deg/s}$$

$$V_{corner} = \sqrt{\frac{2n_{max_{struc}}(W/S)}{\rho_{SL}\sigma C_{L_{max}}}} = 657 \text{ ft/s}$$

$$\dot{\chi}_{corner} = g\frac{\sqrt{n_{max_{struc}}^2 - 1}}{V_{corner}} = 0.4384 \text{ rad/s} = 25.12 \text{ deg/s}$$

$$D_{corner} = 1/2\rho_{SL}\sigma V_{corner}^2 S\left(C_{D_0} + KC_{L_{max}}^2\right) = 32975.9 \text{ lb}$$

$$\frac{dV}{dt} = g\frac{(T_{avail_{max}} - D_{corner})}{W} = -37.1 \text{ ft/s}^2$$

The information provided in Table 8.2 indicates that, at different airspeeds, aircraft C has an advantage in turn rate and radius of turn, both in sustained turn performance (fastest turn) and instantaneous turn performance (corner). But this does not reveal the whole picture. What if both aircraft are flying at the same velocity, which is different from the ones listed in the table? Using Eq. (8.7), a

Table 8.2 Results of conventional turn performance example

Performance	Aircraft C	Aircraft D
V_{corner}, ft/s	657	705
Turn rate, deg	25.12	23.41
Radius of turn, ft	1499	1726
Deceleration, ft/s	−37.1	−35.1
V_{FT}, ft/s	441	497
Turn rate, deg/s	16.76	15.75
Radius of turn, ft	1,555	1,808

graph of turn rate performance at different specific energy level for both aircraft was plotted in Fig. 8.4.

From Fig. 8.4, it is possible to see that aircraft C has a turning advantage over aircraft D at maximum sustainable turning performance ($P_s = 0$ ft/s) at velocities below approximately 800 ft/s. In the instantaneous turn regime ($P_s = -200$ ft/s), aircraft C has the advantage over a larger velocity range. It is seen, however, that for turns with $P_s = 200$ ft/s, aircraft D has the turn rate advantage over the complete velocity range illustrated here. Even though aircraft D has a larger T/W and maximum L/D, its induced drag coefficient is larger than the one of aircraft C, which means that as the lift is increased during a turn because of the increasing load factor, the drag of aircraft D increases more rapidly than that of aircraft C. In the high-velocity range, the lower minimum drag coefficient of aircraft D combined

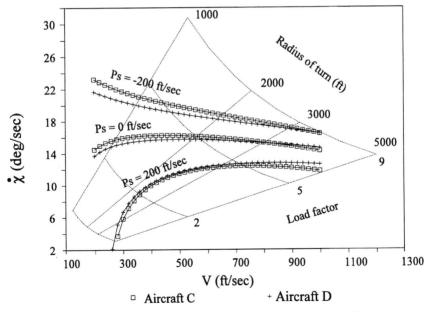

Fig. 8.4 Turning performance of aircraft C and aircraft D.

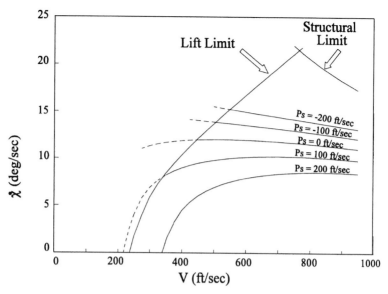

Fig. 8.5   Stall and structural limits of aircraft C combined with its energy maneuverability.

with its larger T/W results in a definite advantage over aircraft C. The curves in Fig. 8.4 do not show all of the structural and aerodynamic limitations of the aircraft, such as those imposed by the maximum lift coefficient and the aircraft's design load factor. Such limits can be added as shown in Fig. 8.5 for aircraft C only. Also, by combining the graph of Fig. 8.4, the limits of the aircraft in terms of maximum lift, and Fig. 5.2, a fairly detailed idea of the general performance of both aircraft can be obtained. This form of presentation would prove to be very confusing due to the number of lines on each graph unless curves could be shown in different colors.

8.4   Energy Method in Flight Testing

The energy method can be a powerful tool to determine the climb performance of an aircraft during flight testing of the prototypes. Examining Eq. (8.3), it can be seen that if the velocity is maintained constant ($dV/dt = 0$), the excess power of an aircraft can be converted into a rate of climb. The climb is initiated 1,000–2,000 ft below the desired testing altitude to give time to the pilot to establish a constant velocity climb. At 500 ft below the testing altitude, the pilot starts to time its climb and stops 500 ft above the test altitude. The engine setting should be checked as the aircraft goes through the test altitude. It should be noted that the weight of the aircraft will decrease as testing progresses due to the fuel being burned. This translates into a larger rate of climb for a given throttle setting, testing altitude, and velocity as testing proceeds.

Another way to estimate the climb performance of an aircraft is to perform constant altitude acceleration runs. To perform such a test, an aircraft adopts a low velocity (close to stall or minimum velocity) at the testing altitude. The pilot then moves the throttle to the desired setting and the aircraft starts accelerating at

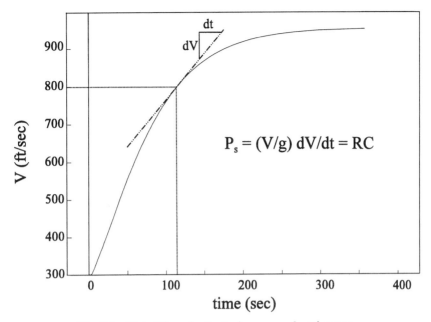

$$P_s = (V/g) \, dV/dt = RC$$

Fig. 8.6   Plot of the velocity during an acceleration run.

a constant altitude. The aircraft will continue its acceleration until it reaches near maximum velocity. The velocity is recorded at regular intervals of time during the acceleration runs (Fig. 8.6).

From Eq. (8.3), it is seen that for a given excess power and velocity, the aircraft can accelerate by a given dV/dt, or climb at a constant velocity, or a combination of both. Thus, if the constant altitude acceleration is known at a given velocity and throttle setting, the maximum steady-state rate of climb for that altitude and velocity can be found.

Problems

8.1.   A CF-18A, flying at Mach 1.3 and 25,000 ft, encounters a pair of MiG-29s (see Appendix C), one at 35,000 ft and Mach 0.9 and the other at 30,000 ft and Mach 1.1. Which aircraft has the energy advantage?

8.2.   A CF-18A is flying at Mach 1.2 and 20,000 ft. It detects an SU-27 flying at the same altitude and at Mach 1.15 in the opposite direction. The detection range is 25 mile (distance b on Fig. P8.2). The pilot of the CF-18A wants to initiate a constant altitude turn, without losing energy, to arrive on the same flight path as the SU-27, 1 mile behind. Note that the minimum drag coefficient at Mach 1.2 is 0.045 and the induced drag coefficient (K) is 0.1316. What is the maximum bank angle the pilot of the CF-18A can adopt to maintain the aircraft energy (while applying maximum thrust)? To what radius of turn and rate of turn does this correspond?

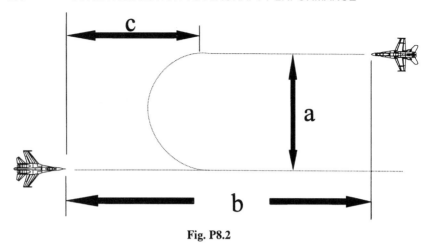

Fig. P8.2

What are the required lateral separation and the horizontal separation (distances **a** and **c**, respectively, on Fig. P8.2) required to achieve the 1-mile trail separation?

9.1 Introduction

MANY assumptions were made in the preceding chapters as to the behavior of aircraft aerodynamics and engines performance with altitude and speed to ease the prediction of the basic performance of an aircraft. In this chapter, the effects of some of these assumptions on the calculated performance are investigated.

9.2 Effect of AoA on Level Flight Performance

To this point, the direction of the thrust vector with respect to the flight path, which is related to the AoA of the aircraft, has been neglected. In fact, for most parts of a typical mission the effects will be negligible because the aircraft would normally operate at velocities where the lift coefficient (thus the AoA) is quite small. In this section we investigate the effect of the AoA on level flight velocities. The concepts of dash AoA and cruise AoA are also introduced.

During level flight, the forces acting on an aircraft are as per Fig. 3.2. The balance of the forces equations are

$$T \cos(\alpha_T) - D = (W/g)a_x \qquad (9.1)$$

$$L + T \sin(\alpha_T) - W = 0 \qquad (9.2)$$

It should be noted here that the acceleration in the x direction is retained to keep the general form of the equation, but that the acceleration in the y direction is zero since the aircraft is in level flight. From Eq. (9.2), it is seen that the throttle setting, thus, the thrust, now influences the equation in the vertical plane by counteracting a portion of the weight of the aircraft. Therefore, a smaller lift force is required to maintain level flight. This in turn will affect the aircraft AoA, which is proportional to C_L and, therefore, will be smaller than a comparable AoA with no thrust component in the vertical plane.

Consider a constant velocity level flight where the aircraft is flying at part throttle with the thrust providing a force component in the direction of the lift. The aircraft is then in steady-state flight at a steady AoA, which is proportional to the lift coefficient. The thrust AoA (α_T) is now also related to the AoA of the aircraft,

$$\alpha = \alpha_0 + C_L \bigg/ \left(\frac{\partial C_L}{\partial \alpha}\right)$$

$$\alpha_T = \alpha + \text{const} \qquad (9.3)$$

This flight condition, with no forward acceleration, will yield the cruise AoA. When the throttle is advanced, the initial aircraft reaction will be to start climbing because of the excess thrust lift, and the aircraft will likely accelerate horizontally as well. To maintain this altitude, the pilot must lower the nose of the aircraft. This will reduce the aircraft's AoA, which in turn translates into a greater horizontal acceleration due to the reduction in α_T and in induced drag (smaller C_L). The final result is a smaller aircraft AoA for a given velocity during acceleration compared to the cruise AoA at the same velocity. This latest AoA is called the dash AoA.

An example with typical numbers may be of some help here. To simplify the analysis, it will be assumed that the lift curve is a straight line up to the stall AoA. We will also assume that the slope ($\partial C_L/\partial\alpha$) will not be affected by the Mach number and that the AoA of the thrust line (α_T) is the same as the AoA of the wing (α). Finally, we will use aircraft C because of its large thrust-to-weight ratio and AoA range. For aircraft C with a maximum lift coefficient of 1.9 at 30 deg and a lift coefficient of 0 at 0-deg AoA, the following lift slope is obtained:

$$\frac{\partial C_L}{\partial\alpha} = \frac{1.9}{30} = (15.789 \text{ deg})^{-1} \tag{9.4}$$

Starting under steady-state (no horizontal acceleration) conditions, the AoA at 36,000-ft altitude and 400 kn true airspeed (Mach 0.7) can be determined. The aircraft weight is 16,000 lb. The following equations thus apply:

$$q = \tfrac{1}{2}\rho_{\text{SL}}\sigma V^2 = 161.4 \text{ psf}$$

$$T\cos(\alpha_T) - qSC_D = 0$$

$$qSC_L + T\sin(\alpha_T) - W = 0$$

$$C_D = C_{D_0} + KC_L^2$$

$$\alpha_T = \alpha$$

$$\alpha = C_L \bigg/ \left(\frac{\partial C_L}{\partial\alpha}\right)$$

which can be rewritten in the form

$$T\cos\left(\frac{C_L}{\partial C_L/\partial\alpha}\right) - qS\left(C_{D_0} + KC_L^2\right) = 0$$

$$qSC_L + T\sin\left(\frac{C_L}{\partial C_L/\partial\alpha}\right) - W = 0$$

This leaves two unknowns, the thrust and the lift coefficient. Through further reduction,

$$T = qS\left(C_{D_0} + KC_L^2\right)\bigg/\cos\left(\frac{C_L}{\partial C_L/\partial\alpha}\right)$$

and

$$qSC_L + qS(C_{D_0} + KC_L^2)\tan\left(\frac{C_L}{\partial C_L/\partial\alpha}\right) - W = 0$$

C_L can now be found by iteration, using Eq. (3.2) to provide an initial estimate

$$C_{L_1} = W/qS = 0.495 \Rightarrow \text{AoA}_1 = 7.82 \text{ deg}$$

$$C_{L_2} = \left[W - qS(C_{D_0} + KC_{L_1}^2)\tan\left(\frac{C_{L_1}}{\partial C_L/\partial\alpha}\right)\right]\bigg/ qS = 0.488$$

$$C_{L_3} = 0.489$$

$$C_{L_4} = 0.489$$

$$\therefore \text{ cruise AoA} = 7.71 \text{ deg}$$

$$T = 1700\,\text{lb} \equiv 22\% \text{ of maximum thrust at altitude}$$

Thus, for aircraft C, at the stated flight conditions, the cruise AoA is 7.71 deg, which differs only by about 1.5% from the estimates using Eqs. (3.2) and (9.4). If, now, the thrust were increased to 100%, it would be found that, before the velocity began to increase, the dash AoA would be 7.34 deg ($C_L = 0.465$) and the instantaneous level acceleration 12.16 ft/s^2 (0.378 g).

It is then apparent from these calculations that the use of the simplified Eqs. (3.1) and (3.2) introduces only a relatively small error during cruise conditions (medium to high speeds) and is a lot simpler. Furthermore, a knowledge of the lift coefficient curve is not required. Therefore, the equations are to be used to estimate the performance of an aircraft during level flight as long as the AoA is smaller than about 15 deg.

Unfortunately, the same cannot be said while the aircraft is flying in the low-speed regime (slow flight) where the aircraft can reach a high AoA, as shown in Fig. 9.1. This situation is typical of high-performance military aircraft such as aircraft C where very slow flight is possible due to its large T/W ratio and the high AoA at $C_{L_{\max}}$.

The thrust contribution to lift at these high AoAs can become very important. In the case of aircraft C at a weight of 16,000 lb and flying at sea level, the stall speed is reduced from 111 kn in power-off stall condition [Eq. (3.11)] to 74 kn at maximum thrust condition [Eq. (3.12)], a reduction of 34%. In this case, the thrust provides 56% and the wing 44% of the vertical force. It should be noted, however, that this is not a steady-state condition since the aircraft would be accelerating at a rate of 28.1 ft/s^2 (0.87 g) because of the horizontal thrust component, which exceeds the drag produced by the aircraft. The power-off stall condition is still the one used because of the flight safety implications in case of an engine failure in flight. Level flight high AoA is not a normal flight condition but, because it is quite spectacular, it is frequently displayed at airshows. The F/A-18 is an excellent aircraft to display this flight profile because of its large maximum lift AoA of around 35 deg and its retention of good handling qualities during slow flight.

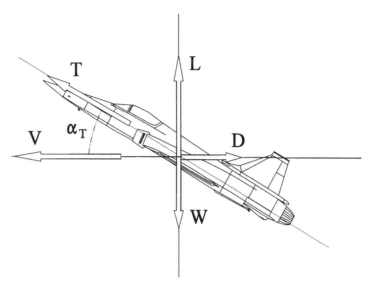

Fig. 9.1   Level flight at high AoA.

A final comment on slow flight. It should be apparent that while flying at AoAs below the minimum drag velocity that a variation in thrust will control the altitude; a variation in pitch, thus AoA, will control speed.

9.3   Unrestricted Turning Flight

In Chapter 5, the aircraft was flying in a coordinated unaccelerated turn with no loss of altitude. Let us now evaluate the effect of pointing the thrust vector toward the inside of the turn as well as taking into account the effect of AoA on the thrust vector. This last point is particularly important for modern fighters where flights at high AoA are an important feature during ACM. Figure 9.2 shows the forces and angles during a yawed turn.

Summing the forces along the flight path we get

$$T \cos(\alpha_T) \cos(\beta) = D \qquad (9.5)$$

where β is the sideslip angle. The forces acting in the vertical direction are

$$L \cos(\phi) + T \sin(\alpha_T) \cos(\phi) = W \qquad (9.6)$$

and the forces along the radius of the turn are

$$L \sin(\phi) + T \sin(\alpha_T) \sin(\phi) + T \cos(\alpha_T) \sin(\beta) = (W/g)V\dot{x} \qquad (9.7)$$

Solving for the bank angle,

$$\tan(\phi) = \frac{V^2}{gr} - \frac{T}{W} \cos(\alpha_T) \sin(\beta) \qquad (9.8)$$

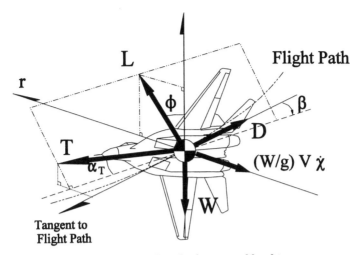

Fig. 9.2   Forces and angles in a yawed level turn.

and for the radius of turn,

$$r = \frac{2W[1 - (T/W)\sin(\alpha_T)\cos(\phi)]}{g\rho C_L S[\sin(\phi) + (T/W)\cos(\phi)\cos(\alpha_T)\sin(\beta)]} \tag{9.9}$$

This last equation indicates that the radius of turn will be minimized by a large thrust-to-weight ratio and maximum lift coefficient. Using aircraft C at a typical combat weight of 16,000 lb, with a T/W ratio of 1.125, and flying at its maximum lift coefficient at sea level, which then provides a thrust AoA of about 30 deg, the preceding equations were solved to provide the radius of turn and the vertical velocity of the aircraft as a function of its bank and sideslip angles. The results are shown in graphical form in Fig. 9.3. Note that the velocity is adjusted at every bank angle so as to have zero acceleration along the flight path, although a rate of climb or descent results. Here, it was assumed that the drag coefficient increased with sideslip angle by a factor $(1 + \beta)$ with β in radians.

Another way of finding the minimum radius of turn for a given sideslip angle would be to differentiate Eq. (9.9) with respect to the bank angle and setting the result to zero. Then some iterations are required to determine the lowest value of ϕ, which provides the smallest radius of turn,

$$\frac{\partial r}{\partial \phi} = 0$$

$$T\sin(\alpha_T) - W\cos(\phi) + T\sin(\phi)\cos(\alpha_T)\sin(\beta) = 0 \tag{9.10}$$

$$\text{iterate to find } \phi \text{ for min } r$$

Let us use a specific example to demonstrate some of the complexity encountered in trying to determine the radius of turn during an unrestricted turning

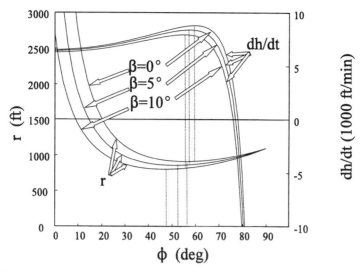

Fig. 9.3 **Radius of turn as a function of ϕ and β for maximum lift coefficient and thrust at combat weight, aircraft C.**

flight. The minimum radius of turn for three specific conditions will be examined 1) during a coordinated turn at constant altitude, 2) during a coordinated turn with altitude change, and 3) during a turn with altitude change and 15-deg sideslip. Here we will use aircraft C at the same condition used to produce Fig. 9.3 (sea level, $W = 16,000$ lb, $T/W = 1.125$, and $\alpha_T = 30$ deg).

1) For a level coordinated turn, using Eq. (5.26) (for the tightest turn),

$$n_{TT} = 1.41$$

and

$$r_{TT} = 425 \text{ ft} \qquad \text{at} \qquad V_{TT} = 116.6 \text{ ft/s}$$

A check must be made to verify that this theoretical velocity is not smaller than the stall velocity (remembering that the maximum lift coefficient is 1.9), using Eq. (5.28),

$$V_{ST} = 419.8 \text{ ft/s} \qquad \text{and} \qquad n_{ST} = 4.98$$

which gives $r_{ST} = 1123$ ft and $\dot{\chi} = 21.44$ deg/s. These values do not take into account the thrust angle, which is known in this case and is large enough not to be negligible. Correcting the preceding value will give

$$V_{ST} = \sqrt{\frac{2T \cos(\alpha_T)}{\rho C_{D_{\text{stall}}} S}} = 390.7 \text{ ft/s}$$

Remembering that the aircraft would be flying at its maximum lift coefficient, the new load factor will be

$$n_{ST} = [L + T \sin(\alpha_T)]/W = 4.87$$

$$\cos(\phi) = 1/n_{ST} \Rightarrow \phi = 78.15 \, \text{deg}$$

which give

$$r = \frac{2W}{g\rho S \sin(\phi)} \frac{1}{[C_L + C_D \tan(\alpha_T)]} = 994 \, \text{ft}$$

and

$$\dot{\chi} = 22.5 \, \text{deg/s}$$

Thus, the minimum radius of turn at sea level for aircraft C in a level coordinated turn is 994 ft and it occurs while flying at the stall turn condition.

2) For the minimum radius of turn with altitude change, Eq. (9.10) applies. With a sideslip angle of zero degrees,

$$\cos(\phi) = (T/W) \sin(\alpha_T) \Rightarrow \phi = 55.77 \, \text{deg}$$

and then

$$r = \frac{2[W - T \sin(\alpha_T) \cos(\phi)]}{g\rho C_L S \sin(\phi)} = 909.7 \, \text{ft}$$

$$V = \sqrt{gr \tan(\phi)} = 207.5 \, \text{ft/s}$$

$$n = 1/\cos(\phi) = 1.78$$

$$\dot{\chi} = \frac{g\sqrt{n^2 - 1}}{V} = 13.1 \, \text{deg/s}$$

$$\frac{dh}{dt} = \frac{V}{W}\left[T - \frac{D}{\cos(\alpha_T)}\right] = 167.6 \, \text{ft/s} \Rightarrow \gamma = +53.9 \, \text{deg}$$

This rate of climb shows that the aircraft's flight path is in fact a climbing spiral and that the helix angle is 53.9 deg.

Because this is a high-climb angle, the preceding equation can only give an approximation of the actual radius of turn. It is, however, evident from these results that when an aircraft is not restricted to a constant altitude turn, a smaller radius of turn can be achieved.

3) For the same flight conditions, but with 15 deg of sideslip toward the inside of the turn, we use Eq. (9.10) and iterate to find the smallest radius of turn possible. We start with a bank angle of 60 deg, which is close to the value of answer 2,

$$T \sin(\alpha_T) - W \cos(\phi) + T \sin(\phi) \cos(\alpha_T) \sin(\beta) = 0$$

$$\cos(\phi_2) = (T/W)[\sin(\alpha_T) + \sin(\phi_1) \cos(\alpha_T) \sin(\beta)]$$

$$\phi_1 = 60 \rightarrow \phi_2 = 38.66 \rightarrow \phi_3 = 43.94 \rightarrow \cdots \rightarrow \phi_8 = 42.79 \, \text{deg}$$

This gives the following results:

$$r = 747\,\text{ft}$$

$$L = [W/\cos(\phi)] - T\sin(\alpha_T) = 12{,}803\,\text{lb}$$

$$V = \sqrt{\frac{2L}{\rho C_{L_{\max}} S}} = 168.4\,\text{ft/s}$$

$$C_D = C_{D_{\text{stall}}}(1 + \beta) = 0.5421$$

$$C_L/C_D = 3.5 \Rightarrow D = L/3.5$$

$$T_r = \frac{D}{\cos(\alpha_T)\cos(\beta)} = 4367\,\text{lb}$$

$$\frac{dh}{dt} = \frac{V}{W}(T_a - T_r) = 143.5\,\text{ft/s}$$

$$\gamma = 58.4\,\text{deg}$$

$$n = [L + T\sin(\alpha_T)]/W = 1.36$$

$$\dot{\chi} = 10.14\,\text{deg/s}$$

These results, summarized in Table 9.1, show that a verification of the climb angle and yaw angle during a turn can be very important since these two parameters have a serious impact on the possible turn performance of a high-performance aircraft. For angles of climb or descent, during turns, less than about 15 deg, the simpler constant altitude coordinated turn equations provide adequate performance predictions. For aircraft C it may be necessary to include the climb angle in the calculations to get a better estimate of the radius of turn.

In the example, the maximum sustainable load factor (g loading, $n = 10.55$) was greater than the structural limit ($n = 9.0$). It was only used to determine n_{TT}, not as a flight condition. It turned out that minimum radius of turn was obtained while flying at stall conditions (for level coordinated turn) with a load factor well under the structural limit.

With the introduction of highly agile aircraft (such as the F-16), the pilot rather than the structural limit may be the limiting element for ACM. Pilots, on the average, are able to withstand sustained g loadings of between five and six, beyond which they experience tunnel vision/grayout, and, eventually, blackout (if high-g

Table 9.1   Example to demonstrate determination of rate of turn

Condition	V, ft/s	n	ϕ, deg	r, ft	$\dot{\chi}$, deg/s	γ, deg	dh/dt, ft/s
1	390.7	4.87	78.15	994	22.5	0	0
2	207.5	1.78	55.77	910	13.1	53.9	167.6
3	168.4	1.36	42.79	747	10.14	58.4	143.5

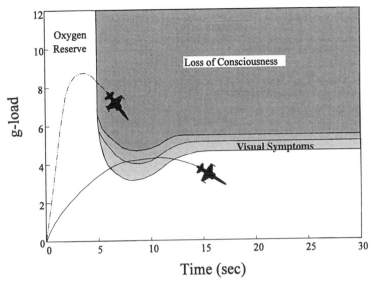

Time (sec)

Fig. 9.4 Typical G-LOC envelope, no antigravity suit or straining maneuvers (adapted from Ref. 9).

loads were maintained) as the heart cannot maintain the necessary flow of blood (and oxygen) to the eyes and brain. The newer generation of combat aircraft, which now have the necessary thrust-to-weight ratio to permit steady-state maneuvers at g loadings beyond human tolerance, have brought on a new aeromedical problem called gravity-induced loss of consciousness (G-LOC).

In older aircraft the pilot usually pulled gs at moderate g onset rates (dg/dt), because of the aircraft limits, until they experience visual symptoms. The pilot could then perform antigravity straining of muscles or unload the aircraft until the symptoms disappeared. In new aircraft, with the rapid g onset rates now available, the pilot would not experience the visual symptoms [because of the reserve of oxygen in the brain, about 5 s worth (see Fig. 9.4, adapted from Ref. 9) before loss of consciousness. In centrifuge tests, it was determined that the average incapacitation as a result of G-LOC is about 15 s after removal of the g load with retrograde amnesia for several seconds more. In all, there are approximately 25 s where the pilot would not have control of the aircraft. Antigravity straining rises these curves by about 1 g. If the pilot talks or stops antigravity straining during a high-g maneuver the pilot may be propelled into the G-LOC envelope (one point against voice-operated cockpits).

More inclined seats (F-15 at 13 deg, F-16 at 35 deg, and Martin–Baker's experimental articulated seat at 35 deg for ejection and crew entry and at 65 deg for high-g loads) may also help the pilot during high-g loads maneuvers but this option could restrict the ejection envelope of the seat. Another option, which is getting more attention these days, is the improved gravity suit. The Canadian Forces are working on a system called sustained tolerance and increased g and the U.S. Air Force is working on a system called combat edge. Both suits are designed

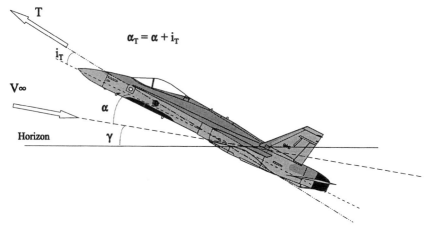

Fig. 9.5 AoA influence on climb performance.

to eliminate G-LOC within the current flight envelope of the aircraft to be fitted with the respective systems.

9.4 Influence of AoA on Climb Performance

It has already been shown that the AoA has relatively little effect on performance while flying at medium to high velocities except while maneuvering. The same is true for climb performance where in steady climb (or descent) the effect of AoA will be negligible in the high spectrum of the velocity regime of a particular aircraft. The basic steady-state equations for climb performance (Fig. 9.5), taking the AoA into account, are

$$T \cos(\alpha_T) - D - W \sin(\gamma) = 0$$
$$L + T \sin(\alpha_T) - W \cos(\gamma) = 0$$

(9.11)

During a steady climb or descent at medium to high speeds $\alpha_T \approx 0$, but during maneuvering (dive pullout) α_T may be high due to the high-lift coefficient requirements. The aircraft pitch angle Θ is the sum of the flight-path angle and the AoA

$$\Theta = \gamma + \alpha$$

(9.12)

Note that previously we did not make any distinction between the pitch angle and flight-path angle because it was assumed that the AoA was small. However, aircraft such as aircraft C have a large AoA range prior to stall. For example, consider aircraft C at a weight of 16,500 lb flying at a 20,000-ft altitude. It is possible for the pilot to establish a steady-state flight condition at maximum AoA (30 deg) and 150 kn (Fig. 9.6). Using Eq. (9.11), it can be established that there will result a steady-state descent angle of approximately 6 deg. This means that

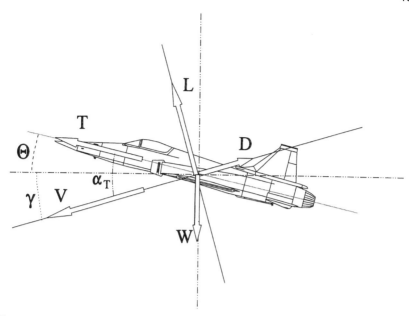

Fig. 9.6 Aircraft C (W = 16,500 lb) in steady-state descent while flying at α_{max}, 150 kn and 20,000-ft altitude.

the nose of the aircraft will be pointing 24 deg above the horizon even though the aircraft is descending at approximately 27 ft/s ($\approx$1600 ft/min).

9.5 Gliding Turn

In Sec. 4.8 of Chapter 4, the concept of gliding flight with no bank angle was introduced. In this section, the effect of the bank angle on the glide angle and sink rate is investigated.

The steady-state gliding turn equations are a combination of Eqs. (4.25), (4.26), and (5.1) to (5.3) as follows:

$$D = -W \sin(\gamma) \tag{9.13}$$

$$\dot{\chi} = \frac{g}{V} \frac{L}{W} \sin(\phi) = \frac{g}{V} n \sin(\phi) \tag{9.14}$$

$$L \cos(\phi) = W \cos(\gamma) \tag{9.15}$$

Note that the bank angle is an independent variable in the preceding equations; it is solely a function of the roll control input from the pilot. For a zero bank angle, the turn rate [Eq. (9.14)] equals zero and Eqs. (9.13) and (9.15) become identical to Eqs. (4.25) and (4.26), the constant heading steady-state gliding condition.

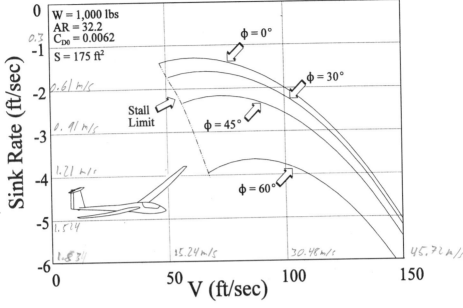

Fig. 9.7   Effect of the bank angle on the sink rate.

The steady-state glide angle, for a given bank angle, can be determined by combining Eqs. (9.13) and (9.15).

$$\frac{L}{D} = \frac{-1}{\tan(\gamma)\cos(\phi)}$$

$$\gamma = -\arctan\left(\frac{1}{E\cos(\phi)}\right) \tag{9.16}$$

Note the marked effects of the bank angle on both the sink rate and the glide angle (Fig. 9.7). The ratio of the sink rate to the aircraft velocity equals $\sin(\gamma)$, which is approximately equal to the glide angle when γ is small. The increased sink rate is a function of the increased specific power required by the aircraft in a bank due to the increased load factor. The load factor in a bank is determined from Eq. (9.15) and is equal to

$$n = \frac{L}{W} = \frac{\cos(\gamma)}{\cos(\phi)} \tag{9.17}$$

The sink rate is, thus,

$$RD = -\frac{dh}{dt} = \frac{DV}{W}$$

$$\therefore RD = \frac{1}{2}\rho_{SL}\sigma V^3 S C_{D_0} + \frac{KW^2}{\frac{1}{2}\rho_{SL}\sigma V S}\left[\frac{\cos(\gamma)}{\cos(\phi)}\right]^2 \tag{9.18}$$

Note that a positive rate of descent corresponds to a loss of altitude with time (i.e., a negative rate of climb). Note that the increase in sink rate is solely due to the increase in induced drag because the minimum drag is not affected by the bank angle. The turn rate [Eq. (9.14)], can be rewritten as follows:

$$\dot{\chi} = (g/V) \cos(\gamma) \tan(\phi) \qquad (9.19)$$

and the turn radius is

$$r = \frac{V}{\dot{\chi}} = \frac{V^2}{g} \frac{1}{\cos(\gamma) \tan(\phi)} \qquad (9.20)$$

For a given bank angle and velocity, the turn rate will decrease and the turn radius will increase with an increase in glide angle. Increasing the bank angle will have the opposite effect. For the glider performance illustrated in Fig. 9.7 flying at a velocity of 67 ft/s, the sink rate more than triples when the bank angle is increased from 30 to 60 deg (going from 1.3 to 3.95 ft/s), but the radius of turn is reduced from 241.5 to 80.6 ft and the turn rate is increased from 15.9 to 47.6 deg/s. At 67 ft/s and 60-deg bank, the sailplane of Fig. 9.7 is almost at its stall speed.

9.6 Effects of Using Afterburners

Thus far, the only effect of using afterburners that was discussed was the increase in thrust, which was especially important for high-climb rate, turn rate, and velocity. On the other hand, the use of afterburners, as will be seen in this section, has a detrimental effect on performance such as range and endurance.

As mentioned in Chapter 2, Table 2.4, the use of the afterburner for the J79 engine increased the engine thrust by 64% compared to military setting (maximum no afterburner). This is a substantial increase in thrust (from 10,900 to 17,900 lb). If an aircraft using this engine was flying at conditions of zero excess power at military setting, using the afterburner will now provide an extra 7,000 lb of thrust, which can be used to accelerate and/or climb. The drawback of afterburner use is the large increase in fuel consumption. In the case of the J79, this represents a factor of 2.83 (from 9,156 to 35,084 lb/h) even though the SFC only increased by a factor of 1.33 (from 0.84 to 1.96 lb/lb/h). An aircraft having 10,000 lb of fuel available for the cruise segment would take 1 h and 5.5 min to use all its cruise fuel while flying at military setting and only 17.1 min if flying with afterburner on, which is a 74% reduction.

To illustrate the effects of using afterburners, a simple flight profile was used to quantify those effects. The initial conditions were aircraft C in takeoff configuration (and maximum clean takeoff weight) on a runway at sea level and zero forward velocity. At time equal to zero, maximum thrust is applied [afterburner thrust in one case and military thrust (maximum dry) in the other (see Table 3.1)]. The aircraft accelerates to takeoff conditions ($1.3 V_{stall}$ at 50-ft altitude) then cleans up (retracts flaps and landing gear) and climbs to 1,000 ft while maintaining constant velocity and maximum thrust. Once at 1,000 ft, the aircraft accelerates to its fastest climb velocity (including compressibility effects) at which time it initiates

Table 9.2   Summary of climb profiles

Condition	h, ft	Afterburners thrust				Military thrust		
		t, s	V, ft/s	ΔW_f, lb	X, ft	t, s	V, ft/s	ΔW_f, lb	X, ft
Liftoff	0	8	243	77.9	1,024	13.4	243	33.2	1,690
50 ft	50	9	263	87.7	1,275	15.4	263	37	2,169
1,000 ft	1,000	13.4	263	129	1,915	23.4	263	57.7	4,063
V_{FC}	1,000	42.3	1,001	398.5	20,708	63.3	832	153.1	26,281
30,000 ft	30,000	102.4	895	791.9	68,953	191.4	895	369.7	141,108
Mach 1.2	30,000	134.2	1,193	943.5	102,666	315.6	1,193	522.1	276,993

a climb at fastest climb condition up to 30,000 ft. Finally, upon reaching 30,000 ft, the aircraft accelerates to Mach 1.2 (1,193 ft/s). The results are included in Table 9.2 and illustrated in Fig. 9.8.

From this small exercise it is clear that the use of afterburners will considerably reduce the time required to reach a given altitude/airspeed combination. The penalty for this improved acceleration/climb performance is the greater fuel consumption. Thus, afterburners are only used when required such as during takeoff, rapid accelerations to minimize the time required to get on station, and supersonic flights. The latest generation of fighters (such as the F-22), with their large thrust-to-weight ratios, has the capability to cruise at low to moderate supersonic Mach numbers (1.2–1.6) without using afterburners. This flight condition is termed supercruise (Fig. 9.9).

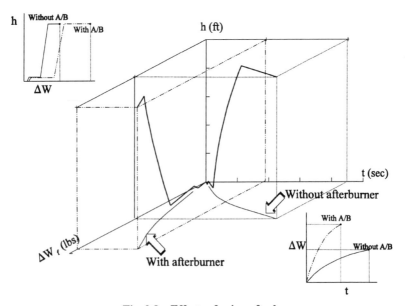

Fig. 9.8   Effects of using afterburners.

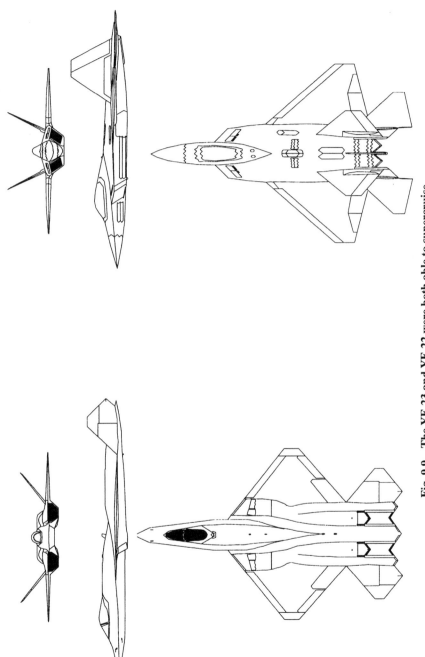

Fig. 9.9   The YF-23 and YF-22 were both able to supercruise.

9.7   Tailoring of the Drag Polar

Thus far, we have used the parabolic drag polar [Eq. (1.11)], for estimating the performance of different aircraft. This form of the drag polar [rewritten here as Eq. (9.21)], assumes that the minimum drag coefficient occurs at a lift coefficient equal to zero. This is achieved by setting a wing, with no or near zero camber, at as small as possible angle of incidence with respect to the fuselage (i.e., aligning the wing minimum drag angle with the fuselage minimum drag angle):

$$C_D = C_{D_0} + K C_L^2 \tag{9.21}$$

However, minimum drag at zero lift coefficient may not be ideal for many aircraft missions where high speed (required for low-lift coefficient) is not a requirement. It is possible to tailor the drag polar to get a minimum drag coefficient at a lift coefficient that is nonzero. This can be done by using a combination of wing camber, wing incidence, and wing geometric/aerodynamic twist. The drag polar, for this case, takes the following form:

$$C_D = C_{D_{\min}} + K \left(C_L - C_{L_0} \right)^2 \tag{9.22}$$

where C_{L_0} is the lift coefficient for minimum drag ($C_{D_{\min}}$). The difference between the two curves is illustrated in Fig. 9.10. In this case, both drag polars, Eqs. (9.21) and (9.22), had equal minimum drag coefficients of 0.02 and induced drag coefficient K of 0.1.

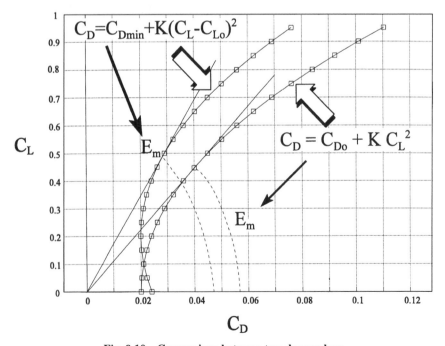

Fig. 9.10   Comparison between two drag polars.

At a glance it can be seen that the drag polar with nonzero lift coefficient for minimum drag [Eq. (9.22)] has a greater lift-to-drag ratio than the drag polar of Eq. (9.21). For two aircraft of similar dimensions, one having the drag polar of Eq. (9.21) (aircraft A1) and the other of Eq. (9.22) (aircraft A2), aircraft A2 has a smaller drag coefficient than aircraft A1 for equal lift coefficients greater than 0.1 (for this case). This translates into greater excess thrust in the low to medium speed regime that can be converted into greater acceleration or rate of climb. It also means that aircraft A2 will use less fuel than aircraft A1 while both aircraft are flying at the same lift coefficient greater than 0.1. At high speed, where the lift coefficient is less than 0.1, aircraft A1 would have the greater excess thrust.

The exact value for the maximum lift-to-drag ratio can be obtained in a way similar to that used to obtain Eq. (3.7),

$$E_m = \frac{1}{2\left(\sqrt{KC_{D_{\min}} + (KC_{L_0})^2} - KC_{L_0}\right)} \tag{9.23}$$

Note the similarity between this last equation and Eq. (3.7). This last equation is Eq. (3.7) corrected for nonzero lift coefficient for minimum drag.

Let us assume that both aircraft have the following characteristics: aircraft weight 5,000 lb, wing area 250 ft^2, and maximum thrust at sea level of 2,000 lb. Figure 9.11 for sea level conditions was created from this basic information.

The velocities for minimum drag (maximum endurance) and best range conditions are indicated in Fig. 9.11. Table 9.3 contains additional information.

It can be noticed that the velocity for minimum drag is slightly lower ($\approx 2\%$) for aircraft A2. The aerodynamic efficiency, on the other hand, has increased by nearly

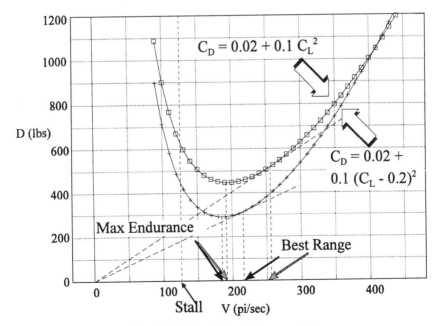

Fig. 9.11 Two different drag curves.

Table 9.3   Summary of drag polar comparison

Conditions	V, ft/s		E		VE, ft/s	
	$C_{D_0}+$ KC_L^2	$C_{D_{min}}+$ $K(C_L-C_{L_0})^2$	$C_{D_0}+$ KC_L^2	$C_{D_{min}}+$ $K(C_L-C_{L_0})^2$	$C_{D_0}+$ KC_L^2	$C_{D_{min}}+$ $K(C_L-C_{L_0})^2$
Minimum drag	194	190 (−2%)	11.28	17.24 (+54%)	2188	3276 (+50%)
Best range	255	220 (−14%)	9.68	16.14 (+67%)	2468	2551 (+44%)

54%. Since a jet aircraft endurance is directly proportional to the aerodynamic efficiency, this corresponds to a 54% increase in endurance (for the same engine SFC and fuel-weight fraction, Chapter 3, Sec. 3.5) compared to aircraft A1.

To obtain best range conditions (Sec. VI.B of Chapter 3), a jet aircraft must fly at a ratio of drag over velocity that is minimum, i.e., maximizing the product of velocity and aerodynamic efficiency VE. From Fig. 9.11 and Table 9.3, it can be seen that, even though aircraft A2 has a best range velocity that is 14% lower than that of aircraft A1, the product of $(VE)_{BR}$ is 44% greater. Therefore, in this case, aircraft A2 has 44% greater range than aircraft A1 or, for the same ratio of drag over velocity (thus same range), aircraft A2 could cruise at approximately 340 ft/s (see intersection of the drag over velocity line for aircraft A1 with the drag curve of aircraft A2), a 33% increase, which would translate into a reduced flight time from one point to another. This can be a significant advantage for a commercial aircraft.

For velocities below approximately 400 ft/s (for sea level conditions), the point where the two drag curves meet (Fig. 9.11), aircraft A2 has less drag than aircraft A1 for the same flight velocity. Thus, in the low-velocity range, aircraft A2 will have better flight performance than aircraft A1, including greater range, endurance, climb angle, and rate of climb. For the high-velocity regime, aircraft A1 would have the advantage, including a greater maximum velocity of 576.5 ft/s compared to the 552 ft/s of aircraft A2, a 4% increase for sea level conditions.

The preceding example was used to show that it is possible to optimize the drag polar to best suit the aircraft primary mission. It must be remembered that changing one parameter (such as wing incidence with respect to the fuselage), to try to tailor the drag polar will, in most cases, modify other parameters and affect the drag polar shape. The best engineering compromise must be sought.

9.8   Turbojet, Turbofan, Turboprop, and Piston-Prop Performance

Referring to Chapter 2, one can appreciate the distinctions between piston-prop, turboprop, turbofan, and turbojet engines. For simplicity during aircraft performance analysis, turboprops were regrouped with piston-prop aircraft while turbofan equipped aircraft were associated with turbojet aircraft. Furthermore, it was assumed that the thrust developed by a turbojet engine was independent of airspeed. The same was true for the power developed by a piston-prop engine for Mach numbers above 0.1. This led to a decreasing thrust available with increasing airspeed for piston-prop aircraft (Figs. 9.12 and 9.13).

Turbojet engines are somewhat inefficient at low speeds (see Chapter 2), whereas piston-prop engines are quite efficient in that speed regime, but their performance decreases rapidly with increasing airspeed. Turboprops and turbofans are gas turbine engines designed to provide the best compromise for the intermediate-speed regime. Turboprop engines remove most of the energy remaining in the

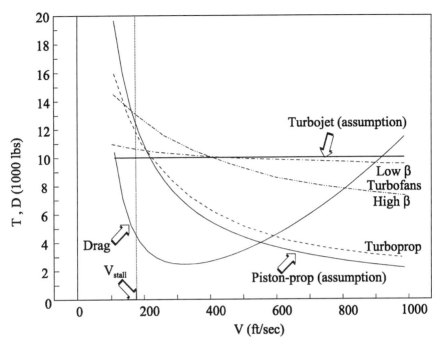

Fig. 9.12   Thrust as a function of airspeed for different propulsion systems.

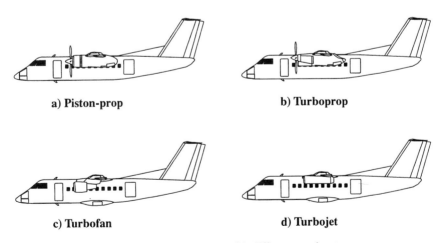

a) Piston-prop   b) Turboprop

c) Turbofan   d) Turbojet

Fig. 9.13   Same basic layout with different engine types.

exhaust flow and provide this energy, through a reduction gear box, to a propeller. Unlike piston-prop aircraft, the turboprop aircraft expands its exhaust gases in a nozzle to produce thrust (approximately 15% of maximum static thrust). This thrust, which remains essentially constant throughout the speed regime, will allow a turboprop aircraft to reach higher speeds than a piston-prop aircraft of the same aerodynamic characteristics, weight, and engine power. Turboprops still suffer from propeller efficiency degradation at high speeds, but the rate of decrease of propulsive efficiency at high speed is somewhat less than that of a piston-prop aircraft. Another advantage of turboprop engines is their large power-to-weight ratio (low-power loading) and low-frontal area, turboprop engines being much lighter and smaller than a reciprocating engine of the same power. This translates into greater lift-to-drag ratio for a turboprop aircraft compared to a piston-prop aircraft of the same general layout. The only present disadvantage of turboprops is their larger SFC, but ongoing research and development will bring these values closer to the ones of piston props.

To improve the very high TSFC of early turbojet engines, bypass flow was introduced (Fig. 9.14). At first, only a small amount of bypass (approximately equal to 1) was used. This flow came either from a front fan (e.g., JTD8) or a rear fan (e.g., CJ805-23). These first-generation turbofans increased the takeoff thrust of the engine and lowered the TSFC compared to the turbojet engine they were often derived from. The fan was acting like a small propeller driving a cold flow (not going through the combustion chamber), thus a slight decrease in thrust occurred with increasing airspeed. This layout increased the engine frontal area slightly. Modern fighters still use turbofans with bypass ratio of approximately 0.3 to provide acceptable SFC in the normal (nonafterburning) flight conditions.

Then came turbofan engines with larger bypass (JTD9, $\beta = 5$). This resulted in a large decrease in TSFC but it also significantly increased the frontal area of the engines, limiting their use to Mach numbers smaller than 1.0. These engines are basically large turboprops (without a reduction gear box) with ducted propellers.

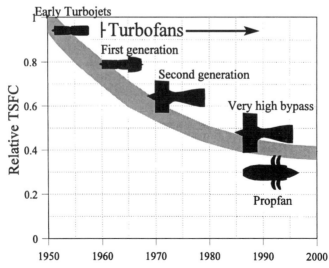

Fig. 9.14   Reduction in the TSFC of turbofan engines.

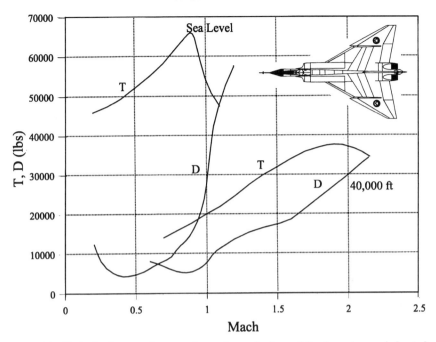

Fig. 9.15   Full afterburner thrust and aerodynamic drag of the Avro Arrow (adapted from Ref. 10).

These engines have large takeoff thrust compared to turbojet engines using the same core, but this thrust tends to decrease with increasing airspeed (Fig. 9.12).

The assumption of constant thrust for turbojet (airspeed independent) may not always be the case. Ram effects, for properly designed propulsive system (from intake to exhaust) over their operating Mach number range and engine mass flow can actually increase the available thrust with increasing airspeed. This is dramatically illustrated in Fig. 9.15 for the Avro Arrow.

Properly designed intakes for a piston-prop aircraft can also increase the ram pressure available to the engine, which would increase the MEP (see Chapter 2, Sec. 2.3). This translates into a larger shaft power with increasing airspeed for piston prop, which tends to flatten the thrust-available curve.

When one studies the performance of an aircraft, a first estimate can be obtained by using the assumptions of the preceding chapters. Specific performance analysis requires a better knowledge of the actual behavior of the propulsive system at specific speeds and altitudes.

9.9   Real Aircraft Performance

To obtain simple-to-use equations to estimate aircraft performance, some assumptions were made. Among them, the SFC remained constant throughout the flight, i.e., independent of aircraft velocity, altitude, and throttle setting. Other assumptions included constant thrust for turbojet equipped aircraft (including turbofans) and constant power for piston-prop aircraft (including turboprops). In this section, some of these assumptions are removed, which somewhat complicates the analysis of aircraft performance, but it illustrates that the basic principles used in

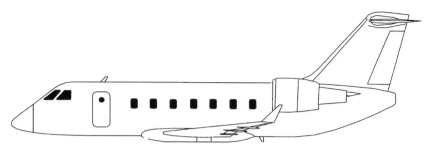

Fig. 9.16   Business jet aircraft.

the preceding chapters, such as maximum excess thrust for maximum climb angle, are still valid. For this analysis, a typical business jet aircraft is used (Fig. 9.16).

This business jet has an operational empty weight (W_{OE}) of 26,000 lb and a maximum takeoff weight of 45,000 lb. It can carry up to 20,000 lb of fuel and 5,000 lb of payload (the sum of fuel and payload not exceeding the maximum takeoff weight). It can be configured to carry up to 19 passengers. Its wing has an AR of 8.5 and an area of 490 ft². Its trimmed drag polar clean configuration (flaps and landing gear retracted), as a function of Mach number, is shown in Fig. 9.17.

The aircraft maximum lift coefficient is approximately 1.65 at low Mach numbers. This business jet is powered by two turbofans with a bypass ratio of 3. The installed maximum thrust for each engine is shown in Fig. 9.18. Note that the thrust decreases while the TSFC increases with increasing Mach number.

As a first exercise, one can establish what the flight envelope may look like for an aircraft weight of 40,000 lb. Aircraft of this category usually have a maximum

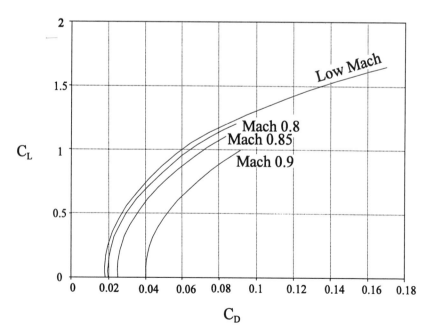

Fig. 9.17   Drag polar of business jet.

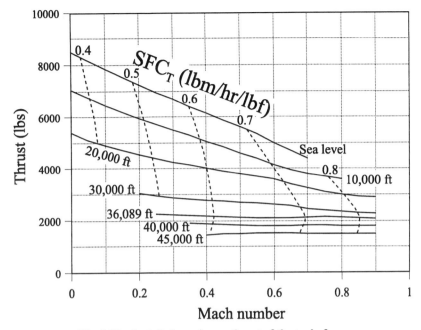

Fig. 9.18   Installed maximum thrust of the turbofans.

altitude of 45,000 ft due to cabin pressurization structural limit. The maximum speed of the aircraft will be limited by either the maximum Mach number (never to exceed) of 0.9, or a maximum dynamic pressure equivalent to Mach 0.7 at sea level in a standard atmosphere. At 30,000 ft, for example, the drag and maximum thrust curves would be as shown in Fig. 9.19.

Note that the maximum excess thrust does not coincide with the aircraft's minimum drag condition due to the decreasing thrust with increasing Mach number. Performing the same exercise for different altitudes, one can determine the boundaries of the flight envelope (Fig. 9.20).

Next, one can determine the best range conditions (see Chapter 3, Sec. 3.3). When it was assumed that the TSFC was constant, it was found that the range could be maximized when flying at the minimum ratio of drag over velocity [i.e., maximizing the specific instantaneous range equation (3.24)]. This same equation must still be maximized to find the best range conditions, but this time the TSFC must be accounted for. Now, to maximize the specific instantaneous range, one must maximize

$$V/SFC_T D \tag{9.24}$$

No partial throttle TSFC is given in this section; it will be assumed that the TSFC is independent of throttle setting. For the business jet, flying at 30,000 ft with a weight of 40,000 lb, the variation of specific instantaneous range with velocity is shown in Fig. 9.21.

Thus, the best range conditions are obtained while flying at approximately 650 ft/s (Mach 0.65) where the TSFC would be approximately 0.69 lb/lb/h. Using Eq. (3.24) with a constant TSFC of 0.7 lb/lb/h, one finds that the estimate for best

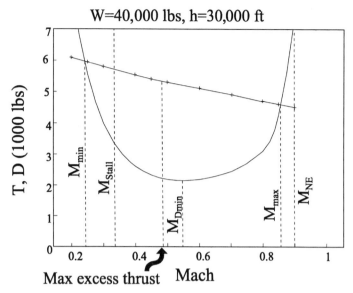

Fig. 9.19   Drag and thrust curve.

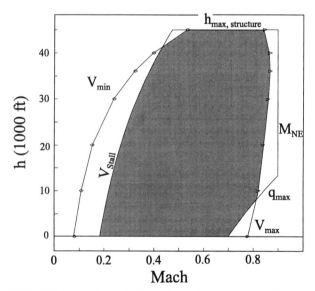

Fig. 9.20   Flight envelope of the business jet at a weight of 40,000 lb.

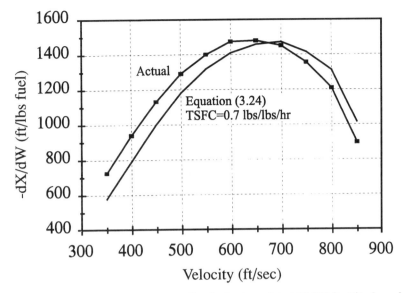

Fig. 9.21 Specific instantaneous range for the business jet at 30,000-ft altitude and a weight of 40,000 lb.

range conditions would occur at a slightly higher speed (Mach 0.7 in this case). The use of a constant TSFC would underestimate the specific instantaneous range for Mach numbers below 0.67 and overestimate it when flying at Mach number superior to 0.67. Overall, the use of a constant TSFC facilitates that analysis of aircraft range while introducing only small errors in specific instantaneous range values, provided one uses a reasonable estimate for the cruise TSFC.

To determine the flight conditions for maximum endurance, one must maximize the ratio of aerodynamic efficiency to the TSFC (Fig. 9.22).

It can be seen that this ratio occurs in the region of slow flight (above V_{stall} and below $V_{D_{min}}$), see Chapter 3, Sec. 3.2. It might be preferable for the pilot to fly slightly faster than $V_{D_{min}}$ to avoid the region of reverse command.

To determine the performance of the aircraft during a climb, one can use the velocity hodograph (Chapter 4, Sec. 4.4). In this case, since the maximum thrust available decreases with increasing velocity for low altitudes (below approximately 30,000 ft), it will be determined that the maximum climb angle will fall at a velocity that is slightly lower than the velocity for minimum drag (maximum lift-to-drag ratio). This point still corresponds to the maximum excess thrust condition (see Fig. 9.19). The maximum rate of climb will still occur at the velocity for maximum excess power.

For the takeoff case at sea level conditions for a takeoff weight of 45,000 lb, the takeoff thrust will decrease during the ground roll and subsequent climb. At zero forward velocity, the maximum takeoff thrust will be 17,000 lb (two engines giving 8500 lb each) and will decrease to 14,000 lb at liftoff ($V_{LO} = 236$ ft/s $= 140$ kn, Mach 0.21), a decrease of 17.6%. This will translate into a decreasing acceleration during takeoff and a slightly longer takeoff distance compared to the constant thrust assumption.

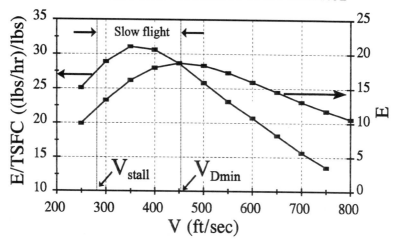

Fig. 9.22   Lift-to-drag ratio E and E/TSFC variation with velocity, h = 20,000 ft, W = 40,000 lb.

Problems

9.1.   A CF-18A (see Appendix E), flying at 19,000 ft, encounters a MiG-29 at the same altitude. Which aircraft has the minimum radius of turn (constant altitude)? Which one has the minimum radius of turn without altitude restriction? Assume 0 deg of sideslip angle in all cases.

9.2.   A CF-18A (see Appendix E) is doing a low-speed, low-altitude (100-ft) flight pass during an airshow. Its AoA is 30 deg (C_L = 1.65) (Fig. P9.2). What are the aircraft velocity and thrust required to do such a pass?

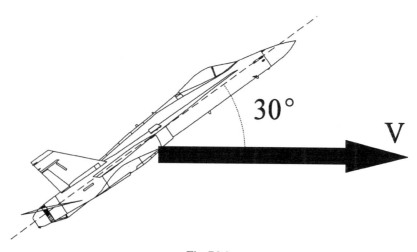

Fig. P9.2

10
Stability and Control

10.1 Introduction

U P to now we have been dealing with the aerodynamics and performance of aircraft under steady-state conditions. It was assumed that the aircraft was stable and controllable so that the pilot could achieve the calculated performance. If an aircraft was unstable, every disturbance encountered, no matter how small, would have to be countered by the pilot to maintain the desired flight path. This would be possible only if the aircraft was controllable. This brings us to define controllability and stability.

Controllability is the ability or ease (in terms of forces and moments applied) with which an aircraft responds to control surfaces displacement and achieves a new desired condition of flight. It is a measure of the responsiveness to inputs.

Stability, on the other hand, is the ability of an aircraft to return to some particular condition of flight after having been slightly disturbed from that condition without control input from the pilot. It is a measure of the resistance to disturbance. A stable aircraft will reduce the workload of the pilot.

There are two categories of stability: static stability, which is the tendency a body shows after it has been displaced from equilibrium, and dynamic stability, which is the time history of the tendency. A statically neutral aircraft is one which, once disturbed from its equilibrium position, will remain in this new position unless some control surfaces are deflected to bring the aircraft back to its original position. A statically unstable aircraft is one which, once disturbed, will tend to further deviate from its original position. This condition requires constant inputs from the pilot to maintain a proper flight path and in some cases may simply be uncontrollable. Whereas statically neutral aircraft could be acceptable for short-duration flights, the workload imposed on the pilot for a long trip could be unacceptable. On the other hand, a statically unstable aircraft is unacceptable in any condition. A statically stable aircraft will tend to return to its equilibrium position after being disturbed (Fig. 10.1).

The coordinate system used in stability and control analysis varies from author to author. Some use the body-axis system, which is fixed to the aircraft (the longitudinal, lateral, and vertical axes as described in Appendix A); others use the stability axis system, which is similar to the wind axis system but neglects the effects of the yaw angle β. The body-axis system will be used here for simplicity. The moments about the three axes are, respectively, the pitching moment M, the yawing moment N, and the rolling moment $\mathcal{L}$:

$$M = \tfrac{1}{2}\rho_{\text{SL}}\sigma V^2 S\bar{c}C_M$$

$$N = \tfrac{1}{2}\rho_{\text{SL}}\sigma V^2 SbC_N \qquad (10.1)$$

$$\mathcal{L} = \tfrac{1}{2}\rho_{\text{SL}}\sigma V^2 SbC_{\mathcal{L}}$$

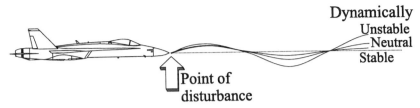

Dynamically
Unstable
Neutral
Stable

Point of
disturbance

Fig. 10.1   Dynamic response of statically stable aircraft.

Note the use of a moment arm in these equations (the mean aerodynamic chord $\bar{c}$ and the wing span b).

A full stability and control analysis requires the evaluation of the six degrees of freedom of an aircraft (three mutually perpendicular displacements and three moments) and many textbooks have been written on the subject. This chapter only intends to show how some stability and control considerations affect aircraft performance. No in-depth analysis of static and dynamic stability is done in the following sections.

10.2 Longitudinal Static Stability

The longitudinal static stability involves moments (M) and rotations (pitch angle, θ) about the lateral axis. The center of that rotation is the center of gravity (c.g.) of the aircraft. A positive pitching moment will move the nose of the aircraft up as shown in Fig. 10.2. Figure 10.3 shows the major contributor to the pitching moment.

The sum of the pitching moments about the c.g. is

$$\sum M_{cg} = M_{ac} + M_{fus} - L(X_{acW} - X_{cg}) - L_H(X_{acH} - X_{cg})$$
$$- TZ_T - F_i(X_i - X_{cg}) \tag{10.2}$$

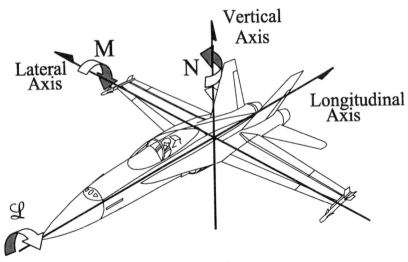

Fig. 10.2   Axes and moments.

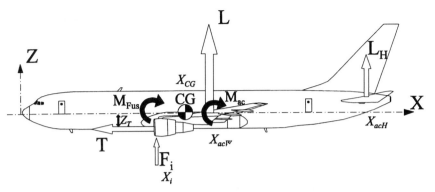

Fig. 10.3 Moments and forces acting on an aircraft in flight.

Assuming that the dynamic pressure is approximately constant from one end of the aircraft to the other, the sum of the pitching moments can be written in coefficient form as follows:

$$\sum C_{M_{cg}} = C_{M_{ac}} + C_{M_{fus}} - C_L \left(\frac{X_{ac} - X_{cg}}{\bar{c}} \right) - C_{L_H} \frac{S_H}{S} \left(\frac{X_{acH} - X_{cg}}{\bar{c}} \right)$$

$$- \frac{T}{\frac{1}{2}\rho_{SL}\sigma V^2 S} \frac{Z_T}{\bar{c}} - \frac{F_i}{\frac{1}{2}\rho_{SL}\sigma V^2 S} \left(\frac{X_i - X_{cg}}{\bar{c}} \right) \qquad (10.3)$$

A wing with a positive camber contributes a nose-down pitching moment about the aerodynamic center ($C_{M ac}$) that remains constant up to stall. This aerodynamic center is located approximately at 25% of the MAC in subsonic flight and moves back to approximately 50% of MAC at supersonic speeds. The nose-down pitching moment increases significantly when the flaps are lowered for takeoff and landing. The wing also contributes a lift force to the pitching moment. This lift force can be considered as a concentrated force acting through the aerodynamic center. Since the forces and moments produced by the wing are the most important contributors to the pitching moment, the aerodynamic center is usually located close to the c.g. to reduce the moment arm of the lift force. The position of the c.g. is usually expressed in terms of % MAC. For example, the safe range for the location of the c.g. of a particular aircraft is between 18 and 23% MAC. This aircraft has a MAC of 9 ft, which means that the c.g. is located between 1.62 and 2.07 ft of the LE of the MAC, an allowable range of a little less than 6 in.

The shape of the fuselage (its camber) will dictate the 0-deg incidence pitching moment. The pitching moment for the fuselage ($C_{M fus}$) increases with increasing fuselage AoA (nose-up pitching moment) providing a destabilizing effect on the overall aircraft longitudinal static stability.

The engine contribution to the pitching moment is twofold. First, the engine thrust line location will provide a nose-up, destabilizing moment for engines located below the c.g. and a nose-down, stabilizing moment for engines mounted above the c.g. (Fig. 10.4).

Second, the incoming flow entering the intake will turn from their incoming angle to the engine thrust line angle (Fig. 10.5). This causes a change in momentum of the airflow proportional to the mass flow and the change of the velocity vector.

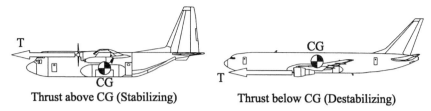

Thrust above CG (Stabilizing)   Thrust below CG (Destabilizing)

Fig. 10.4   Engine thrust line contribution to pitch.

Engines with inlets located in front of the c.g. will provide a destabilizing moment to the pitching moment at positive AoA.

For the static trim condition to exist, the sum of the pitching moment must equal zero (i.e., $\sum C_{Mcg} = 0$). This is the main function of the horizontal tail. It must contribute a pitching moment, through its lift and moment arm, large enough to compensate for the other components of pitch equation (10.3). For an aft-mounted tail, this moment contribution is usually negative lift (a downward force) because the wing aerodynamic center is usually located behind the c.g. contributing to a nose-down pitching moment, which is further increased by the wing's aerodynamic moment (M_{ac}). The critical conditions for tail sizing are during takeoff and landing where the flaps and landing gear are extended (increased nose-down pitching moment) and the dynamic pressure is low, and during flights at transonic and supersonic speeds where the efficiency of the tail decreases and the aerodynamic center moves back from its 25% MAC position.

For an aircraft to be statically stable in the pitching plane, it must generate a compensating moment opposed to the one generated by a disturbance. This means that for an increasing AoA (wing lift coefficient), there must be a negative moment (nose-down pitching moment) generated by the rest of the airframe. Thus, the slope of the pitching moment coefficient [Eq. (10.3)] must be negative, i.e., the derivative of Eq. (10.3) with respect to the wing lift coefficient must be negative.

Changing the lift of the horizontal tail, by either deflecting the elevator (it acts as a flap) or changing the incidence of the horizontal tail (all moving tail) will not change the slope of the horizontal tail moment contribution to the overall pitching moment (see Fig. 10.6). It will change the intersect value at zero wing lift coefficient, thus allowing for a new trim point at a different wing lift coefficient when the flight condition changes.

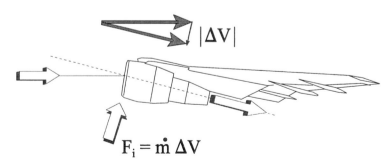

$$F_i = \dot{m}\,\Delta V$$

Fig. 10.5   Force acting on the intake due to the turning of the flow.

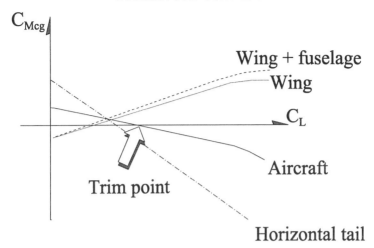

Fig. 10.6   Main contributors to the aircraft pitching moment and aircraft trim point.

10.2.1 Wing Flowfield Influence

The wing strongly affects the flow around the entire aircraft. The generation of lift generates a circulation around the wing. This circulation tends to bend the incoming flow upward (upwash) and bend the flow downward (downwash) after going past the wing. The upwash will push up on anything that is located in front of the wing, such as the front fuselage, while the downwash will push down on everything located aft of the wing, such as the rear fuselage and tail.

This combination of upwash and downwash on the fuselage further contributes to the destabilizing (nose-up) pitching moment effects of the fuselage. As well, the tail is now in a flow that has a smaller AoA than the AoA of the wing. Also, the dynamic pressure around the horizontal tail is usually less than that of the wing by about 10% (Fig. 10.7).

10.2.2 C.G. Limits

During a mission, the weight of the aircraft will change, leading to a change in the location of the c.g. This change can be slow (fuel burned), very fast (payload dropped), or the weight can even increase in-flight (in-flight refueling) (Fig. 10.8).

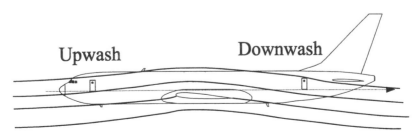

Fig. 10.7   Wing flowfield influence on the AoA of the flow at other stations.

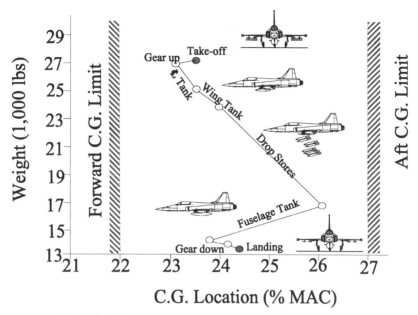

C.G. Location (% MAC)

Fig. 10.8   Displacement of the c.g. location during a mission.

The displacement of the c.g. during the mission will affect the trim setting of the aircraft. If the c.g. moves forward, a greater downward force from the tail will be required, therefore creating more trim drag and increasing the static stability of the aircraft (i.e., larger control deflections are required). If the c.g. moves backward, the trim drag will be reduced and the stability decreases. The most forward c.g. location is limited by the stick force required to move the elevator and by the maximum up elevator deflection at all flight speeds, with takeoff and landing being the critical conditions. An aircraft with its c.g. at the most forward position will feel nose heavy to the pilot. If the c.g. moves farther forward, the pilot may not be able to lift the nose wheel during takeoff. The aft c.g. location is limited by longitudinal stability and control sensitivity. If the c.g. is located at the maximum aft position, the aircraft will feel tail heavy. If the c.g. moves past the maximum aft position in flight, the pilot may not be able to compensate for the large nose-up pitching moment; the aircraft will stall and may be unrecoverable.

Thus, an aircraft that is heavily loaded in the forward part of the airframe will have more drag, which translates into reduced lift-to-drag ratio. This reduced lift-to-drag ratio will have a direct effect on performance such as range, endurance, and climb. On the other hand, an aft c.g. will increase the lift-to-drag ratio (Fig. 10.9).

As the c.g. moves aft, the slope of the aircraft pitching moment will increase. At a given position of the c.g., a change of AoA will not change the pitching moment (see Fig. 10.10). This is the neutral point for the aircraft. At this position of the c.g., the aircraft has neutral static stability. If the c.g. moves any farther aft, the aircraft will become unstable. The distance between the c.g. and the aircraft neutral point is called the static margin. A positive static margin means that the aircraft's c.g. is

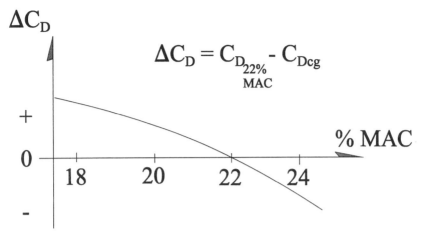

$$\Delta C_D$$

$$\Delta C_D = C_{D_{22\%}} - C_{Dcg}$$
MAC

% MAC

+

0

18   20   22   24

−

Fig. 10.9   Effect of change of location of c.g. on aircraft drag, from a reference drag coefficient when the c.g. is at 22% MAC.

located in front of the neutral point and that the aircraft is statically stable. Typical static margins for the most aft c.g. location vary between 5 and 10%.

Modern fighters, such as the F-16, were designed with relaxed static stability (RSS) with a negative static margin (by as much as −15%) to increase the aircraft responsiveness in the pitch plane (Fig. 10.11). To be able to control such an aircraft, it must be equipped with active control technology systems that actively monitor atmospheric conditions and disturbances and deflect the control surfaces several times per second to maintain the desired flight path. RSS aircraft will usually have smaller wings and horizontal tail than aircraft designed with a positive static margin because the horizontal tail now produces positive lift to trim the aircraft rather than negative lift. This results in less induced drag and skin friction drag leading to higher lift-to-drag ratio.

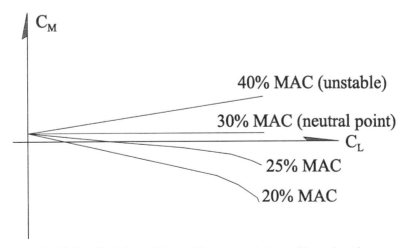

$$C_M$$

40% MAC (unstable)

30% MAC (neutral point)
$$C_L$$

25% MAC

20% MAC

Fig. 10.10   Variation of the pitching moment slope with c.g. location.

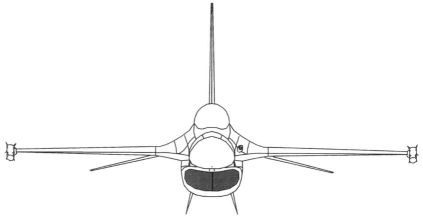

Fig. 10.11   F-16 has RSS.

10.3   Lateral-Directional Static Stability

The analyses of lateral and directional static stability are closely coupled because both are affected by the aircraft yaw angle β. The deflection of ailerons or rudder will produce some form of yawing and rolling moment. In normal steady flight, the yaw angle is zero and there are no yawing or rolling moments.

The most important contributor to lateral stability (about the longitudinal axis) is the wing. The wing's dihedral angle has a positive effect on lateral stability when the angle is positive, i.e., the dihedral of the wing rolls the aircraft in a direction away from the sideslip angle. The vertical location of the wing also has an effect on lateral stability with a high wing providing a stabilizing effect. This stabilizing effect is usually expressed in terms of equivalent dihedral angle (Fig. 10.12).

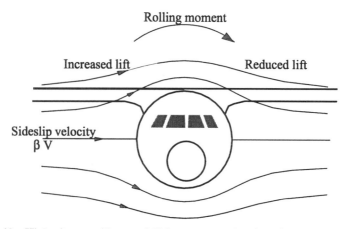

Fig. 10.12   High wing provides a stabilizing moment, the aircraft rolls away from the sideslip.

A wing with sweepback (positive sweep angle) will provide a stabilizing effect. The vertical tail contributes positively to the static lateral stability, being above the fuselage, while a ventral fin is destabilizing.

The primary roll control surfaces are the ailerons: one aileron goes down producing more lift on that wing while the aileron on the other wing goes up reducing the lift on that side. There is, however, a cross effect when deflecting ailerons. The wing producing more lift also produces more drag, which will create a yawing moment in the opposite direction to the roll. This effect is known as adverse yaw.

Another control surface used for roll is the spoiler. This control device, installed on some aircraft, deflects up on the upper wing surface, decreasing the lift and increasing the drag; thus, the wing yaws in the same direction as the roll. Modern fighters can use their all-moving horizontal tail differentially to provide roll control without the adverse yaw effect.

The major contributor to directional stability (about the vertical axis) is the vertical tail. The vertical tail must provide sufficient directional stability to compensate for the destabilizing effects of the fuselage section ahead of the c.g. when the sideslip angle is nonzero (Fig. 10.13).

The primary directional control surface is the rudder. The rudder basically acts like a flap to provide increased lift at the tail. Because the center of lift of the vertical tail is above the c.g., a rolling moment is created when the rudder is

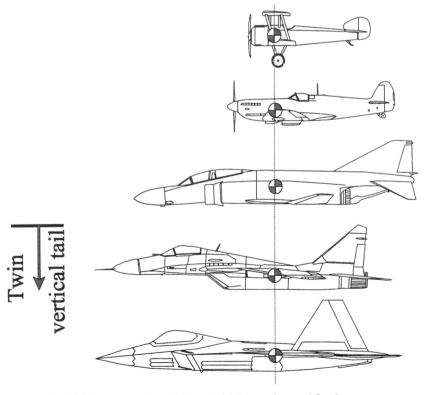

Fig. 10.13   Increased vertical tail with larger forward fuselage area.

deflected. This rolling moment is in the same direction as the intended yawing moment and is termed proverse roll.

10.4 Engine Out Problems

When one engine from a multiengine aircraft stops working in-flight, the pilot must obviously correct for the loss of thrust available. However, to maintain a moment equilibrium condition about all three axes, the pilot must also deflect the aircraft control surfaces. For example, a low-wing-mounted, twin-engine aircraft (Fig. 10.14) loses the thrust from the right engine. This engine will now be either windmilling or, worse, seized up and acting like a nonstreamlined solid object. In either case, it will be creating extra drag below and to the right of the c.g. This asymmetric thrust case about the vertical axis will create a yawing moment, which must be balanced by the deflection of the rudder. In fact, the sizing requirements for the vertical stabilizer/rudder are usually dictated by the takeoff one engine out situation where the thrust is very high and the dynamic pressure around the vertical tail is very low.

The center of lift of the vertical tail is above the c.g. (distance z_{vt} in Fig. 10.14). The required force from the vertical tail to compensate for the asymmetric thrust will create a rolling moment about the longitudinal axis. This moment must be compensated by the deflection of ailerons. The right wing, the one with the dead engine, having to produce more lift than the left wing to compensate for the rolling moment, will now produce more induced drag than the left wing, thus a yawing moment, which will have to be compensated for by a further deflection of the rudder.

The deflection of the rudder and the ailerons will create drag, which must be added to the drag of the dead engine. This requires that the remaining engine(s) have enough maximum continuous thrust (the maximum thrust that can safely be maintained over long periods of time) to compensate for such drag. If not, the aircraft must slow down and/or lose altitude. Assuming there is enough thrust, the

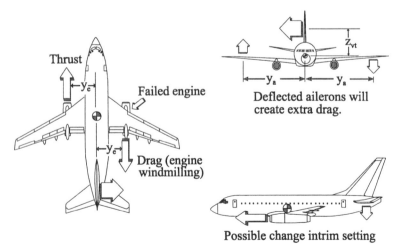

Fig. 10.14   Asymmetric thrust case.

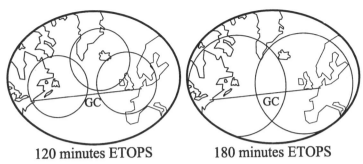

120 minutes ETOPS   180 minutes ETOPS

Fig. 10.15   Area of operation over the North Atlantic for two different ETOPS certified aircraft.

new force distribution about the c.g. will require a new trim setting, which may
produce additional drag.

Overall, the thrust-to-weight ratio of the aircraft decreases by as much as 50%
(for a twin-engine aircraft), and the lift-to-drag ratio will also decrease due to the
additional drag sources. The decrease of thrust-to-weight ratio translates directly
into smaller climb angle, rate of climb, and/or accelerations, whereas a decrease
in lift-to-drag ratio will decrease the aircraft range and endurance.

For twin-engine airliners operating under ETOPS regulations [FAA AC 120-
42A (see Chapter 3, Sec. 3.4.10)], the possible loss of an engine imposes significant
constraints on long-range operations. These aircraft must remain within a specified
flight time (up to 180 min) from an adequate airport at all times during cruise. For
an aircraft crossing the North Atlantic, for example, 180 min ETOPS capability

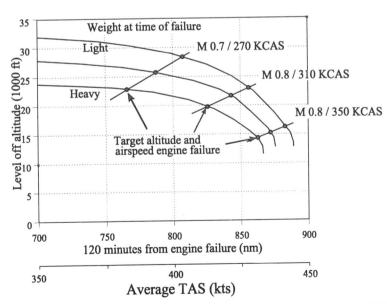

Fig. 10.16   Target altitude and airspeed after engine failure for a given ETOPS
certified aircraft.

would allow the aircraft to follow the Great Circle route (the shortest distance between two points on the globe), with corrections for the prevailing winds, whereas an aircraft certified for 120-min ETOPS would have to deviate from that route to remain within 120-min flying time from an adequate airport (see Fig. 10.15). This means that the 120-min ETOPS certified aircraft would require more fuel to travel between two given airports than an 180-min ETOPS certified aircraft (for the case illustrated in Fig. 10.15).

The area covered by the circles of Fig. 10.15 is directly related to the aircraft flight altitude and velocity after engine failure. Upon engine failure, the aircraft will start decelerating due to the decrease in available thrust, increased drag from the dead engine, and the control surfaces deflection. The pilot must then select a target altitude and airspeed (established by the airline, in conjunction with the aircraft manufacturer), which determines the area of operation (area covered by the circles) for the particular aircraft configuration and weight (see Fig. 10.16).

This very short chapter on stability and control was used to highlight the fact that an aircraft must be controllable before one can hope to verify the predicted performance in flight. Stability and control considerations (such as empennage size and location) will have a large effect on the aircraft trimmed drag polar, which in turn will affect its performance.

11
Elements of Aircraft Design

11.1 Introduction

T HE aim of this chapter is to provide the reader with an idea of the level of effort involved in designing a new aircraft to meet specified performance criteria. It is also to provide the reader with a practical application to what was learned in previous chapters. This chapter shows how aircraft performance is accounted for during conceptual design, then introduces such subjects as Stealth technology (unclassified information only), drag reduction research, and high-AoA flights, which will impose some restrictions to an aircraft designer during the design process.

11.2 Design Phases in Aircraft Development

Every aircraft is designed to meet an existing operational requirement or a contractor perceived future market requirement. This is essential because the development of a new aircraft can add up to billions of dollars and development costs must be recovered in one form or another for a company to survive and prosper. Once the requirements are defined by trade studies and market analysis, the design process can start. This design process can be broken down into three major phases, which are described in the following (Fig. 11.1).

1) The conceptual design phase is the phase during which the initial sizing is done and the general configuration is laid out to meet the performance as set out in the SOR. It is also during this phase that an initial cost analysis is performed to determine if the proposed designs are affordable or if the requirements have to be relaxed.

2) The transition between the conceptual design and the preliminary design phases is not black and white but occurs approximately when no more major modifications to the aircraft layout are done. It is a time to do some fine tuning (minor modifications) to the layout and to start designing the associated systems of the aircraft (avionics, structure, environmental conditioning system, etc.). At one point though, the design has to be frozen. Lofting (the mathematical modeling of the aircraft parts, skin, and components) is performed to verify that everything fits together. During the Boeing 777 development this was carried to the point where even the technicians were modeled to ensure they could reach any part of the aircraft for repairs. Testing in the fields of aerodynamics, propulsion, and structure is also done to verify that the configuration chosen will actually meet the diverse military or civilian specifications, as well as the requirements of the customer.

3) The detail design phase starts with the decision to go to full-scale development. It is during this phase that the actual pieces to be built are designed, as well as the tools that will be required to build them. Testing is intensified and performance estimates are finalized. This phase ends when fabrication of the aircraft is started (again, not black and white).

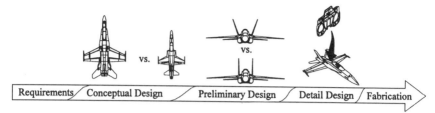

Fig. 11.1   Aircraft design phases.

The emphasis of this section of Chapter 11 is on how aircraft performance criteria are taken into account during the conceptual design phase so that the knowledge gained in previous chapters can be applied.

11.2.1   Weight Buildup

The aircraft design takeoff weight (W_{TO}) is the total weight of the aircraft before takeoff for the mission it was designed to perform. This may not necessarily be the maximum takeoff weight of the aircraft as more and more military aircraft are designed for multirole missions and can carry a variety of external and/or internal loads. The design takeoff weight is the summation of the weight of the crew W_{crew}, the payload weight W_{PL}, the fuel weight W_f, and the aircraft empty weight W_E

$$W_{TO} = W_{crew} + W_{PL} + W_f + W_E \qquad (11.1)$$

The crew of an aircraft is the personnel required to safely accomplish the mission it was designed to do or any other mission that it may be asked to perform. A weight of 200 lb per crew member,[11, 12] including luggage and/or survival gear (military pilots) is used for initial weight sizing.

The payloads consist of passengers and their luggage, cargo, weapons (bombs, missiles, etc.), and/or special mission specific equipment. In short, everything that is not usually fixed or attached to the aircraft. A weight of 210 lb is used for passengers with their luggage. Note that both the crew weight and the payload weight are known before initial sizing because they are specified in the SOR.

The fuel weight is the weight of all of the fuel required to accomplish the mission plus some reserve fuel as determined by regulations and specifications, as well as trapped fuel that cannot be pumped from the tanks or that remains in the fuel lines. Typically, unless specified in the SOR, 5% of the takeoff weight is reserve fuel whereas 0.5–1% is trapped fuel, the rest being mission fuel (W_{mf}). Thus, the total fuel weight can be written as (with 0.5% trapped fuel)

$$\frac{W_f}{W_{TO}} = 1.055\left(1 - \frac{W_x}{W_{TO}}\right) = 1.055\left(1 - \frac{1}{MR}\right) \qquad (11.2)$$

where W_x is the weight of the aircraft at the end of the mission and MR is the mass ratio for the mission as defined by Eq. (3.25).

Finally, the empty weight of the aircraft includes such things as the weight of the structure, the avionics, the engines, the landing gear, any fixed equipment (for a fighter aircraft, this also includes any internally mounted gun), and some ballast if required (to maintain the aircraft's c.g. within acceptable limits). The operating empty weight W_{OE} is the sum of the aircraft empty weight plus the weight of the crew and the weight of the trapped fuel and oil W_{tf}. Therefore, Eq. (11.1) can be rewritten as

$$W_{TO} = W_{OE} + W_{mf} + W_{PL} \tag{11.3}$$

where

$$W_{OE} = W_E + W_{crew} + W_{tf} \tag{11.4}$$

11.2.2 Aircraft's Primary Mission

To determine the takeoff weight of the aircraft, one has to know what mission it is intended to perform. This is specified in the SOR (military) or market survey/customer requirements. As an example, a short specimen SOR for a military aircraft trainer (MAT) is provided (Fig. 11.2). Note that the propulsion system is not specified here, and it is the designers' responsibility to determine the best propulsion system to fulfill the requirements. In these days of budget restraint, meeting the bare minimum requirements rather than offering a gold-plated aircraft is often the winning combination.

11.2.3 Initial Weight Sizing

The first step toward designing a new aircraft is to determine the approximate takeoff weight, which will then be used to select the propulsion system and initial layout of the aircraft. The mission described in the SOR is divided into multiple segments that can be analyzed individually. The segments used in this section of Chapter 11 will be takeoff, climb to mission altitude, cruise, endurance/loiter, descent, and landing (including taxi and shutting down the engines). This way, equations developed in the preceding chapters can be used to determine the weight ratio of each segment of the mission. Where no equations can be used, typical weight ratio values will be provided. These values are based on a statistical analysis of existing aircraft. The reader is encouraged to use experience and good judgment while using these values. Simulations using a similar type of aircraft could produce better approximations of the weight ratio for every segment.

Takeoff. The takeoff portion of the mission includes the following steps: engine startup, taxi and engine warmup, ground roll on runway, and liftoff to 35/50 ft above the runway. A typical value of weight ratio for this phase is

$$W_{i+1}/W_i \approx 0.98 \tag{11.5}$$

In the example of Chapter 7, a weight ratio of 0.998 was obtained for the ground roll and liftoff to 50 ft. With the addition of the engine startup and warmup plus taxi to the runway, one may expect the weight ratio to be around 0.98 as indicated in Eq. (11.5).

SPECIMEN
Statement of Operational Requirement
for a
Military Aircraft Trainer (MAT)

A.   Background

By the end of the 1990's, many countries will have military aircraft trainer that are 20-30 years old and may need replacement. , A trainer aircraft that could be used to do pilot training from just after the basic training (piston-prop trainer aircraft) to just before type specific training (fighter, bomber, transport, etc.) is required. There is a need for a low cost trainer.

B.   Mission

The MAT shall be capable of performing the mission described below:

Take-off   From a runway at 2,000 ft pressure altitude with air temperature of 90°F, with 5 min fuel allowance at idle power for engine start up and taxi, the MAT shall lift-off in less than 1,200 ft on a wet concrete runway (μ_g=0.05). V_{LO}=1.2 V_{stall}.

Climb   The MAT shall climb to best cruise altitude using a minimum time to altitude climb profile.

Cruise   The MAT shall perform a best range cruise-climb profile until the total distance covered during climb and cruise is 150 nm.

Descent   Descend to 15,000 ft. No range/fuel/time credit for descent.

Training   Training shall be simulated by an equivalent 2.0 hrs flight in loiter mode. No range credit for training.

Return   The MAT shall perform a best range cruise-climb profile covering a distance of 150 nm to return to base.

Descent   Descend to 5,000 ft. No range/fuel/time credit for descent.

Reserve   Loiter 20 min at 5,000 ft at best endurance conditions.

Landing   Descend and land on a wet concrete runway (μ_g=0.2) at a pressure altitude of 2,000 ft in less than 1,200 ft of ground roll. V_{TD} = 1.1 V_{stall}.

C.   Other Requirements

The cockpit shall accommodate of crew of two with dual control and the back seat elevated to enable the backseater unobstructed forward vision (including 11° downward). The cockpit shall also be equipped with ejection seats.

Fig. 11.2   Specimen SOR.

Climb. The weight ratio for a climb from a lower altitude to a higher one can be approximated using the weight ratio range provided next. Obviously, the larger the altitude difference, the smaller the weight ratio. As well, depending on the climb profile, this value can change considerably

$$W_{i+1}/W_i \approx 0.96\text{–}0.995 \tag{11.6}$$

One may wish to estimate the weight ratio for the climb segment by assuming a climb profile that yields simple-to-use equations. For a jet aircraft, the maximum angle of climb profile is done at maximum excess thrust conditions (minimum drag). If the altitude increment is known, then one may approximate the weight ratio for the climb segment by using Eq. (3.67) and replacing the time variable (t) by the ratio of the altitude to climb to the estimated average rate of climb ($\Delta h / RC_{avg}$). The following equation results:

$$\frac{W_{i+1}}{W_i} = \exp\left(\frac{-\Delta h\, SFC_T}{RC_{avg}E}\right) \tag{11.7}$$

Doing the same for a propeller driven aircraft, the following equation for the climb segment results:

$$\frac{W_{i+1}}{W_i} = \exp\left(\frac{-\Delta h\, V SFC_P}{RC_{avg}\eta_p E}\right) \tag{11.8}$$

One should ensure that the units used in these equations are consistent.

Cruise. The weight ratio of the weight at the end of the cruise segment to the weight at the beginning of the cruise leg can be estimated for jets by using Eq. (3.42):

$$\frac{W_{i+1}}{W_i} = \exp\left(\frac{-X\, SFC_T}{EV}\right) \tag{11.9}$$

Thus, the weight ratio for the cruise segment can be found by knowing the range X (usually specified in the SOR) and the cruise speed V (can be specified in the SOR), the engine TSFC or SFC_T and the lift-to-drag ratio E of the cruise segment. The TSFC during cruise can be estimated from existing or in-development engines, whereas the L/D ratio can be found by using typical values from historical trends. If it is a cruise for best range, Chapter 3 mentions that the value of the lift-to-drag ratio for a jet aircraft will be approximately 86.6% of the aircraft maximum L/D ratio. If the aircraft is doing a high-speed dash, then the lift-to-drag ratio is expected to be less than the value for best range. Figure 11.3 illustrates the variation of the L/D ratio, for typical aircraft C (compressibility effects included), as a function of Mach number. This figure is to be used as a guidance for the variation of the L/D ratio with Mach number noting that the maximum value and the actual shape of the curve are dependent on aircraft aerodynamic characteristics.

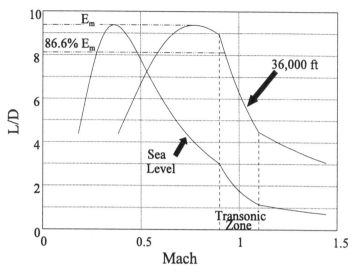

Fig. 11.3   L/D as a function of Mach number for aircraft C at maximum takeoff weight.

For a propeller driven aircraft, the weight ratio for the cruise segment can be estimated using Eq. (3.59),

$$\frac{W_{i+1}}{W_i} = \exp\left(\frac{-X\,SFC_P}{\eta_p E}\right) \tag{11.10}$$

Here, the range X is required as well as the engine PSFC (SFC_P), the propeller efficiency η_p, and the aerodynamic efficiency E. For a propeller driven aircraft, to achieve best range, it must fly at its maximum lift-to-drag ratio conditions (at E_m).

Loiter.   The weight ratio, for a loiter segment for a jet aircraft, can be found from Eq. (3.67),

$$\frac{W_{i+1}}{W_i} = \exp\left(\frac{-t\,SFC_T}{E}\right) \tag{11.11}$$

where t is the length of the loiter segment (units of time). To achieve maximum endurance, a jet aircraft must fly at its maximum lift-to-drag conditions (E_m).

The weight ratio, for a loiter segment for a propeller driven aircraft, can be found from Eq. (3.73),

$$\frac{W_{i+1}}{W_i} = \exp\left(\frac{-t\,V\,SFC_P}{\eta_p E}\right) \tag{11.12}$$

To achieve maximum endurance, a propeller driven aircraft must fly at its minimum power-required conditions ($E \approx 0.866 E_m$). If the velocity V during the loiter is not specified in the SOR, one must assume the velocity for minimum power, which corresponds to about 76% of the velocity for minimum drag (see Chapter 3).

Combat/training. Combat or training usually involve changes of heading, altitude, and speed. Thus, the aircraft will not be flying straight and level at a constant throttle setting. The throttle movement will probably mean that the engine will not be operating at its maximum efficiency for the thrust required. The weight ratio for this segment can be estimated either by specifying the time duration at maximum thrust (typically, 3 min for fighter combat) or by specifying a certain number of combat turns at a given altitude and Mach number.

If the time duration is known, then the weight ratio will be

$$W_{i+1}/W_i = 1 - SFC_T (T/W)_{\text{combat}} \, t_{\text{combat}} \qquad (11.13)$$

On the other hand, if the number of turns is specified, one can estimate the time required to complete the number of turns by using Eq. (5.5), rewritten here,

$$t_{\text{turns}} = \frac{2\pi}{\dot{\chi}} N_{\text{turns}} = \frac{2\pi V}{g\sqrt{n^2 - 1}} N_{\text{turns}} \qquad (11.14)$$

where N_{turns} is the number of turns. The value of the load factor can be estimated by using Eq. (5.4),

$$n \le \frac{q C_{L_{\max}}}{W/S} \qquad \text{and} \qquad n \le n_{\text{max}_{\text{struc}}}$$

This value should not exceed the maximum structural load factor of the aircraft or cause the aircraft to stall.

Descent. Typical weight ratios for a descent segment are

$$W_{i+1}/W_i \approx 0.985\text{--}0.995 \qquad (11.15)$$

depending on the altitude range to descend.

Landing. The landing segment includes the landing from 50 ft above the runway to complete stop plus taxi to the ramp and shutdown of the engine(s). A typical value for this segment is

$$W_{i+1}/W_i \approx 0.99 \qquad (11.16)$$

Reserve fuel. The reserve fuel requirement for a mission is usually specified in terms of loiter time just prior to landing. But it could also be specified in terms of a range to divert to another airport plus the loiter time prior to landing. In any case, the equations developed can be used to estimate the weight ratio for the reserve fuel segment. If not specified in the SOR, then it should be assumed to be equal to a certain percentage of the takeoff weight, as indicated in Eq. (11.2).

Mission weight ratio estimate. The aircraft empty weight can be estimated once each of the mission segments are established. Using simple multiplication, the total mission weight ratio can be determined in the following way:

$$\frac{W_x}{W_{\mathrm{TO}}} = \frac{W_1}{W_{\mathrm{TO}}} \frac{W_2}{W_1} \frac{W_3}{W_2} \cdots \frac{W_x}{W_{x-1}} = \prod_{i=\mathrm{TO}}^{x-1} \frac{W_{i+1}}{W_i} \qquad (11.17)$$

where (W_x/W_{TO}) is the mission weight ratio ($\mathrm{MR}_{\mathrm{mission}}$). This mission weight ratio can also be expressed in the following way:

$$\frac{W_x}{W_{\mathrm{TO}}} = \frac{W_{\mathrm{TO}} - W_{\mathrm{mf}}}{W_{\mathrm{TO}}} = 1 - \frac{W_{\mathrm{mf}}}{W_{\mathrm{TO}}} \qquad (11.18)$$

One should note here that the only loss in weight during the mission was that of the fuel used to produce thrust. If any payload was dropped during the mission, such as bombs and missiles, then the equation would be written as

$$\frac{W_x}{W_{\mathrm{TO}}} = \frac{W_{\mathrm{TO}} - W_{\mathrm{mf}} - W_{\mathrm{PL_{dropped}}}}{W_{\mathrm{TO}}} \qquad (11.19)$$

Once the mission weight ratio is determined, one may proceed to the next step, determining the takeoff weight and aircraft empty weight. Historical trends have shown that there exists a linear relation between the logarithm of the empty weight and the logarithm of the takeoff weight. This can be written as follows:

$$\ell n(W_E) = y + x\,\ell n(W_{\mathrm{TO}}) \qquad (11.20)$$

To determine the actual values of the variables x and y, one must determine the category of the aircraft to be sized. The selection of the categories may seem arbitrary and varies from author to author, but it has proved to be a valuable tool for initial sizing. This author uses the categories listed in Table 11.1 for aircraft weight sizing.

To perform the actual weight sizing, Eq. (11.20) may be rewritten without the logarithmic terms as follows:

$$W_E = A W_{\mathrm{TO}}^x \qquad (11.21)$$

where $A = e^y$. A curve fit for military turboprop trainers is shown in Fig. 11.4 so that the reader can see the relation between the equivalent weight distribution curve compared to a few existing military turboprop trainers (the year of each respective first flight is indicated within parentheses).

To relate the mission weight ratio to the ratio of the empty weight to the takeoff weight, Eq. (11.1) can be written in the following way:

$$W_{\mathrm{TO}} = \left[(W_{\mathrm{PL}} + W_{\mathrm{crew}}) \middle/ \left(1 - \frac{W_f}{W_{\mathrm{TO}}} - \frac{W_E}{W_{\mathrm{TO}}} \right) \right] \qquad (11.22)$$

Table 11.1 Coefficients of Eq. (11.20)

Categories, gross takeoff weight	y	x	Categories	y	x
Gliders	−0.1659	0.9475	Military trainers:		
Homebuilt (prop)	−0.6541	1.0213	Jet (design)	0.9466	0.8595
Single-engine prop	0.3282	0.8933	Jet (gross TOW)	1.2691	0.8019
Twin-engine prop	−0.3181	0.9814	Turboprop (design)	−1.2583	1.1085
Regional TBP	−0.8470	1.0371	Piston-prop (design)	−1.2754	1.1151
Regional jets	0.8576	0.8724	Fighter/attack:		
Business jets	−0.2949	0.9713	Jet with ext stores	0.21874	0.91290
Transport jets	0.1751	0.9363	Jet clean	0.19276	0.94206
			Military transport patrol and bomber:		
			Jet (gross TOW)	0.0632	0.9303
			TBP (gross TOW)	1.3727	0.8281

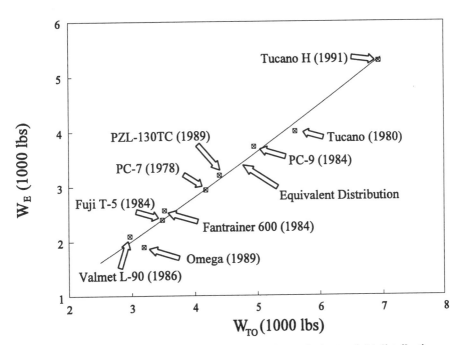

Fig. 11.4   Some turboprop military trainers and the equivalent weight distribution.

Table 11.2   Total fuel coefficient

B	Description
1.005	0.5% trapped fuel, no reserve (or reserve included in mission weight ratio)
1.055	0.5% trapped fuel plus 5% reserve fuel

where

$$\frac{W_f}{W_{TO}} = \frac{W_{mf}}{W_{TO}} + \frac{W_{tf}}{W_{TO}} \tag{11.23}$$

(W_{mf}/W_{TO}) is calculated with the help of Eq. (11.18); remember, no payload drop is simulated. Depending on the SOR description, (W_{tf}/W_{TO}) may or may not contain the reserve fuel. In any case, Eq. (11.23) may be written as

$$\frac{W_f}{W_{TO}} = B\left(1 - \frac{W_x}{W_{TO}}\right) \tag{11.24}$$

where B takes on the values in Table 11.2.

Finally, Eq. (11.21) can be rewritten in a way that is compatible with Eq. (11.22)

$$W_E/W_{TO} = AW_{TO}^{(x-1)} \tag{11.25}$$

Remember that $(A = e^y)$. Equations (11.22), (11.24), (11.25) and Tables 11.1 and 11.2 are used to do the initial weight sizing for a new aircraft design.

11.2.4   Weight Sizing for the Specimen SOR

In this section, the specimen SOR will be used to illustrate the theory just described. Two cases will be evaluated to determine which would best meet the requirements set in the specimen SOR. The first case will be a light, turbofan equipped, military trainer and the second case will be a light, turboprop equipped, military trainer. Figure 11.5 summarizes the mission for which the military trainers will be sized.

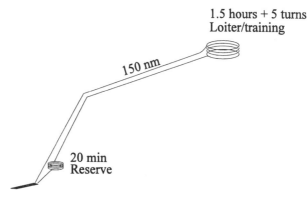

Fig. 11.5   Example of mission requirement for a new military aircraft trainer.

Table 11.3   First estimate for MAT

Initial physical characteristics	Turbofan trainer	Turboprop trainer
Maximum lift coeffficients		
Clean	1.5 (1.2–1.8)	1.5 (1.2–1.8)
Takeoff	1.8 (1.4–2.0)	1.8 (1.4–2.0)
Landing	2.0 (1.6–2.2)	2.0 (1.6–2.2)
AR	6.0 (4.5–7.0)	6.5 (5.5–7.0)
(SFC_T) or (SFC_P)	0.7 (0.4–1.0) lb/h/lb	0.5 (0.4–0.7) lb/h/HP
Propeller efficiency η_p	—	0.82 (0.75–0.85)
Oswald coefficient e		
Clean	0.8	0.8
Takeoff	0.76	0.76
Landing	0.72	0.72
Minimum drag coefficient C_{D_0}		
Clean	0.020	0.022
Takeoff (flaps and gear)	0.045	0.045
Landing (flaps and gear)	0.055	0.055
$K = 1/(\pi ARe)$		
Clean	0.06631	0.06121
Takeoff	0.06980	0.06443
Landing	0.07368	0.06801
$(L/D)_{max}$ [from Eq. (3.7)]		
Clean	13.73	13.63
Takeoff	8.92	9.29
Landing	7.85	8.17

The initial assumptions for each trainer are specified in Table 11.3 and some initial sketches are provided (Fig. 11.6). The assumptions were made using data from existing military trainers (range of typical values indicated inside parenthesis). Remember that design is an iterative process; the values chosen here will have to be revisited at a later stage of design to verify their validity.

Takeoff.   For the takeoff portion, the weight ratio will be assumed to be 0.99 for both types of propulsion system. Thus,

$$W_1/W_{TO} = 0.99$$

Climb.   The climb portion of the mission does not specify the final climb altitude, and so an initial assumption will be used. For the turbofan powered trainer (TFMAT), it will be assumed that the aircraft will climb from 2,000 (runway altitude) to 20,000 ft (where it should achieve best range conditions for this segment) at an average climb rate of 2,000 ft/min. The resulting weight ratio will

Fig. 11.6   Preliminary sketches for the turbofan powered MAT (TFMAT) and the turboprop powered MAT (TPMAT).

be, using Eq. (11.7) and noting that the units were made to match,

$$\frac{W_2}{W_1} = \exp\left[-\left(\frac{(18,000 \text{ ft})(0.01167 \text{ lb/min/lb})}{(2,000 \text{ ft/min})(13.73)}\right)\right] = 0.9924$$

For the turboprop powered trainer, since best range does not depend on the cruising altitude, it will be assumed that the TPMAT will climb to 15,000 ft, the training altitude. The average rate of climb will be 1,000 ft/min and the climbing speed will be 150 kn. The weight ratio is then, using Eq. (11.8),

$$\frac{W_2}{W_1} = \exp\left[-\left(\frac{(13,000 \text{ ft})(253 \text{ ft/s})(2.525 \times 10^{-7} \text{ ft}^{-1})}{(16.67 \text{ ft/s})(0.82)(0.866 \times 13.63)}\right)\right] = 0.9949$$

Cruise.   The SOR mentions that the some cruise credit can be taken during the climb segment. For the TFMAT, the altitude climbed is 18,000 ft at an average climb rate of 2,000 ft/min, which means that it took 9 min to reach cruising altitude. Assuming that the aircraft average velocity during the climb was 200 kn, the average climb angle is 5.66 deg, the horizontal distance covered would be 29.85 n mile. Thus, the total cruise distance remaining to cover is 120.15 n mile. Assuming a cruise speed of 300 kn and using Eq. (11.9), the weight ratio is

$$\frac{W_3}{W_2} = \exp\left[-\left(\frac{(120.15 \text{ n mile})(0.7 \text{ lb/h/lb})}{(0.866 \times 13.73)(300 \text{ kn})}\right)\right] = 0.9767$$

Again, it may seem that a lot of assumptions are used to find the weight ratios. One must keep in mind that at this point the designer has no real aircraft configuration to work from; best engineering estimates must be used. The preceding values will be rechecked at a later stage of the design process to confirm their validity or to update them.

For the TPMAT, it took 5.2 min to climb to the cruise altitude at an average speed of 150 kn; thus, the average climb angle was 3.76 deg, and the horizontal distance covered was 32.53 n mile. The remaining distance for the cruise leg is 117.47 n mile. The weight ratio can be estimated by using Eq. (11.10)

$$\frac{W_3}{W_2} = \exp\left[-\left(\frac{(117.47 \text{ n mile} \times 6,080 \text{ ft/n mile})(2.525 \times 10^{-7} \text{ ft}^{-1})}{(0.82)(13.63)}\right)\right]$$
$$= 0.9840$$

Descent. It was assumed that the TFMAT would cruise at 20,000 ft, and so it must descend 5,000 ft to get to the training altitude. The weight ratio during this short descent could be estimated to be

$$W_4/W_3 = 0.995$$

For the TPMAT, the cruise altitude to get to the training area is the same as the training altitude and so the weight ratio is

$$W_4/W_3 = 1.0$$

Training. The training assumes a 2.0-h flight in loiter mode. Since the aircraft probably will not be in a straight and level flight during the training, it is assumed that the lift-to-drag ratio will be less than optimal for this phase. A value of about 50% of E_m for each aircraft is assumed here. The weight ratio for the TFMAT, thus, will be, from Eq. (11.11),

$$\frac{W_5}{W_4} = \exp\left[-\left(\frac{(120 \text{ min})(0.01167 \text{ min}^{-1})}{(0.5 \times 13.73)}\right)\right] = 0.8155$$

For the TPMAT, it will be assumed that the training will take place at a velocity of about 200 kn, and so the weight ratio is, from Eq. (11.12),

$$\frac{W_5}{W_4} = \exp\left[-\left(\frac{(120 \text{ min})(20{,}267 \text{ ft/min})(2.525 \times 10^{-7} \text{ ft}^{-1})}{(0.82)(0.5 \times 13.63)}\right)\right] = 0.8959$$

Cruise back. For the cruise back, the full 150-n mile distance must be used as stated in the SOR. The weight ratios are as follows (assuming both aircraft come back at 15,000 ft). For TFMAT

$$W_6/W_5 = 0.9710$$

and for TPMAT

$$W_6/W_5 = 0.9796$$

Descent to 5000 feet. This is a relatively small descent and so both aircraft will have a weight ratio of approximately

$$W_7/W_6 = 0.995$$

Reserve. A 20-min loiter at best endurance conditions is used to simulate the reserve required for this mission. The weight ratios, thus, for TFMAT

$$W_8/W_7 = 0.9831$$

and for TPMAT

$$W_8/W_7 = 0.9869$$

Landing.   Finally, the landing part will be approximately the same for both aircraft at a value of

$$W_9/W_8 = 0.99$$

Mission weight ratio.   The mission weight ratio for the aircraft will be for TFMAT

$$W_9/W_{TO} = (0.99)(0.9924)(0.9767)(0.995)(0.8155)$$

$$\times (0.9710)(0.995)(0.9831)(0.99) = 0.7322$$

and for TPMAT

$$W_9/W_{TO} = (0.99)(0.9949)(0.9840)(1.0)(0.8959)$$

$$\times (0.9796)(0.995)(0.9869)(0.99) = 0.8269$$

For the same mission, it can be seen that the TPMAT will have a larger mission weight ratio; thus, it will have used less fuel to perform the same mission compared to the TFMAT. This is representative of the fact that turboprop aircraft are generally smaller and have a better fuel consumption than turbofan aircraft of the same category. However, they also have a lower maximum speed (this does not seem to be a problem for this SOR because no cruising speed or maximum speed were mentioned).

The crew weight will be approximately 400 lb [two crew members at 200 lb each (see Sec. 11.2.1)] and there will be no payload. The fuel reserve is already included in the mission profile so a value of 1.005 for B [Eq. (11.24)] will be used. One can now estimate the aircraft takeoff weight, empty weight, and fuel weight required for the mission by using data from Table 11.1 and Eqs. (11.22), (11.24), and (11.25). An initial guess must be used to start the iterative process. A typical takeoff weight for the TFMAT is approximately 10,000 lb whereas for the TPMAT a takeoff weight of approximately 4,500 lb will be used. For the TPMAT,

$$\frac{W_f}{W_{TO}} = 1.005\left(1 - \frac{W_9}{W_{TO}}\right) = 0.1740 = \text{const}$$

$$\left[\frac{W_E}{W_{TO}}\right]_1 = \left(e^{-1.2583}W_{TO}^{(1.1085-1)}\right) = 0.7078$$

$$W_{TO} = W_{\text{crew}}\Big/\left(1 - \frac{W_f}{W_{TO}} - \frac{W_E}{W_{TO}}\right) = 3,384\text{lb}$$

A new value for the takeoff weight that is in between the 4,500 lb first estimated and the new value of 3,384 lb is used to continue the iterative process until the takeoff weight estimate does not change more than 0.5%. In this case, the final results for the TPMAT are shown in Table 11.4. The same process is used for the TFMAT to get the results of Table 11.5.

Thus, if the only constraint imposed on the design of this new military trainer was the mission specified in the SOR, one would probably prefer to build a TPMAT, which would use less fuel for the specified mission (lower operating cost)

Table 11.4 TPMAT

$W_E/W_{TO} = 0.6630$	$W_{TO} = 2{,}460$ lb
$W_f/W_{TO} = 0.1740$	$W_E = 1{,}631$ lb
	$W_f = 428$ lb

and would most likely be cheaper to buy and maintain. The low-takeoff weight indicated in Table 11.4 might be larger in reality if special equipment (such as ejection seats, which are not usually carried by turboprop military trainers) is required because that weight will have to be accounted for in the form of payload.

The preceding example was used to demonstrate how performance equations can be used to estimate the takeoff weight of an aircraft for a given mission. However, this is only an initial estimate. As the design phase progresses and more and more details are accounted for, this initial estimate will change to better reflect the end product.

11.2.5 Mission Segment Sizing

Other performance equations can be used to estimate the thrust-to-weight (or power loading) and wing loading of the aircraft to be built. This section provides an overview of this process.

Takeoff. For the takeoff segment, a maximum takeoff distance will usually be specified (in the specimen SOR, 1,200 ft on wet runway, $\mu_G = 0.05$, 2,000-ft pressure altitude, and 90°F → $\sigma = 0.8775$). From this, using Eqs. (7.5) and (7.7), one can determine the relationship between the takeoff thrust-to-weight ratio and wing loading. Combining the two equations gives

$$\left(\frac{T}{W}\right) = \left[\frac{(1.2)^2}{\rho_{SL}\sigma g X_{GR} C_{L_{max_{TO}}}}\right]\left(\frac{W}{S}\right) \tag{11.26}$$

Cruise or maximum speed. If a given cruise speed or maximum speed is required, one can estimate the thrust-to-weight ratio required to fly at that speed by using Eq. (3.9) in the following form:

$$\left(\frac{T}{W}\right) = \frac{qC_{D_0}}{(W/S)} + \frac{K}{q}\left(\frac{W}{S}\right) \qquad q = \frac{1}{2}\rho_{SL}\sigma V^2 \tag{11.27}$$

Table 11.5 TFMAT

$W_E/W_{TO} = 0.6953$	$W_{TO} = 11{,}200$ lb
$W_f/W_{TO} = 0.2691$	$W_E = 7{,}788$ lb
	$W_f = 3{,}014$ lb

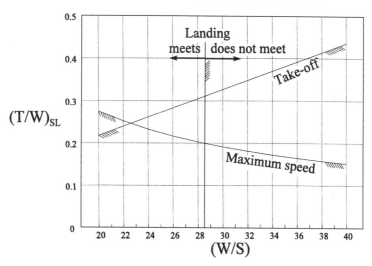

Fig. 11.7   Relationship between the T/W and W/S for the TFMAT to meet the specimen SOR.

Landing.   For the landing segment, one can estimate the required wing loading to meet the maximum landing distance as specified by the requirements (for the specimen SOR, less than 1200 ft on wet runway, $\mu_G = 0.2$, at 2000-ft pressure altitude $\rightarrow \sigma = 0.9427$). Using Eq. (7.34) and assuming zero residual thrust, one gets the wing loading required to meet the landing distance,

$$\left(\frac{W}{S}\right) = \frac{\rho_{SL}\sigma g \mu_G X_{GR} C_{L_L}}{(1.1)^2} \tag{11.28}$$

An aircraft designed to meet the requirements of the specimen SOR and, for example, a maximum speed of 300 kn at 15,000-ft standard altitude, would require a takeoff thrust-to-weight ratio and wing loading similar to the one shown in Fig. 11.7. For the TFMAT, the minimum thrust-to-weight ratio required to meet the specimen SOR would be approximately 0.25 for a wing loading of about 22.5 lb/ft². From the results of Table 11.5, this means that a thrust of 2,800 lb and a wing surface of 498 ft² are required. This is a first estimate of the size of the TFMAT using known performance equations.

11.3   Stealth Technology

Stealth or low observable (LO) technology encompasses every means to defeat the systems by which an aircraft can be detected. The aim of stealth technology is to render the aircraft undetectable by any means [radar, infrared (IR), audio, visual (optical), or ultraviolet] by reducing the aircraft signature to a level so low that it will be lost in the background noise of a possible detecting system.

11.3.1   Defeating Radars

The radar is the means to detect an aircraft at very long range (some radar are capable of beyond the horizon detection). The radar signature of the aircraft [also

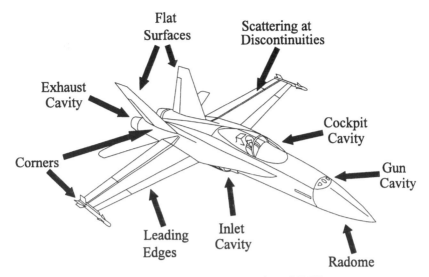

Fig. 11.8   Contributors to the aircraft RCS.

called the radar cross section (RCS)] will vary depending on which side faces
the radar. The basic law of physics for radar is that the angle of reflection of the
radar wave is equal to the angle of incidence (just as light on a mirror). To prevent
radar energy from bouncing straight back to its source, it must be absorbed or
reflected in a direction other than that of the source. Thus, the RCS of the aircraft
is influenced by such parameters as the engine intake and exhaust cavities, leading-
and trailing-edges orientation, cockpit cavity, flat fuselage and vertical tail sides,
edges of plates and doors, surface discontinuities, etc. (Fig. 11.8).

The strength of the radar echoes that are reflected back toward the source
is inversely proportional to the range of the target to the fourth power. Other
factors also influence the strength of the echoes: characteristics of the source radar
(power of the transmitter, the size of the antenna, wavelength of the radio waves),
characteristics of the surrounding environment (strength of the background noise
or clutter), the profile of the aircraft mission (the number of search scans in which
the target appears and the length of time the target is in the antenna beam during
each search scan) and the reflecting characteristics of the target (RCS)

$$\frac{\text{Strength}}{\text{echoes}} \propto \frac{\text{RCS}}{\text{Range}^4} \qquad (11.29)$$

$$\frac{\text{Detection}}{\text{range}} \propto \sqrt[4]{\frac{\text{RCS}}{\frac{\text{Strength}}{\text{echoes}}}} \qquad (11.30)$$

Usually a low-frontal and underside RCS is desired for aircraft designed to
penetrate enemy air defense radar system at medium-to-high altitudes. A low-
frontal hemisphere RCS is desired if the aircraft is to confront enemy fighters at
beyond visual range (BVR). Figure 11.9 gives an idea of the relative detection

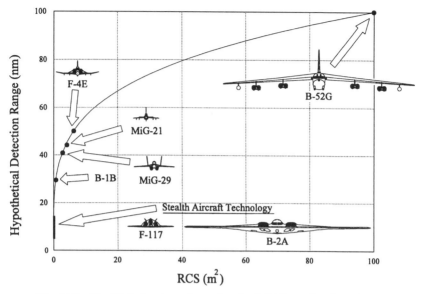

Fig. 11.9   Stealth technology decouples the RCS from the aircraft size.

range as a function of the aircraft frontal RCS based on a radar that can detect a B-52G at 100 n miles.

Most of today's combat aircraft have a significant RCS for their size and, to avoid radar detection, have to fly below the radar horizon of ground-based station and low enough that their radar signature is hidden from air radars in the ground clutter. The disadvantage of flying low is that the aircraft range is reduced and the aircraft becomes more vulnerable to ground fire.

The use of radar absorbent paint and radar absorbent material (RAM) does reduce the RCS significantly. But the RAMs are usually very heavy, which is again a penalty for range and endurance. On the other hand, a lower RCS does reduce the distance at which the aircraft can be detected and, in turn, gives less time to the enemy to react. One aircraft that uses RAM is the B-1B, and it has a radar signature approximately 100 times smaller than that of the B-52 (or about the size of a large bird). The SR-71 (as well as its predecessor, the Lockheed A-12) was also covered with RAM, and its small size compared to the B-1B gave it a lower RCS but the large exhaust plumes from its two engines were excellent radar reflectors (the intensity of the radar return being dependent on the ion density in the wake, which is itself driven by the jet temperature).

Another way to reduce the RCS is to shape the aircraft so that the radar energy that is not absorbed by the RAM is reflected away from the source (or with minimum return to the source). The F-117 represents the first generation of operational combat aircraft to use this approach. It had a flat bottom and faceted upper surface. The engine intake and exhaust are hidden from ground radar by the wing. As well, the intake lips are fitted with screens to attenuate radar returns (the grids are made of 0.6×0.6 in. mesh, which is half the wavelength of X-band radars). The wing LE sweep angle is 67.5 deg and there is no horizontal control surface. Radio

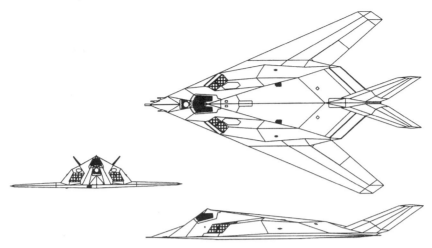

Fig. 11.10   Lockheed F-117A.

antennas are retractable and weapons (two 2,000-lb bombs/missiles) are carried internally to maintain LO characteristics (Fig. 11.10).

Faceting, which may seem a good idea for lowering RCS has its disadvantages. For one thing, it increases the drag of the aircraft, which in turn reduces the achievable range with a given amount of fuel. This can be countered by air-to-air refueling, for which the F-117 is equipped, making high lift-to-drag ratio (which probably does not exceed 5) needed for long range less important.

Later generations of stealth aircraft (such as the B-2A, YF-22, and YF-23) did away with total aircraft faceting and adopted smoother curves and better aerodynamic shaping by using selected angles of reflection. The idea is to concentrate the radar reflections in directions that have low probability of being intercepted for any meaningful length of time, mainly away from the aircraft flight path and side (Fig. 11.11).

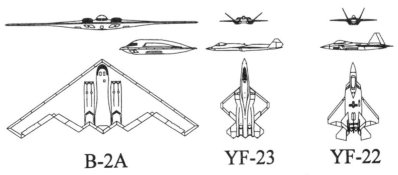

B-2A   YF-23   YF-22

Fig. 11.11   Latest generation stealth aircraft.

11.3.2   IR Sources

The IR signature comes from the heat sources on the aircraft with the exhaust pipes producing the largest contribution. The IR signature is proportional to the object temperature to the fourth power minus the ambient temperature (IR noise) to the fourth power

$$IR_{sign} \propto \left(T_{obj}^4 - T_{amb}^4 \right) \tag{11.31}$$

From this equation it is clear that the use of afterburners will greatly increase the IR signature of an aircraft. Note that when the temperature is high enough, the IR signature comes into the visible light range. This signature, which cannot be eliminated (friction drag produces heat on the aircraft skin), can be reduced by squeezing the exhaust wakes into thin, planar sheets (F-117A); by causing hot gases to flow over heat absorbing tiles (YF-23A, B-2A); by shielding wakes at large angles off behind fins, rudders, and other aircraft surfaces (F-117A, B-2A, YF-22A, YF-23A); by increasing ambient airflow/hot gases mixing before the exhaust gases are expelled from the aircraft (RAH-66); or by any combination of the foregoing.

Normal exhaust flow for a turbojet engine with no afterburner can be in the neighborhood of 400 K, whereas using afterburners can bring the temperature to around 1,000 K. The ability to supercruise (to fly supersonically without the use of afterburners) will reduce the IR detectability of the aircraft at those speeds, as well as increase the aircraft range due to the lower SFC when not using afterburners.

The sun or its reflection off clouds, bodies of water, etc., is an example of a source of IR noise that can disrupt IR guidance and tracking systems. To avoid IR noise sources, IR guidance systems are usually configured to monitor specific IR wavelength ranges such as between 3 and 4 μm (as shown in Fig. 11.12).

11.3.3   Other Means of Detection

The audio signature is usually less important because the efficiency of even the most sophisticated acoustic sensors is degraded by wind noise. Staying subsonic will produce less noise than flying supersonically, but from head on the latter signature will reach the detector only after the aircraft has gone by.

The visual signature can be fairly reduced by the use of camouflage paint, by weather, by reducing smoke emission from engines, and by using the cover of the night. An aircraft of small size (F-20) or of low profile (B-2A) will also have low-visual signatures (Fig. 11.13). While flying at high altitudes, the aircraft engine exhaust jets produce contrails, which can be picked up at great distances on a clear day.

After all of the foregoing points are taken into consideration while designing an aircraft, the designer should not forget that an aircraft can also be detected by the noise it radiates. The navigational, air-to-air, or air-to-ground radars can all be detected at a distance far beyond their useful range, thus alerting the enemy of an incoming aircraft. Electronic countermeasure (ECM) systems will also alert enemies even if they are unsure of the source of ECM. The same is true for radio communication. Depending on the frequency used, the enemy can estimate that the aircraft is within a certain radius from the detector. To avoid being detected while going to or coming from your target, passive systems such as the forward

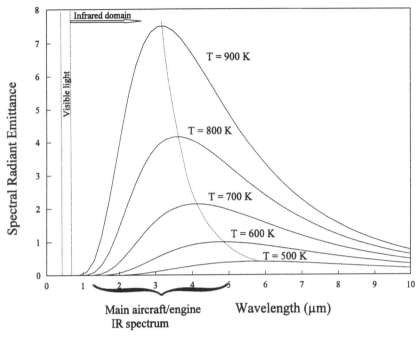

Fig. 11.12   Spectral radiant emittance from a black body at various temperatures.

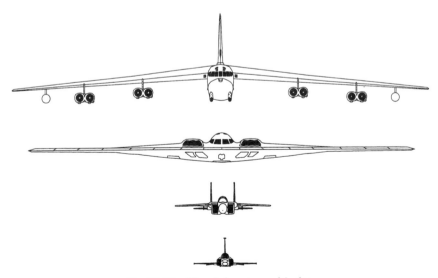

Fig. 11.13   Visual signature with size.

looking IR camera should be used. Low probability of detection radar and radio will also increase the chances of getting to the target undetected.

11.4 Drag Reduction

Most aircraft performance elements require the aircraft to have high-aero-dynamic efficiency (high lift-to-drag ratio) for maximum performance. This was demonstrated in preceding chapters and is summarized in Appendix C. Large excess thrust is also required for fast acceleration and large rate of climb. Most of today's airliners have large excess thrust at their normal cruising altitude and speed. The large thrust is required for acceptable field performance.

The main driver for research in the field of drag reduction over the past 25 years has been the rising cost of fuel (Fig. 11.14). The fuel consumption of an aircraft is directly related to the aircraft lift-to-drag ratio and the engines' SFC. The late 1970s and early 1980s saw the development and entry into service of several twin-engine aircraft with large bypass ratio engines and significantly lower SFC. There was also a push for the development of propfans, but by the end of 1996, only the Antonov company had come up with a flying prototype (the An-70) for a service aircraft, and this aircraft had an uncertain future at the time of publication of this textbook.

To further reduce fuel consumption, the aircraft lift-to-drag ratio must be improved. The various components of drag were described in Chapter 1 with specific information on excrescence drag (Chapter 1, Sec. 1.7.1), interference drag (Chapter 1, Sec. 1.7.2) and trim drag (Chapter 1, Sec. 1.7.3 and Chapter 10). For a jet airliner, 90% of the fuel used during a flight is spent during climb and cruise

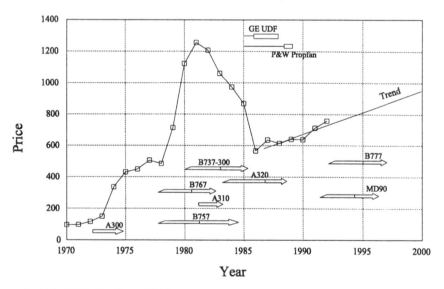

For Jet A1 Fuel: 1,000 US gallons = 6,500 lbs

Fig. 11.14   Price of 1,000 U.S. gallons of Jet A1 fuel (average price, in U.S. dollars, for the month of July).

where the skin-friction drag and the induced drag represent between 85 and 90% of the aircraft total drag. For this section, the drag component of interest will be skin-friction drag. An insight into some of the research being done in the field of skin-friction drag reduction with the most promising technology is given followed by an example on the economic benefit of drag reduction.

Because of the viscosity of the air, the layer of air immediately adjacent to the surface has zero relative velocity (no-slip condition) (Chapter 1). A BL will form within, which the air velocity will rapidly increase from zero at the surface to the freestream value outside the BL. It is within this BL that the skin friction occurs, with a turbulent BL producing more drag than a laminar one.

Active research has been ongoing for over 50 years in the field of skin-friction drag reduction. Early efforts concentrated on finding ways to naturally delay the transition from laminar to turbulent BL by shaping the wing profile. NACA developed the first successful series of such profiles, the NACA 6-series. The shape of the profile pushes the point of maximum suction farther aft of the LE, thus reducing the zone of adverse pressure gradient and delaying the transition of the BL to a point closer to the trailing edge. This form of passive laminar flow control (LFC) is effective for a given design condition (given AoA range) but the performance of those profiles rapidly deteriorates at off-design conditions or in turbulent air (Fig. 11.15). Even the noise and vibrations from the wing-mounted engine can affect the transition point of the BL.

Active LFC involves the use of mechanical system to control the BL. Full wing suction of the BL uses the principle that if the low-energy BL could be partially removed, the transition from laminar to turbulent can be delayed until much farther aft on the wing. Active LFC requires energy from the aircraft, which is extracted

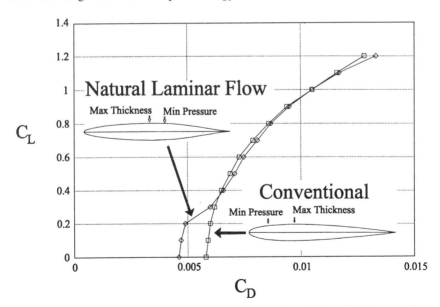

Fig. 11.15   Laminar flow airfoils have an operating lift coefficient (AoA) range for low drag.

from the fuel carried. With an effective distribution of the suction holes, there is an overall decrease in fuel consumption for a given flight. Such systems are heavy and bulky (reducing payload capacity), they might have problems with contamination from the atmosphere (bugs/dust clogging holes), and might prove expensive to maintain.

A compromise between passive and active LFC is being investigated by several nations. It is called hybrid LFC (HLFC). Such a system requires a properly shaped airfoil, using suction only on the first 15–20% of the chord. This can delay transition from laminar to turbulent BL to much higher chord Reynolds numbers (60×10^6 demonstrated compared to the 4×10^6 of a flat plate with no pressure gradient[13]). HLFC has demonstrated promising results and its applications to the wing, as well as vertical and horizontal tails, and engine nacelles are being investigated at the present time.

NASA also has a program to investigate HLFC effectiveness at supersonic speeds. A specially modified F-16XL has been fitted with a suction panel glove on its inner left wing. This panel is perforated by millions of nearly microscopic holes laser drilled in a sheet of titanium. The aim of this program is to demonstrate that transition can be delayed to approximately 50% of the chord even in supersonic flight. This technology could then be applied in the design and development of future high-speed civil transport to provide a 7–10% reduction in aircraft drag and 8–9% reduction in gross takeoff weight.

An example will be used to demonstrate the advantage of reducing an aircraft friction drag. Let typical aircraft A (Chapter 3) use HLFC to reduce skin-friction drag by 10% (here, a 10% reduction of C_{D_0}). For an aircraft weight of 290,000 lb at the beginning of a cruise leg at 30,000 ft where $\zeta_{cruise} = 0.2$, if the HLFC is not used the aircraft would fly 2,291 n mile before it would initiate a descent to land [Eq. (3.42) and Table 3.1]. When the HLFC is turned on, the minimum drag coefficient of the aircraft becomes 0.0162 (instead of 0.018), the maximum lift-to-drag ratio increases to 17.5 (5.4% increase), and the aircraft best range velocity increases to 476 kn (Mach 0.81), a 2.6% increase. If the aircraft uses the same fuel quantity for cruise ($\zeta_{cruise} = 0.2$), it will cover a distance of 2479 n mile before initiating descent (an 8.2% increase in range). If, instead of going farther, the aircraft initiates descent after 2,291 n mile as in the first case, the cruise fuel fraction is reduced to 0.186 from 0.2 (6.8% decrease). This results in the saving of approximately 4000 lb of fuel for the same distance covered or about $500 in 1995 U.S. dollars according to Fig. 11.14. There is, therefore, a potential for substantial savings in direct operating costs for airlines in adopting HLFC for their aircraft fleets.

This very short section was used to demonstrate the advantages of drag reduction on aircraft performance and economics. Every designer must strive to arrive at a design that will minimize fuel consumption to either increase an aircraft's range, reduce direct operating cost, or reduce aircraft weight and size for a given mission.

11.5   Store Carriage: A Design Challenge

Military aircraft, compared to civilian ones, can be a nightmare for an aerodynamicist. An aircraft can be designed to perform at its topmost efficiency throughout its flight envelope just to be loaded afterward with pylons and stores with little consideration for basic aerodynamics. The resultant drag increase (especially in the transonic region) can significantly decrease the aircraft performance. Sometimes

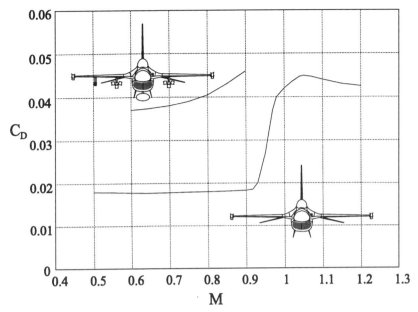

Fig. 11.16   Stores effect on F-16 minimum drag (adapted from Ref. 14).

the carriage of stores produces drag increases of the same order of magnitude as the aircraft minimum drag. For example, the minimum drag of an F-16 in the clean configuration (2 × Aim-9) is compared with an air-to-ground configuration (12 × MK-82, 2 × Aim-9, 1 × 300 gal, 1 × ALQ-119 pod) (Fig. 11.16).

Military aircraft must also operate outside of cruise condition when loaded, if only to maneuver to avoid threats or engage an aggressor without jettisoning their weapon load, thus scrubbing the mission. During Operation Desert Storm in 1991, two F/A-18 aircraft en route to attack a ground target were intercepted by two enemy fighters. They successfully engaged and destroyed the enemy interceptors before proceeding to their assigned target and returning to their carrier in the Red Sea.

Weapon loads will not only severely increase drag but will also limit the usable lift, adversely affect stability, and lower buffet limits. Interference between stores may even affect their separation trajectories. All this must be accounted for in some way during the design stages.

Most military combat aircraft are designed to carry external stores on pylons and racks. This provides flexibility in loading different types of weapons, minimizing aircraft volume (and weight) when not doing a mission requiring maximum payload/volume. Table 11.6 is included to provide a rough order of magnitude of weight and drag for different stores that could be carried on fighters such as typical aircraft C (see Chapter 3).

Aircraft C represents a military aircraft that can be used as a fighter as well as an attack aircraft; thus, a large variety of external stores can be fitted on pylons and racks in a wide variation of arrangements depending on the particular mission to be carried out. As a baseline, the values of the clean aircraft (no external stores) at maximum takeoff weight and at combat weight are included in Tables 11.7 and 11.8.

Table 11.6   Weight and drag of various weapons

Type	Subtype	Weight, lb	Drag count[a]	Comments
Fuel tank	330 gal	300	10.0	Can carry approximately 2,000 lb of fuel
Bombs	MK-82 LD	531	4.0	Low drag
	MK-82 Snakeye	565	6.0	Retarded fall, speed brakes deploy after release
	MK-84 LD	1970	7.0	Low drag
	MK-84 LGB	2082	15.0	Laser guided
	Rockeye II	760	8.0	Cluster bomb
Rocket pod	LAU-61	522	8.0–31.5	Fairing breaks away when rockets fired, which increases the drag
Air-to-surface missiles	AGM-65	634	12.0	Several types of guidance
Air-to-air missiles	AIM-9	195	6.0	IR guidance
	AIM-7	510	6.0	Semiactive radar guidance
	AIM-120	333	5.0	Active radar guidance
Pylons and racks	Wing	200	8.0	Wing pylon
	₵	100	3.0	Centerline pylon
	VER	172	9.0	Versatile ejector rack
	MER	220	15.0	Multiple ejector rack

[a]Note: 1 drag count $\equiv C_D = 0.0001$.

Table 11.7   Performance comparison, fighter configuration

Parameters	Combat weight	Maximum takeoff weight	Fighter config.	Δ Fighter/maximum takeoff %
W, lb	16,000	18,540	21,330	+15.0
C_{D_0}	0.025	0.025	0.02525	+1.0
Max ceiling, ft	67,648	64,582	61,562	−4.7
$X_{BR(VCL),20K}$, n mile	700	1,411	1,896	+34.4
t_{max}, h	2.01	3.75	4.71	+25.6
$X_{TO,50ft}$, ft	959	1,275	1,648	+29.3
$RC_{max,SL}$, ft/min	45,283	39,167	33,984	−13.2
$\dot{\chi}_{SL}$, deg/s	21.4	18.3	15.8	−13.7

Table 11.8   Ground attack configuration, performance comparison

Parameters	Combat weight	Maximum takeoff weight	Ground attack config.	Δ Ground attack/ maximum takeoff, %
W, lb	16,000	18,540	27,338	+47.4
C_{D_0}	0.025	0.025	0.0461	+84.4
Max ceiling, ft	67,648	64,582	50,135	−22.4
$X_{BR(V,CL),20K}$, n mile	700	1,411	1,016	−28.0
t_{max}, h	2.01	3.75	2.59	−30.9
$X_{TO,50ft}$, ft	959	1,275	2,802	+119.8
$RC_{max,SL}$, ft/min	45,283	39,167	19,631	−49.9
$\dot{\chi}_{SL}$, deg/s	21.4	18.3	11.7	−36.1

The first configuration to be evaluated is the fighter configuration, which consists of two wing-tip-mounted AIM-9 Sidewinder heat-seeking missiles and one external fuel tank on a centerline ($\mathbb{C}$) pylon. This results in an increase in maximum takeoff weight of 2,790 lb (2,000 of which is extra fuel). The minimum increase in drag due to external stores is equal to the sum of individual store drag and is 25 drag counts (see Table 11.6). The minimum drag coefficient is now 0.02525. Spacing between stores is such (see Fig. 11.17) that interference drag (in this configuration) is negligible. It is assumed that there is no effect on induced drag, although wing-tip-mounted missiles do increase the effective AR slightly. Finally, it is assumed that there is no change in wave drag.

The immediate advantage of this configuration is the increase in range and endurance due to the external fuel tank, which (when full) increases the fuel weight fraction from 0.2724 to 0.3305. The penalties in other performance factors are relatively small and are mainly due to the increased weight (the increase in minimum drag being only 1.0%). Every aircraft that carries external fuel tanks

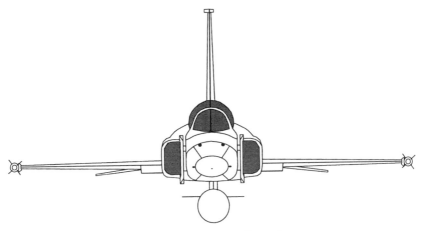

Fig. 11.17   Fighter configuration.

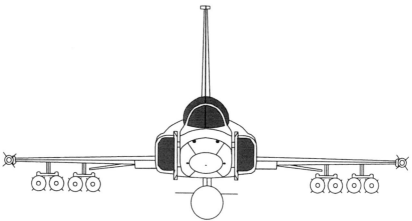

Fig. 11.18   Ground attack configuration.

will usually use the fuel in those tanks first and eject the tanks before going into combat (wartime scenario, in peace time the tanks are retained to reduce operational expenditures). Once the external fuel tank is ejected, the performance of the fighter would be somewhere between that for maximum takeoff weight and combat weight.

The next configuration is that of a ground attack aircraft. The aircraft will carry the same basic load as the fighter configuration plus four wing pylons, four versatile ejector racks with two MK-82 Snakeyes each (configuration shown in Fig. 11.18). The takeoff weight is now 27,338 lb, and the minimum drag increase due to the stores is 141 drag counts. The close proximity of the stores with each other (especially the bombs) will cause interference drag. A value of 70 drag counts due to the interference drag is assumed for all flight conditions except for takeoff, where the value is assumed to be negligible (low velocity). The aircraft will also be limited to Mach 1.0. The changes in performance that result are shown in Table 11.8.

Here, it is clear that external stores seriously reduce all performance of aircraft C compared to the clean configuration. Part of the degradation comes from the large increase in aircraft takeoff weight, but the most serious degradation comes from the large increase in drag of the aircraft. An aircraft loaded with this weapons load is usually limited to Mach numbers below 1.0. Also, the turning performance is severely degraded and the aircraft is at a serious disadvantage against a more lightly loaded aircraft during a dogfight. Its only chance of surviving in air combat is to detect the enemy and fire its missile first or it must drop its bomb load and external fuel tank before engaging the enemy. The mission is scrubbed and the aircraft can be lost.

Not all stores interference is detrimental to aircraft performance. For example, adding missiles and launchers to the wing tip of a fighter effectively increases the aircraft wing span (thus the AR), which will reduce the drag in the maneuvering range (high C_L). This configuration does increase the minimum drag but so will the installation of the missiles and launchers under the wing, and this last configuration does not provide the high-lift drag decrease. In Chapter 1, Sec. 1.7.2, it was seen that properly shaped wing tip tanks on an F-5 fighter actually reduced the aircraft cruise drag compared to a clean configuration.

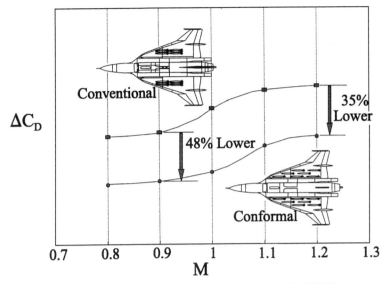

Fig. 11.19 Drag comparison between conformal carriage of 14 MK-82 and conventional carriage of 12 MK-82 (adapted from Ref. 9).

One way to minimize store drag is to stow stores internally, such as was done on most bombers and on 1950s fighters (F-106). This was also adopted on more recent fighters such as the advanced tactical fighter candidates (YF-23 and YF-22). This effectively eliminates weapons drag (except during jettison) and also has the advantage of reducing the RCS of the aircraft. The main disadvantages are the limited weapon flexibility (the weapon bay size and configuration) as well as increased fuselage volume, which translates into a heavier fuselage with less fuel volume than another aircraft of similar size.

Another approach to minimizing the drag while carrying multiple bombs is conformal carrying. Weapons are mounted conformally on short individual ejector pylons and positioned to take full advantage of tandem shielding and stagger. For example, the conformal carriage of 14 MK-82 bombs on an F-16XL produced less drag at all Mach numbers than the conventional carriage of 12 MK-82 on pylon/MER. Figure 11.9 indicates that there is a significant advantage of using conformal carriage. Lower drag results in increased range, higher penetration speed, and, in this case, a higher number of weapons carried.

One interesting approach to the problem of increased stores drag was used by McDonnell Douglas to take advantage of the F-15 growth potential (Fig. 11.20).

Fig. 11.20 CFT equipped F-15 compared to one equipped with two external fuel tanks.

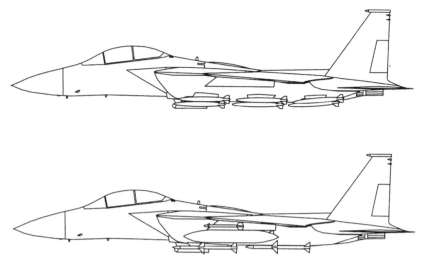

Fig. 11.21   F-15E with CFT, air-to-ground and air-to-air configurations.

Conformal fuel tanks (CFT), also called fastpacks were designed with both air-to-air and air-to-ground stations. Two CFT (one on either side of the fuselage) contain a total of 9360 lb of fuel. An F-15 equipped with 4 AIM-7, 4 AIM-9 and a centerline fuel tank (3965 lb of fuel) has a drag index (for the stores) of about 58 compared to the drag index of about 50 for a CFT equipped F-15. In level flight, the total subsonic drag of a CFT equipped F-15 is less than one carrying a centerline external tank. But above Mach 1.0 this tendency is reversed. To carry the same amount of fuel as a CFT equipped F-15, a conventional F-15 would have to carry three external tanks, which produces much more drag than two CFTs. Thus, the CFT provide for an increased range (Fig. 11.21). In the air-to-ground mode, the stores are carried tangentially to decrease frontal area.

11.6   Toward Higher Maneuverability

Close-in combat has been present in aerial warfare since World War I. Aircraft designers have always tried to incorporate in their design the latest technology (aerodynamics, avionics, armament, etc.) that they thought would give the new aircraft the edge in air combat. Back in the days of straight wing aircraft, maximum lift coefficient would be reached at AoA between 12 and 17 deg; beyond that point, stall and, possibly, a spin would occur. Then, fuselage pointing and aiming was achieved by changing the velocity vector orientation and maneuverability was measured in terms of maximum sustained turn rate and minimum radius of turn (see Chapter 5).

The advent of turbojet engines offered the aircraft designers the possibility of creating aircraft with ever-increasing maximum speeds (Fig. 11.22). To reach those speeds, the wing had to be swept back and the AR had to be decreased from a typical 6.5 to a maximum of approximately 3.5 for supersonic aircraft. These changes decreased the lift coefficient slope and increased the stall AoA significantly. Provided the aircraft is still controllable at the large AoAs required

WW I

WW II

1960s

1980s

Early
21st century

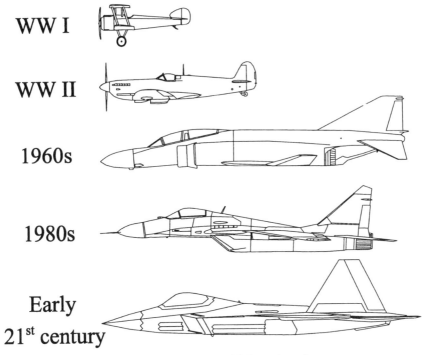

Fig. 11.22   Evolution of fighter aircraft.

to reach maximum lift coefficient, the turn performance of that aircraft will be improved (see Chapter 9). At these large AoAs, the flow around the aircraft is largely turbulent and detached. Fighters of the late 1950s and early 1960s, which were designed with the philosophy that aerial combat would be won at long range with the help of missiles, were prone to loss of controllability at these large AoAs due to the degrading lateral/directional stability and reduced control surface power leading to uncontrolled maneuvers, departures, and spins. One of these aircraft was the McDonnell Douglas F-4 Phantom. During the Vietnam War, driven by the need to correctly identify a possible target, F-4s had to get closer to the detected aircraft, which often resulted into dogfights against smaller, more maneuverable aircraft such as the MiG-17, and many F-4s were lost because of loss of control at high AoAs. Extensive research resulted in the development of LE slats, which improved the departure characteristics of the aircraft. With the addition of LE slats, the aircraft was allowed to reach its maximum AoA before departure occurred, but the fixes were done after the aircraft entered service (Fig. 11.23).

Further studies in the early 1970s highlighted the destabilizing contribution of asymmetric vortex shedding by the nose of the aircraft to the degradation of lateral/directional stability. This led to the development of a series of aircraft equipped with LE extensions (LEXs) such as the F-16, the F/A-18, the MiG-29, and the Su-27 (Fig. 11.24). Utilization of vortex flow allows a designer to partially control the separated flowfield and to use additional energy from the vortex flow to obtain higher lift coefficients at higher AoAs.

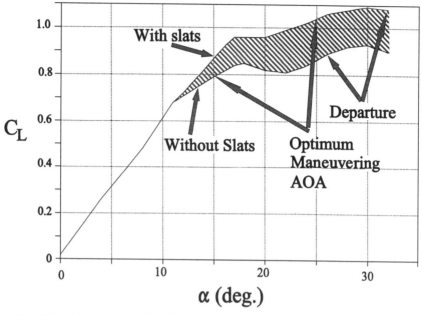

Fig. 11.23   Improvement in lift curve with the addition of LE slats on an F-4.

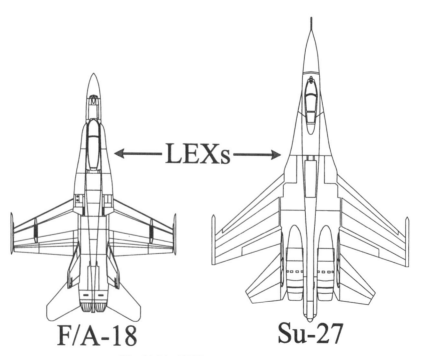

F/A-18   Su-27

Fig. 11.24   LEX equipped aircraft.

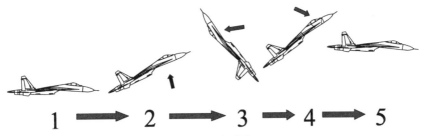

Fig. 11.25   Pougachev's Cobra maneuver.

During aggressive maneuvering some fighters, such as the U.S. teen fighters (F-14, F-15, F/A-18), can reach transient AoAs of up to 50 deg or more. The soviet Su-27 and MiG-29 aircraft demonstrated the ability to perform maneuvers at AoAs exceeding 90 deg (Pougachev's Cobra maneuver and tail slide). During the Cobra maneuver, the pilot rapidly pitches the aircraft up from level flight at around 230 kn to reach an AoA of approximately 100–130 deg, then brings the nose back down (Fig. 11.25). For a ground observer, the impression given by this maneuver is one of a cobra rising its head to strike. During this maneuver, the aircraft can decelerate at up to 60 kn/s and reach a maximum load factor of approximately 3.5–4 g.

Although the Cobra maneuver may not prove to be a valuable tactical maneuver, the ability to perform it implies great agility in most of the flight envelope, which may translate into a tactical advantage over other aircraft not able to perform such a maneuver. But tactical advantage in air combat is not just a measure of airframe agility and maneuverability, it is a question of weapon system agility (the integration of the following: pilot detection of the enemy, airframe pointing to within the weapons acquisition, and firing envelope). At this point it may be useful to define airframe maneuverability and agility.

1) Maneuverability is the ability and limitation of an aircraft to change its flight-path vector. Energy maneuverability is a steady-state quantity that can best be described by plotting the aircraft turn rate against its speed (or Mach number) or the specific excess power against the turn rate (Chapter 8).

2) Agility is the ability to change rapidly and precisely the maneuver state of the aircraft. It is proportional to the inverse of time for transition from one maneuver to another (or to the time rate of change of acceleration) and addresses exclusively the translation of a moving object in three dimensions.

3) Supermaneuverability is agility at stall and poststall AoAs.

A very agile airframe may be of little use, except for defensive maneuvers, if the pilot cannot timely detect the enemy aircraft or cannot use the onboard weapons at the extreme maneuvering conditions. Research is underway in the fields of avionics/radar/IR (including helmet-mounted sight), airframe agility (including thrust vectoring), and weapons capability such as large angle off-boresight acquisition and firing. Some short-range IR guided missile presently in service (the Russian Vympel R-73E, NATO code name AA-11 Archer) can acquire targets at more than 60-deg off boresight, with the use of a helmet-mounted sight. Several other missiles presently in development or near service entry (the British ASRAAM and the U.S. AIM-9X) will be able to acquire a target at 90-deg off boresight, maneuvering at more than 30 g and AoAs in excess of 60 deg either with or without thrust vectoring.

Airframe agility is, nevertheless, very important. Modern fighters will have to rely more and more on agility and will have to fly at high AoAs (or high alphas) to generate the angular accelerations and positional maneuver advantages to win within-visual-range (WVR) engagements. Even stealth aircraft, which now rely mostly on BVR missiles to defeat conventional enemies, may be faced with other stealth aircraft and be unable to detect them before they are so close to each other that a dogfight is inevitable. Another similar scenario would be that the pilot of a stealth aircraft would come within visual or IR range before detecting the enemy because it did not want to alert the latter by radiating the radar. The focus of the remainder of this section is on airframe agility and maneuverability, and its effects on aircraft performance.

High maneuverability requires a combination of high rate of turn and low radius of turn. In Chapter 5, it was demonstrated that both the aircraft turn rate and radius of turn are strongly affected by the aircraft's velocity–load factor combination. In fact, the radius of turn increases proportionally to the square of the velocity, whereas the rate of turn decreases proportionally to the inverse of the velocity for a constant load factor (Fig. 11.26).

In Chapter 9, it was demonstrated that the contribution of the thrust perpendicular to the flight path to decreasing the radius was important. It was mentioned that today's fighters with their high-stall AoA could point the thrust vector toward the inside of the turn to increase the turning load factor, thus reducing the radius of turn and increasing the rate of turn,

$$n = \frac{L}{W} + \frac{T\sin(\alpha_T)}{W} = \frac{\text{wing}}{\text{contribution}} + \frac{\text{thrust}}{\text{contribution}} \qquad (11.32)$$

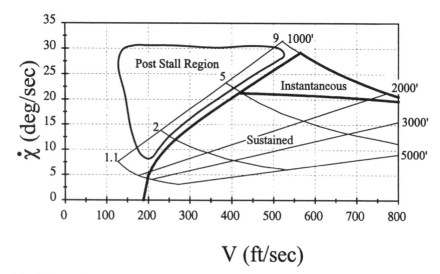

Fig. 11.26   Aircraft rate of turn envelope with the grid of radius of turn and load factor superimposed.

Note that the thrust contribution to the turning load factor is approximately constant for a given altitude (if turning at a fixed α such as that for stall), whereas the contribution from the wing will increase as the square of the velocity, the maximum lift being only limited to the maximum lift coefficient ($L_{max} = \frac{1}{2}\rho V^2 S C_{L,max}$). The load factor must not be greater than the design structural limit of the aircraft. Thus, the thrust vector contribution to the rate of turn [Eq. (5.5)] decreases with increasing velocity (proportional to the inverse of the velocity) whereas the wing contribution to the rate of turn increases with increasing velocity (up to the structural limit of the aircraft).

The Harrier uses thrust vectoring nozzles on either side of the c.g. Such a vectoring system can be used in-flight to increase the turn load factor without causing thrust-induced pitching moment. This way, the wing can be maintained at maximum lift coefficient ($C_{L_{max}}$), and the thrust vector can be maintained perpendicular to the flight path to maximize the turning load factor (again limited by structural limits). Turning all of the thrust perpendicular to the flight path and flying at maximum lift coefficient will create a large deceleration (large drag, no forward thrust). To regain the lost velocity after the maneuver, the aircraft must have large thrust-to-weight ratio, which the Harrier has when flying in a fighter configuration (Fig. 11.27).

As mentioned previously, some fighters are able to maneuver past their AoAs for maximum lift. Although the wing is stalled and produces less and less lift with increasing AoA, the thrust contribution to the turning load factor will increase with increasing AoA. Again here, for high AoAs, the drag will be very high and the forward component of thrust will be low resulting in a rapid deceleration. Maneuvering past the stall barrier (in the poststall region, Fig. 11.26) would significantly reduce the radius of turn of an aircraft, but that aircraft requires good

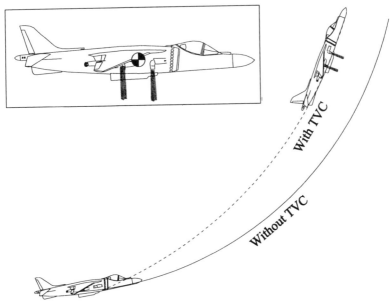

Fig. 11.27   Reduction in radius of turn with the use of thrust vector control (TVC).

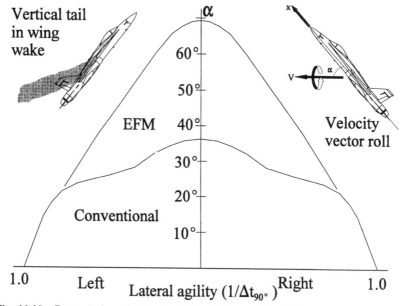

Fig. 11.28   Degradation of lateral agility with increasing AoA, conventional fighter envelope vs EFMs.

roll and yaw control to roll around the velocity vector (this is required to lessen the effects of inertia coupling) at these high AoAs. At 70-deg AoA, a roll around the velocity vector requires a large yawing moment (Fig. 11.28). Most of today's fighters are limited to pitch only maneuvers because of their rapidly degrading lateral agility (inverse of time required to roll to 90 deg of bank) as the AoA is increased up to stall and no rolling power past the stall AoA due to the wing being stalled (loss of aileron effectiveness) and to the vertical tail being in the wake of the wing and fuselage. As well, most fighters have low-nose-down power once they reach high AoAs; thus, it requires excessive time to bring the nose back down.

NASA and other agencies throughout the world are investigating high AoA flights to determine the factor affecting control, agility, and maneuverability at those angles. Thrust vectoring is at the forefront of most of these investigations. NASA's high alpha research vehicle (HARV) is a modified F/A-18, which was fitted with paddles that can deflect into the exhaust jet to deflect the flow both in pitch and yaw. The use of thrust vectoring enhanced the maneuverability and control of the aircraft when conventional aerodynamic controls were ineffective. The HARV demonstrated trimmed flight at 70-deg AoA (maximum of 55 deg for the unmodified F/A-18) and rolling at high rates at 65-deg AoA. This research vehicle is also investigating the use of actuated nose strakes for enhanced rolling to provide improved yaw control at high alphas.

Another aircraft used to demonstrate the effectiveness of thrust vectoring was the F-16 Vista. The conventional engine of this aircraft was replaced with one that had an axisymmetric vectoring engine nozzle that could deflect both in pitch and yaw. This aircraft demonstrated transient AoAs of 110 deg and sustained AoAs

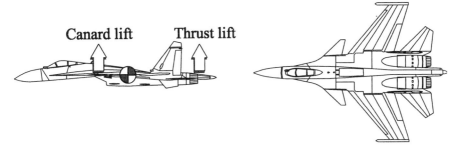

Canard lift   Thrust lift

Fig. 11.29   Sukhoi Su-37.

of 80 deg. The purpose of this program is to investigate combat effectiveness of pitch/yaw thrust vectoring on the combat performance of a fighter.

The only aircraft designed specifically to investigate poststall maneuvering and combat effectiveness is the X-31 enhanced fighter maneuverability (EFM). It is a delta wing aircraft with canard and thrust vectoring. It has a thrust-to-weight ratio slightly greater than 1 on takeoff (approximately 1.1). The aim of this program was to determine how to improve close-in combat fighter effectiveness in both conventional and poststall regimes. This was achieved by designing an aircraft with rapid deceleration, increased negative-g, and improved fuselage pointing capability. The X-31 also demonstrated that turn optimized, gravity assisted poststall maneuvers with a 180-deg heading change (Herbst maneuver, or J-turn when heading change is other than 180 deg) could significantly decrease the aircraft turn radius and increase its rate of turn. On April 29, 1993, the X-31 first demonstrated the Herbst maneuver. At 19,400 ft and 200 kn, the aircraft pulled up to 74-deg AoA, rolled around the velocity vector, and rapidly accelerated in the opposite direction. The 180-deg turn was completed in a radius of approximately 475 ft, and an effective rate of turn of 18.6 deg/s was achieved. In combat simulation against a NASA F/A-18, the X-31 achieved a kill ratio of 31.5:1 while using thrust vectoring. Without thrust vectoring, the maneuvering capability of the X-31 was slightly lower than that of the F/A-18, and the X-31 lost approximately half of the engagement (kill ratio of 1:1).

Using thrust vectoring with aft-mounted engines and aft horizontal tail does not allow the use of full deflection for improved turn load factor, the deflected thrust creating a large nose-down pitching moment. The use of a control surface forward of the c.g., such as a canard, can compensate for the nose-down pitching moment. This is used on the Sukhoi Su-37 (an improved Su-35). The aircraft has a combination of canard-wing-horizontal tail and thrust vectoring in pitch, which enables it to stay controllable down to near zero velocity (Fig. 11.29).

12
Flying in Adverse Weather

12.1 Introduction

T HIS chapter describes how the performance of an aircraft can be affected by weather conditions. More specifically, the following atmospheric conditions are described: aircraft icing, gusts, wind shear, and heavy precipitation.

12.2 Aircraft Icing

The subject of ice formation on an aircraft can be divided into two distinct accumulation phases with their respective influence on aircraft performance: the snow/ice/frost accumulation while the aircraft is on the ground and the ice formation while the aircraft is flying. In both cases, the aircraft performance is altered not so much because of the additional weight due to such accumulation but because of the alteration of the aircraft shape and aerodynamic characteristics, particularly that of the wing. In general, ice accumulation will increase the aircraft drag and lower the maximum lift coefficient, as well as lower the stall angle resulting in an increase in stall speed.

12.2.1 Ice Formation on the Ground

While the aircraft is on the ground it is subjected to the weather. Humidity in the air may produce frost on exposed, cold surfaces. Freezing rain may freeze on impact on the upper part of the aircraft or it may dribble down and freeze on the lower surface and may even block moving parts such as ailerons or flaps. Falling snow or wet snow will accumulate on the upper surface. In every case, there results a more or less uniform accumulation on the upper part of the aircraft with an associated increase in surface roughness. Aviation regulations, such as FAR 121.629, 91.209, and 135.227, prohibit a pilot from taking off when frost, snow or ice is adhering to the wings, control surfaces, or propellers of the aircraft. Aircraft manufacturers support this clean aircraft concept, and FAR does not certify aircraft to takeoff under such conditions.

If the ice/snow/frost accretion (from now on called ice accretion) is not removed prior to taxiing, it may blow completely off during takeoff, but there may be residual traces upon liftoff. The ice accretion that is not shed will alter the aircraft performance primarily because of the modified wing shape. Ice accretion on any other part of the aircraft may modify the c.g. location. The effects on the horizontal tail are also important to consider because it is the horizontal tail that controls the pitch of the aircraft. This will be discussed in more detail in the in-flight icing section.

The upper surface of the wing will be modified by the ice accretion (Fig. 12.1). This ice accretion can range from minute traces of ice to a total coverage of the upper wing. In the latter case, the pilot will notice the buildup and notify ground crews to deice the aircraft, but in the case of small buildups, the pilot may not

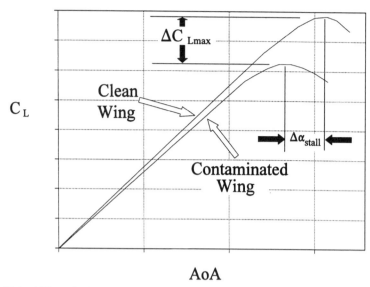

AoA

Fig. 12.1   Effect of wing contamination on maximum lift coefficient and stall angle.

be able to see such buildups (such was the case of USAir Fokker F-28-4000 crash at LaGuardia Airport on March 22, 1992) or may think it will blow off during takeoff (Air Ontario Fokker F-28-1000 crash at Dryden, Ontario, on March 10, 1989). Fokker has conducted wind-tunnel experiments on the upper surface contamination (roughness) and determined that particles of 1–2 mm in diameter, about 1–2 mm deep, and having a density of one particle per square centimeter can reduce an F-28 wing (MAC $\approx$ 3.15 m, 10 ft 4 in.) lift by 22% IGE and 33% OGE. These data combined with others from many flight and wind-tunnel tests by other companies were used to create the Fig. 12.2 graph of lift coefficient loss as a function of wing surface nondimensional roughness height (k/c).

Accumulation on the fuselage will not be as critical because it provides only a small fraction of the lift on takeoff. The main effect of ice accretion here will be the increased drag, which will reduce the aircraft-specific excess power, thus reducing the climb angle.

Falling snow that accumulates on the runway will have an adverse effect on ground acceleration during takeoff by increasing the rolling friction force coefficient. Even when the runways are plowed often, there still remains a residual layer of snow until the precipitation stops and the runways are finally cleaned completely.

A numerical example will be used to illustrate the danger of ice accretion on the aircraft during takeoff (Table 12.1). Using typical aircraft A and assuming the following conditions: maximum lift coefficient reduced by 40% (from Fig. 12.2: $k/c \approx 7 \times 10^{-4}$; from Table 3.1: aircraft A MAC = 19.73 ft, thus $k \approx 0.166$ in.), minimum drag coefficient increased by 50%, rolling friction force coefficient increased from 0.02 (dry concrete runway) to 0.05 (runway with snow accumulation), and Oswald coefficient reduced by 10%. Note that the increase in drag due to the landing gear ($C_{D,LG} = 0.015$) and the takeoff flaps ($C_{D,TOflaps} = 0.01$) are

Table 12.1 Example of danger of ice accretion on typical aircraft A

Aerodynamics	Without accretion		With accretion	
$C_{L_{max}}$	2.0		1.2	
e	0.76		0.684	
C_{D_0}	0.018		0.027	
E_m	16.2		12.5	
μ_g	0.02		0.05	
TO Performance	90% W_{TOmax}	100% W_{TOmax}	90% W_{TOmax}	100% W_{TOmax}
Ground roll, ft	2,051	2,525	3,656	4,619
TO Dist., ft	2,588	3,185	4,484	5,658
BFL, ft	3,619	4,438	8,315	10,553
V_1, ft/s	186	202	208	228
V_{LO}, ft/s	218	230	281	296
V_2, ft/s	236	249	305	321

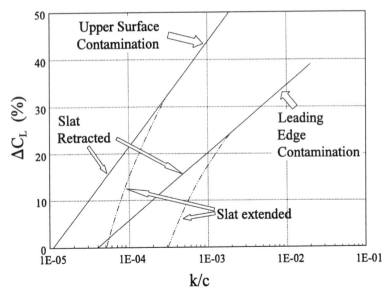

Fig. 12.2 Wing surface roughness effects on maximum lift coefficient loss (adapted from Ref. 15).

not affected in this example. The air temperature will be $-4°F$ ($-20°C$) and the air pressure will be 14.7 psi, which gives an air density equal to 0.00266 slug/ft^3 (an equivalent standard atmosphere altitude of -3893 ft). The braking rolling friction force will be reduced from a typical 0.35 to 0.2 due to the slippery surface (and to maintain control of the aircraft during braking).

In this example, the ground roll distance increased by approximately 80%, the takeoff distance by about 75%, and the BFL by about 134%, depending on the aircraft takeoff weight. In the case of the maximum takeoff weight, the BFL is 10,553 ft, which is longer than most runways presently in service (around 10,000 ft) and not far from the typical maximum length runways of 12,000 ft. During takeoff, the pilot may not notice that more runway length than usual is taken, and liftoff may still be attempted at the predicted liftoff speed. In the case of the maximum takeoff weight, the stall speed with ice accretion is 247 ft/s compared to the 230-ft/s predicted liftoff speed. By the time the pilot realizes that there is something wrong because the aircraft is not lifting off the runway, there may not be enough runway left to stop the aircraft or the aircraft could accelerate past its stall speed and liftoff just to stall and crash a little farther down the runway.

If the accretion is not removed before takeoff, it may present many other dangers to the aircraft, as well as the ones mentioned. One such danger is that eventual shedding of ice during takeoff, or climb out, from the fuselage or inner part of the wing may hit the empennage and damage control surfaces, with the elevator being the most critical because it controls pitch. Aircraft with rear-fuselage-mounted engines have the additional possibility of ice ingestion, which may damage the engines enough to cause total loss of power (Fig. 12.3). The term most widely used to describe such damage is foreign object damage (FOD).

To avoid taking off with ice on the wing of the aircraft, the pilot must ask the ground crew to remove the ice accretion. To deice an aircraft, commercial operators use type I deicing fluids, which are mixed with water. These fluids are applied hot so as to remove the ice formation on the aircraft. Their viscosity is relatively low, and so they form only a thin layer of protection on the aircraft and flow off easily during takeoff. This thin layer of fluid does not provide significant protection against further precipitation and so type II anti-icing fluids must be applied to the clean wing for longer protection. Type II fluids have non-Newtonian viscosity, which decreases rapidly with increasing shear stress. When applied on the wing, they form a thick, gel-like film, which protects the wing for longer periods of time (holdover time) than type I fluids do. The viscosity of these fluids decreases rapidly when they are subjected to the shear of the air flowing over the wing during takeoff; they then flow off easily. These fluids are not recommended for aircraft with rotation speeds below 85 kn and short takeoff run times (about 15 s) because this does not provide enough shear to reduce the viscosity of the fluids and not enough time for the fluids to flow off the wings. Fluids that remain

Fig. 12.3   Possible FOD by shed ice.

on the wing produce a wavy surface, which increases the wing surface roughness in about the same way ice accretion does.

12.2.2 Ice Formation in-Flight

Ice formation in-flight is a different accretion process than the process of accumulation while the aircraft is on the ground. For one thing, the frontal surface rather than the top surface is now the possible accumulation region. The amount of ice that forms on the aircraft is strongly influenced by the liquid water content (LWC) of the air, the air temperature T_∞, and the size of the water droplets, as well as the shape, size, and airspeed of the aircraft. Here we concentrate our attention on the formation of ice on wings mainly because of the adverse effects of ice accretion on the maximum lift coefficient.

While flying, a pilot is more likely to encounter icing conditions when flying through 1) newly formed clouds, especially cumulus, 2) the top part of stratus clouds, and 3) wave clouds due to their high LWC. LWC is generally expressed as grams of water per cubic meter of air. Note that water in the form of vapor, snow, or ice will generally not adhere to the aircraft in flight. Clouds have water droplets of different size, and the LWC varies from one point to the next. Figure 12.4 illustrates the maximum LWC of a typical cloud as a function of the mean effective diameter (MED) of the droplets in the cloud and of the air temperature.

The mass flow rate of water droplets hitting a wing is defined as

$$\dot{m}_{Cap} = \eta V_\infty S_F \text{LWC} \tag{12.1}$$

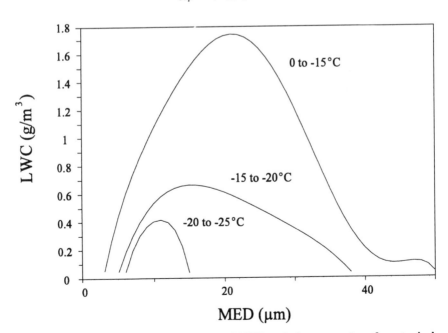

Fig. 12.4 Maximum LWC as a function of MED and air temperature for a typical cloud (from Ref. 16).

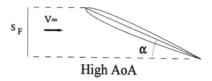

Fig. 12.5   Variation of the frontal area with the wing's AoA.

where η is the collision efficiency of the water droplets. Here, η is a function of the airspeed V_∞, droplets size (MED), wing shape and size, temperature, pressure, and a few other parameters. S_F is the projected frontal area of the wing perpendicular to the incoming air as shown in Fig. 12.5.

Because flying in a given icing condition will yield a specific LWC and MED irrespective of aircraft airspeed, the effects of the variation of these values will not be analyzed here. Note only that for a higher MED, the collision efficiency η of the droplets is higher (see Fig. 12.6, from Ref. 16).

The airspeed has a more than proportional effect on the quantity of water hitting the aircraft. Not only does the maximum mass flow rate ($\dot{m}_{Cap}$ for $\eta = 1.0$) increase with increasing airspeed, as seen in Eq. (12.1), but the collision efficiency also increases with increasing airspeed, all other parameters being constant. Using the

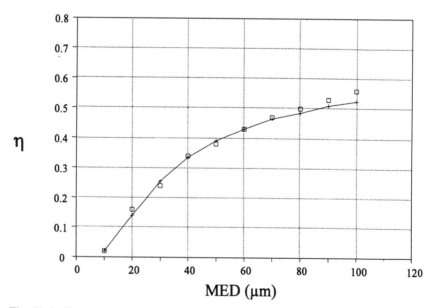

Fig. 12.6   Typical effects of the MED on the collision efficiency of droplets (from Ref. 16).

computer program from Ref. 16, the Fig. 12.7 graph of the collision efficiency as a function of airspeed is obtained.

From this alone, one may wish to fly as slowly as possible to avoid the high mass of ice accretion ($\dot{m}_{Cap}$) at higher speeds. As the aircraft slows down, the AoA will increase, which will most likely increase S_F. An increase in AoA will generally decrease the collision efficiency of the droplets because the flow is affected farther upstream of the LE, which gives more time to the droplets to actually change direction (inertia involved here) and go around the wing. However, the frontal surface S_F will increase somewhat with the AoA, thus leading to an increase in the mass flow rate of droplets hitting the wing, all other parameters being constant.

From Fig. 12.8, it can be seen that to minimize the mass flow rate of droplets hitting the wing, one must keep the AoA as low as possible. It must be remembered that the AoA of the wing is closely tied to the airspeed of the aircraft. In fact, the AoA is proportional to the inverse of the airspeed squared $(1/V_\infty^2)$

$$C_L = \frac{2W}{\rho_\infty V_\infty^2 S} \approx \frac{dC_L}{d\alpha}(\alpha - \alpha_0) \qquad (12.2)$$

$$\therefore \quad \alpha \propto 1/V_\infty^2$$

Thus, unless the exact variation of the collision efficiency with AoA and airspeed is known, a compromise between speed and AoA must be made. From Fig. 12.9, it can be seen that, for velocities above the velocity for maximum lift-to-drag ratio

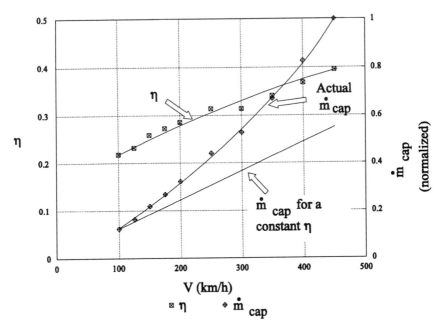

Fig. 12.7 Typical effect of the velocity on the collision efficiency and mass flow rate of water hitting a NACA 4412.

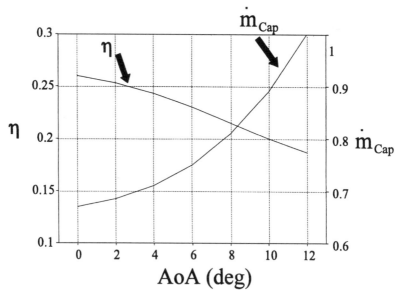

Fig. 12.8 Collision efficiency and normalized mass flow rate of droplets hitting a NACA 0012 as a function of AoA.

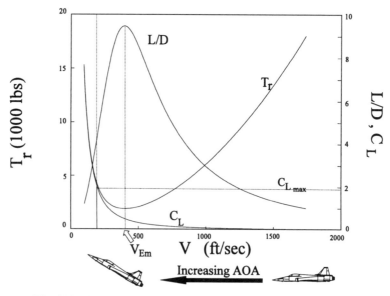

Fig. 12.9 Increasing AoA (through C_L) with decreasing airspeed.

(V_{E_m}), the AoA does not change appreciably with airspeed, while at velocities lower than V_{E_m}, the AoA will increase rapidly. Thus, V_{E_m} seems, at first, to be the best velocity to minimize $\dot{m}_{Cap}$.

Two study cases using a small military jet trainer will be done to determine what is the best velocity to minimize the ice accretion in-flight for that particular aircraft. The two study cases are 1) case A, level flight through icing conditions along a given horizontal distance (X_{icing}), and 2) case B, descent through icing conditions of a given thickness (Δh_{icing}). Both cases are shown in Fig. 12.10.

Case A. For case A, the time required to travel the icing condition distance (X_{icing}) is directly proportional to the inverse of the airspeed (i.e., $\Delta t_{icing} = X_{icing}/V_\infty$); thus, the mass of ice forming on the wing [see Eq. (12.3), second line] should be minimized by minimizing the product (ηS_F). The small jet trainer has the following characteristics: wing aspect ratio 6.0, wing span 36 ft 8 in., and wing surface 220 ft²; wing profile at MAC is a NACA 63A012, a symetrical profile with a 0-deg AoA for zero lift.

The MAC of the jet trainer is approximately

$$\text{MAC} \approx S/b = 6\,\text{ft}$$

The approximate lift coefficient slope is [from Eq. (1.8)]

$$\frac{dC_L}{d\alpha} = a_\infty \bigg/ \left[1 + \frac{a_\infty}{\pi \text{AR}}\right] \approx 0.0822\,\text{deg}^{-1}$$

The profile has a 12% relative maximum thickness, which translates to a maximum thickness of 8.64 in. for a MAC of 6 ft. The actual value of S_F with AoA will be determined using the program of Ref. 16.

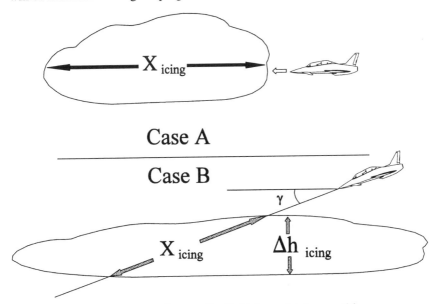

Fig. 12.10   Two study cases for flight through icing conditions.

The atmospheric conditions are: temperature $-20°C$ ($-4°F$) and pressure altitude 20,000 ft. The corresponding air density is 0.001242 slug/ft³. The MED of the water droplets is 20 μm, the LWC is 0.6 g/m³, and the icing conditions extend for 10 n mile (X_{icing})

$$m_{Cap} = \dot{m}_{Cap}\Delta t_{icing} = \eta S_F LWC(V_\infty \Delta t_{icing})$$

$$\therefore \quad m_{Cap} = \eta S_F LWC \, X_{icing}$$

(12.3)

The program of Ref. 16 was used to determine the average collection efficiency as a function of airspeed and AoA. The graph of Fig. 12.11 results. It can be seen that the collection efficiency will increase with increasing airspeed and decreasing S_F, as can be expected. The mass flow rate of water hitting the wing increases with an increase in airspeed as well, but the total mass of ice that hits the wing tends to decrease with an increase in airspeed because the time required to cross the icing conditions decreases with an increase in airspeed. Equation (12.3) showed that, for a given icing distance X_{icing}, the mass of ice hitting the wing is minimized by minimizing the product (ηS_F).

This simple analysis does not take into consideration the effects of total temperature on the amount of water that actually freezes on the wing and on the resulting ice shape. In the preceding analysis, it was assumed that the temperature was cold enough so that the water droplets would freeze on contact. This type of ice is called rime ice; it is opaque and white because air is trapped between frozen droplets. Glaze ice is formed when the water droplets hit the wing and run back a certain

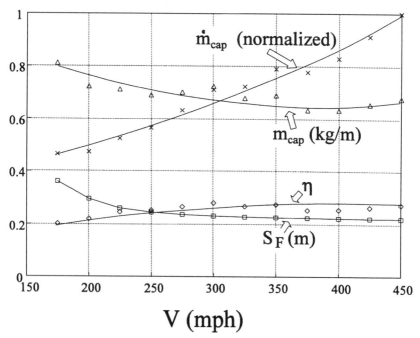

Fig. 12.11   Icing results using the program of Ref. 11 (two-dimensional results).

-26°C   -20°C   -18°C   -15°C   -12°C   -8°C   -5°C   -3°C   -1°C   0°C

Total Temperature

Fig. 12.12   Total temperature effects on ice shape, NACA 0012, 0.53-m chord, 4-deg AoA, 209 km/h, LWC 1.3 g/m³, time 8 min (reproduced from NACA TM-83556, Jan. 1984, Ref. 17).

distance toward the trailing edge. This happens at warmer temperatures. Glaze ice is clear. The actual ice that is formed on the wing is somewhat in between these two extremes. It ranges from the pointy rime ice to the horn-shaped glaze ice as shown in Fig. 12.12.

Thus, an aircraft that flies fast enough will not accrete ice due to the aerodynamic heating of the wing. Fighters do not have any protection systems against icing for this reason, but they are still subject to icing while on approach and during landing when their airspeed is relatively low.

The actual size of the wing also has a major effect on the collision efficiency of the water droplets. In general, the larger the wing, the smaller the relative ice accretion is, as shown in Fig. 12.13. This is because the flow is affected farther in front for a large wing, giving more time to the droplets to change course and go around the wing.

Case B.   For this case, the icing time Δt_{icing} is strongly influenced by the dive angle and aircraft velocity combination (i.e., the rate of descent),

$$X_{\text{icing}} = \frac{\Delta h_{\text{icing}}}{\sin(\gamma)}$$

$$\Delta t_{\text{icing}} = \frac{X_{\text{icing}}}{V_\infty} = \frac{\Delta h_{\text{icing}}}{V_\infty \sin(\gamma)} = \frac{\Delta h_{\text{icing}}}{RD}$$

(12.4)

It can be seen in Fig. 12.14 that the time required to go through the icing condition layer (Δh_{icing}) for a given dive velocity decreases rapidly for small angle of dive variation up to 10 deg, after which the time changes more slowly. As the dive angle increases, the dive velocity will also increase, which will increase the rate of accumulation of ice being formed on the wing. For this reason and following the analysis just given, it is this author's opinion that a dive angle of between 8–10 deg would be the best to minimize the ice formation on the wing, the smaller of the two angles yielding a lower airspeed.

Landing with ice accretion on the wings.   If the pilot suspects that there is some ice on the wings of the aircraft, the approach and landing speed should be increased by 5–20% depending on the estimated ice accretion and shape. This will provide a safe stall margin. If the pilot can fly in a zone where the air temperature is above the freezing point, it should be done so as to burn off the ice formation.

The actual increase in approach and landing speed is a judgement call by the pilot. In general, rime ice will form at very cold temperature and will be relatively

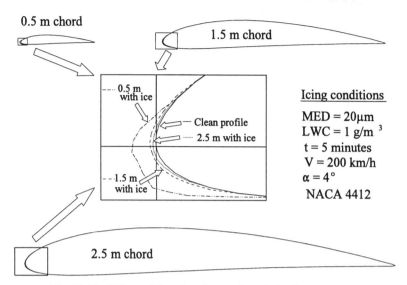

Fig. 12.13   Effect of the wing size on the relative ice accretion.

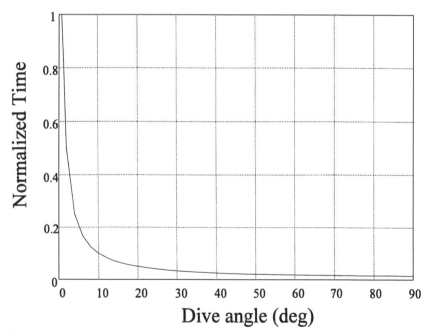

Fig. 12.14   Normalized time of descent at a constant airspeed as a function of the dive angle.

smooth, resulting only in a low degradation of aircraft performance. On the other hand, glaze ice forms at temperatures closer to the freezing point and will result in a significant increase in drag and decrease in maximum lift coefficient and stall AoA.

Pilots should also be aware of icing due to cold soaked fuel. After a long cruise leg at high altitude, the temperature of the fuel in the wing will reach the ambient total air temperature. That fuel may not have time to warm up during the descent for landing. This may cause moisture to condense and freeze to the wing. This results in a degradation in aircraft performance during landing and possible FOD ingestion for aft-mounted engines. This problem is important enough that some aircraft manufacturers, such as McDonnell Douglas for the MD-80 series, are working actively to find solutions to prevent accidents from cold soaked fuel moisture icing.

Tailplanes are usually more sensitive to icing than the wing because they are smaller. The danger of having an iced tailplane is that it could stall, leaving the aircraft with no pitch control. This is particularly important during approach and landing because of the increased pitch control demand on the tailplane due to the use of flaps. Flaps increase the lift coefficient, but they also increase the nose-down pitching moment, thus requiring an increased download from the tail. A tailplane stall will result in an uncontrollable pitch down, which may result in a crash. Such was the case of a Vickers Viscount that crashed in Stockholm in January 1977. If icing is encountered in-flight and there is no means of removing the ice on the tailplane, the pilot should select a lesser amount of flap and land a little faster.

Ice accretion prevention in flight. To prevent the large loss of aircraft performance due to the accretion of ice, the LEs of the aircraft must be protected. The areas most important to protect are: the wing and empennage LEs, the windshield, the propeller blades LEs, engine air inlets, and essential flight instruments. These could be protected with either a deicing system or an anti-icing system. The anti-icing systems usually prevent the formation of ice by applying enough heat to evaporate the water droplets hitting the aircraft surface. In this case a good computer program (such as the one from Ref. 16 or NASA's Lewice program) can be used to determine the potential icing surfaces. These results must be confirmed by flight testing. The disadvantage of an anti-icing system is that it requires a lot of heat energy, which comes from either the engine compressor bleed air or from the electric system. Protecting the entire aircraft with an anti-icing system cannot be done economically. Instead, only the critical parts of the aircraft are protected, and the ice is allowed to build up on less critical components. As well, there exist systems that detect icing conditions and that turn on the anti-icing systems only when required, thus using less energy (which results in less fuel being burned for a given flight profile) than activating the anti-icing system when icing conditions are suspected.

Some aircraft may not have the required power margin to use anti-icing systems; they must then revert to deicing systems. Deicing systems will allow some ice accretion up to a predetermined thickness before removing the ice. Deicing systems can use either heat to remove the ice or pneumatic boots, which break up the ice by expanding. Once the ice has broken up, it is removed by the airflow and the boot deflates back to its original shape. Wings using deicing systems must be designed to tolerate the aerodynamic penalties of the small accumulations.

Following a series of icing related fatal accidents in the 1980s, mostly involving aircraft during takeoff, several commissions were established in the U.S. and Canada to review existing flight regulations concerning aircraft icing. The Dryden Commission Report (Canada), released in March 1992, contained 191 recommendations for improving flight safety. Among the new or revised regulations of the MOT, designed to meet the recommendations of the Dryden Commission, one states that, "No person shall takeoff or attempt to takeoff in an aircraft that has frost, ice, or snow adhering to any of its critical surfaces (clean aircraft concept)." This is a major change compared to previous regulations where it was left to the pilot to determine whether a small amount of accretion would adversely affect flight performance.

12.3   Effects of Gusts on the Load Factor

A gust is a sudden change in the wind velocity. It can be broken down into a horizontal component called wind shear and a vertical component called updraft or downdraft and clear air turbulence. The horizontal component of a gust will alter the relative velocity between the aircraft and the air, whereas the vertical component affects the aircraft's AoA.

Let us consider a horizontal gust first. An aircraft in level flight at a velocity V encounters a gust of intensity $\pm u$, the positive sign being for a gust that increases the relative velocity of the aircraft. The new load factor of the aircraft will be

$$n_{\text{gust}} = \frac{L_{\text{gust}}}{W} = \frac{\frac{1}{2}\rho(V \pm u)^2 S C_L}{W} \tag{12.5}$$

Because the aircraft was in level flight before the gust, the lift L was equal to the weight (i.e., $n = 1.0$). As well, due to the sudden nature of a gust and to the aircraft's inertia, the AoA, thus C_L, remains essentially the same. Equation (12.5), thus, reduces to

$$n_{\text{gust}} = \frac{(V \pm u)^2}{V^2} = \frac{V^2 \pm 2uV + u^2}{V^2} \approx 1 \pm \frac{2u}{V}$$
$$\therefore \quad \Delta n_{\text{gust}} \approx \pm \frac{2u}{V} \tag{12.6}$$

The last term in the preceding equation (u^2/V^2) was neglected because it would usually be a lot smaller than the term (uV). From this last equation it can be seen that the effects of a horizontal gust are proportional to the gust velocity and inversely proportional to the aircraft velocity. Therefore, we can reduce the influence of a horizontal gust on the aircraft's load factor by flying at high velocities. Horizontal gusts, therefore, will have their greatest impact on aircraft during takeoff and landing where the velocity is low.

A vertical gust, on the other hand, will change the AoA (increasing it for an updraft and decreasing it for a downdraft), as well as increasing the aircraft velocity as shown in Fig. 12.15. In this case, the major contributor to the aircraft load factor is the change of AoA, and so it will be assumed that the velocity remains constant for this analysis.

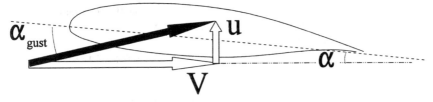

Fig. 12.15   Vertical gust.

It is to be noted that symmetrical vertical gusts in level flight to be used during the design stage of an aircraft are specified in the FARs and in MIL-A-8861. To simplify the analysis, it will be assumed that the aircraft penetrates a sharp-edged vertical gust with an upward vertical velocity of u (feet per second). The AoA will suddenly be increased by an increment $\Delta\alpha$ before the wing begins to react or the airspeed begins to respond to the gust. Thus, the resulting load factor will be

$$n_{\text{gust}} = \frac{C_{L_{\text{gust}}}}{C_L} = \frac{C_L + \Delta C_{L_{\text{gust}}}}{C_L} \tag{12.7}$$

where

$$\Delta C_{L_{\text{gust}}} = \frac{dC_L}{d\alpha}\Delta\alpha_{\text{gust}} \quad \text{and} \quad \frac{dC_L}{d\alpha} = a_0 \bigg/ \left(1 + \frac{a_0}{\pi ARe}\right) \tag{12.8}$$

Therefore, the increase in AoA due to the gust is (here assuming again that the gust velocity is much smaller than the aircraft velocity)

$$\tan(\Delta\alpha_{\text{gust}}) \approx \Delta\alpha_{\text{gust}} = u/V \tag{12.9}$$

Substituting into Eq. (12.7), it is found that

$$n_{\text{gust}} = 1 + \left[\left(\frac{dC_L}{d\alpha}\right)\frac{u}{V}\bigg/\frac{2(W/S)}{\rho V^2}\right]$$

$$= 1 + \left[\left(\frac{dC_L}{d\alpha}\right)\rho u V\bigg/2\left(\frac{W}{S}\right)\right] \tag{12.10}$$

$$\therefore \quad \Delta n_{\text{gust}} = \left[\left(\frac{dC_L}{d\alpha}\right)\rho u V\bigg/2\left(\frac{W}{S}\right)\right]$$

Therefore, the change in load factor due to a vertical gust is proportional to the velocity of the gust u, to the velocity of the aircraft V, to the density of the air ρ, and to the lift slope of the aircraft $(dC_L/d\alpha)$, as well as inversely proportional to the wing loading (W/S). To minimize the effects of such a gust, the pilot should fly relatively slowly (minimize V), fly high (low ρ) or, by design, have a low-lift-curve slope or high-wing loading. Therefore, compromises are required at the design stage because low-level penetration requires a high-wing loading for pilot comfort, precision bombing, and long range, but a low-wing loading is required for fast and tight turns and acceptably short takeoffs and landings.

The sharp-edge gust is evidently a simplification of the actual response of an aircraft to a discrete gust. The actual gust buildup is likely to be gradual and affect the forward part of the aircraft first. This will give time to the aircraft to react, thus reducing the effect of the gust compared to a sharp-edge gust of the same intensity. Current regulations for discrete gusts use a one-minus-cosine pulse to evaluate the aircraft response to a discrete gust. [Discrete gust dynamic loads are not required now by FAR or by the FAA. They are superseded by the continuous turbulence loads requirement of FAR 25.305 (d) and FAR Appendix G (Ref. 18)]. The increase in load due to the gust takes the following form:

$$\Delta n_{\text{gust}} = K_g\left[U_{\text{de}}V_e\left(\frac{\partial C_L}{\partial \alpha}\right)\Big/498\left(\frac{W}{S}\right)\right] \qquad (12.11)$$

where K_g is the alleviation factor (also called gust factor) that accounts for the motion of the aircraft and the time lag in the buildup of aircraft lift. This K_g multiplies an equivalent sharp-edge gust of velocity U_{de} (derived equivalent gust velocity in feet per second). The aircraft velocity in this equation, V_e, is the aircraft equivalent airspeed in knots. Values for U_{de} for altitudes varying from sea level to 20,000 ft are: 66 ft/s at V_B (design rough-air speed of the aircraft), 50 ft/s at V_C (design cruise speed), and 25 ft/s at V_D (design dive speed). For altitudes above 20,000 ft, these values decrease linearly to the following values at 50,000 ft:

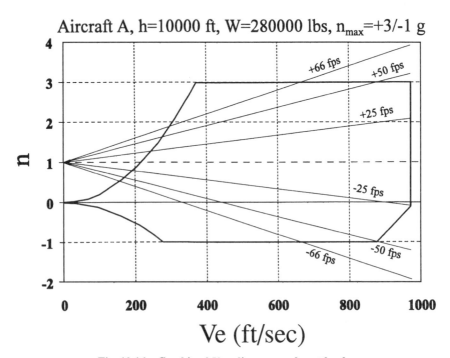

Fig. 12.16   Combined V–n diagram and gust loads.

38 ft/s at V_B, 25 ft/s at V_C, and 12.5 ft/s at V_D. The values of K_g are

$$K_g = \frac{0.88\mu_g}{5.3 + \mu_g} \qquad \text{for } M < M_{\text{crit}}$$

$$K_g = \frac{\mu_g^{1.03}}{6.95 + \mu_g^{1.03}} \qquad \text{for } M > M_{\text{crit}}$$

(12.12)

where μ_g is a mass parameter for the aircraft and is equal to

$$\mu_g = 2\left(\frac{W}{S}\right)\Big/\rho_{SL}\sigma g \bar{c}\left(\frac{dC_L}{d\alpha}\right)$$

(12.13)

An example of the load factor increase for the derived equivalent gust velocity for aircraft A at a weight of 280,000 lb and an altitude of 10,000 ft is shown in Fig. 12.16. The aircraft maximum design rough-air velocity V_B is the velocity at which the gust load line for +66 ft/s meets the line of maximum load factor on the aircraft flight envelope ($\approx$650-ft/s equivalent airspeed in this case).

12.4 Wind Shear

In most cases, a change in wind direction only requires a correction in navigation from the pilot. Turbulence and gusts increase the load factor of the aircraft but, unless the aircraft is flying close to its stall velocity and a gust causes the aircraft to stall, they do not influence the aircraft performance and flight path. A wind shear is defined as: "Any change of wind or downdraught causing a change of flight path that requires significant pilot action for recovery."[19] They are very localized phenomena and often of short durations. Typically, wind shears occur over a distance varying from 150 to 3,000 m and have a life span of approximately 15 min. An aircraft would require between 2–40 s at an approach speed of 150 kn to cross a wind shear region at its largest point.

A wind shear caused by a thunderstorm (microburst) is characterized by a region of large downdraft with the air spreading radially as it gets close to the ground (Fig. 12.17). When encountering a microburst head on, the aircraft will start climbing due to the increasing headwinds, which will increase the lift of the aircraft. The headwinds disappear rapidly and are followed by downdraft (typically, 500–1,200 ft/min) and increasing tailwinds. This sudden change in relative wind (horizontal wind shear of up to 100 kn have been reported[19]) may exceed the climb and acceleration performance of the aircraft, especially if it is flying slowly and low such as during the landing phase.

Prior to the mid-1980s, pilots' techniques for maintaining approach conditions were to use thrust to control flight path and pitch to control airspeed [slow flight, Chapter 3, Sec. 3.4.1]. While encountering a microburst, this technique causes the pilot to first reduce the thrust when encountering headwinds to maintain the flight path. As the aircraft approaches the center of the microburst, the headwinds disappear quickly and the downdraft increases rapidly causing the aircraft to start sinking and descend below the intended flight path. To correct the situation, the pilot applies thrust, but there is a time delay between the throttle movement and the increased thrust by the engines (spool up time due to the inertia of the rotating

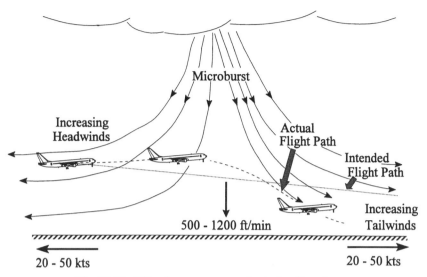

Fig. 12.17 Microburst encounter in the landing phase.

parts). Then, tailwinds start increasing, which lowers the aircraft airspeeds (due to the inertia of the aircraft, its velocity will not react as quickly as the wind velocity change). A pilot normal reaction, due to training, is to lower the pitch angle to increase airspeed, which leads to smaller lift forces, thus increasing the sink rate. Several accidents (26 civil aviation accidents between 1964 and 1985, Ref. 20) where wind shear was a cause factor were attributed to this pilot technique. The first accident to have microburst recorded as the primary cause involved a Boeing 727 on approach to JFK airport in New York on June 24, 1975.[19]

In 1987, following a series of studies, the U.S. FAA published a wind shear training aid, which recommends that, upon pilot recognition of flying through a wind shear, full thrust be applied and that the aircraft should be rotated to an initial 15-deg target pitch angle. This allows the aircraft to remain above or close to its intended flight path and avoid the excessive loss of altitude that may result in the aircraft hitting the ground during approach for landing. Since then, other studies (such as Ref. 20) have identified slightly different techniques to recover from an encounter with a microburst.

12.5 Heavy Precipitation

High-intensity, short-duration rainfall can be a flight safety hazard, especially during takeoff and landing. Rainfall rates R, defined as the linear accumulation depth at ground level per unit time, that can be encountered by aircraft can vary from light rain ($R = 5$–10 mm/h) up to very large rates (the world record being 30.5 mm of rain in 1 min; this is equivalent to 1.83 m/h or 6 ft/h).[21]

Engine thrust is not a problem during heavy rainfall encounters since they are certified to operate at water ingestion rates that exceed the world record rainfall rate. But the impact momentum of heavy rainfall can slow down the aircraft by several knots. Tests done at several laboratories worldwide (including NASA Langley and the Canadian National Research Council Institute for Aerospace

Research) have determined that heavy rainfall will affect the aircraft aerodynamics. One such investigation has determined that there would be an increase of 2–5% in drag, a decrease of 7–29% in maximum lift coefficient, and a decrease of 1–5 deg in the stall AoA for a Boeing 747 encountering rainfall rates varying from 100 to 1,000 mm/h (Ref. 15).

This loss in maximum lift coefficient is not critical for flight performance except during takeoff and landing where the aircraft is flying at a lift coefficient approximately equal to 70% of the maximum available (Chapter 7). Obviously, a 29% loss in maximum lift coefficient would dangerously reduce the stall margin on approach and during landing. As well, heavy rainfall combined with windshear may prove catastrophic.

13
Ground Proximity Warning Systems

13.1 Introduction

T HE advent of avionics integration within new aircraft has allowed aircraft designers to provide the flight crew with only the flight essential information on digital display screens and to display any other information upon request or to alert the crew during an emergency. This has greatly reduced cockpit clutter and enabled the designer to reduce the required flight crew to perform a given mission. Aircraft of the Boeing 767 class, which once required between three and five crew members in the cockpit for safe operation now require only two. Multirole aircraft such as the CF-18 can perform air-to-air missions as well as air-to-ground missions with only one pilot on board. This reduction in crew size has reduced the direct operating cost of aircraft but has somewhat increase the workload of the crew, especially during emergencies.

Controlled flight into terrain (CFIT) is a description for accidents where various human cause factors such as task saturation, distraction, disorientation, or channelized attention cause a pilot to unknowingly fly a serviceable aircraft into the ground. Sufficient information, such as exact height above ground provided by radar altimeter, already exists within the new aircraft, which if properly interpreted by the pilot could prevent such accidents. However, the pilot does not always recognize a critical situation developing in sufficient time to recover the aircraft. This is particularly true while flying low-altitude missions (Fig. 13.1).

A new category of systems, which can use the information provided by the aircraft sensors to predict impending CFIT, is now being introduced to both military and civilian aircraft. These systems are regrouped in a category called ground proximity warning systems (GPWS). These systems can predict an impending CFIT by continuously comparing aerodynamic state and recovery capabilities of an aircraft with its height above ground. When the comparison reveals that a recovery must be initiated to avoid CFIT, audible and visual warnings are issued to the pilot. These systems are completely automatic and require no pilot input other than the selection of a minimum recovery altitude (MRA) that will be a buffer between the aircraft and the ground.

Present International Civil Aviation Organization (ICAO) standards dictate that all aircraft in international commercial operations of a maximum certificated takeoff mass (MCTM) in excess of 15,000 kg or authorized to carry more than 30 passengers, and which were brought into service on or after July 1, 1979, are required to be equipped with GPWS. It is also a recommendation that aircraft brought into commercial service before July 1, 1979 and all those used in international general aviation operations, or a similar MCTM and passenger capacity, should be equipped with GPWS. In March 1995, in Montreal, the ICAO Council adopted amendments to Annex 6, Operation of Aircraft (Part I, International Commercial Air Transport—Aeroplanes and Part II, International General Aviation—Aeroplanes), which provide new requirements for the carriage

Fig. 13.1 Low-altitude flying.

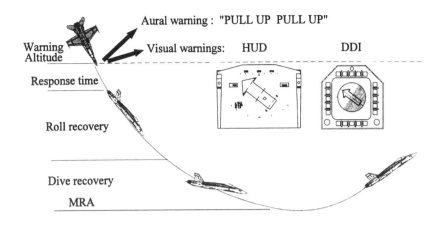

Fig. 13.2 GPWS units broken down in relation to the recovery trajectory.

of GPWS. The new ICAO standards, beginning Dec. 31, 1998, will extend the requirement to carry GPWS to all aircraft, used in both international commercial and general aviation operations, where the MCTM is in excess of 5,700 kg or airplanes are authorized to carry more than nine passengers. The new standards also specify the minimum modes in which the GPWS is required to operate. Several armed forces, such as the U.S. Air Force and Navy and the Canadian Forces, are installing GPWS in fighter and attack aircraft, as well as in their transporters.

This section will highlight the basic aircraft performance equations required to build a basic GPWS algorithm. The flight path trajectory after a CFIT warning will be broken down into three main units (Fig. 13.2): the pilot response time unit where the pilot continues the current maneuver without reacting; the roll recovery unit, where the pilot rolls the aircraft from its current bank angle to nearly wings level; and the dive recovery unit, where the pilot initiates a pull out at a given load onset rate up to a target load factor and then maintains that load factor until the aircraft has recovered (i.e., distance between the aircraft and the ground is increasing and the aircraft is above the MRA). Note that the roll recovery unit will not be described because it deals more with stability and control matters than with aircraft performance. As well, to simplify this discussion on GPWS, it will be assumed that the dive angle, prior to the warning, is constant and that the ground is flat and at sea level.

13.2 Pilot Response Time Unit

This unit calculates the loss of altitude due to the pilot response time after a CFIT warning. It can be assumed that a 1-s pilot response time (Δt_1) after a CFIT warning is issued is a reasonable delay. During such a delay, it is assumed that the pilot is continuing whatever maneuver was occurring at the time of the warning.

13.2.1 Altitude (Mean Sea Level) Loss

The barometric altitude Δh_1 lost during the pilot response time will be a function of its instantaneous vertical velocity V_V and vertical acceleration a_V and can be approximated by

$$\Delta h_1 = \int\!\!\int \left(\frac{\mathrm{d}^2 h}{\mathrm{d}t^2}\right)\mathrm{d}t^2 = V_V \Delta t_1 + \frac{1}{2} a_V \Delta t_1^2 \tag{13.1}$$

where the vertical velocity can be found using Eq. (4.4), reproduced here,

$$V_V = \frac{\mathrm{d}h}{\mathrm{d}t} = V \sin(\gamma) \tag{13.2}$$

The vertical acceleration, assuming a constant dive angle, is

$$a_V = a \sin(\gamma)$$

where a is the acceleration along the flight path.

13.2.2 Ground Altitude (AGL) Loss

The altitude above ground of an aircraft can be determined by using a radar altimeter. Then, the barometric (pressure) altitude of the ground (h_g) is

$$h_g = h_p - h_R \tag{13.3}$$

For our current analysis, the ground will be assumed to be flat and situated at sea level (Fig. 13.3).

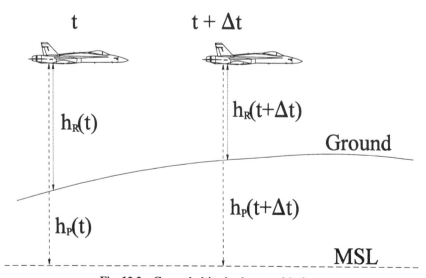

Fig. 13.3   Ground altitude change with time.

13.2.3 New Flight Parameters After Response Time

Assuming that the pilot is continuing the maneuver during the period of response time, the new value for airspeed along the flight path is (assuming constant acceleration)

$$V_1 = V_i + a_i \Delta t_1 \qquad (13.4)$$

The calculation of airspeed variation with time could be better approximated by using the rate of change of acceleration, but this would complicate this discussion on the basics of GPWS equations. Also, a constant dive angle, as well as no roll angle, has been assumed to simplify the equations presented here. Each new level of precision will introduce more variables and more complicated equations. A high level of precision is necessary for any aircraft required to maneuver at low altitudes. To use more precise equations, one must make sure that the readings from the aircraft instruments are also precise or the reading errors will just be compounded and one may get a worse approximation of the recovery profile than by using the preceding more simple equations.

13.3 Roll Recovery Unit

This unit calculates the loss of altitude due to the time it takes for the pilot to roll to nearly wings level (± 5 deg) from its present bank angle (Fig. 13.4). Thus, the angle through which the aircraft has to roll is

$$\Delta \phi = |\phi| - 5 \deg \qquad (13.5)$$

If $\Delta \phi \leq 0$, then the wings are considered level and this subroutine is not executed.

The estimation of the rolling performance can be very complicated due to several factors, such as an aircraft large-AoA range and large Mach number range, as well as the combination of ailerons, flaps, rudders, and elevator deflection to

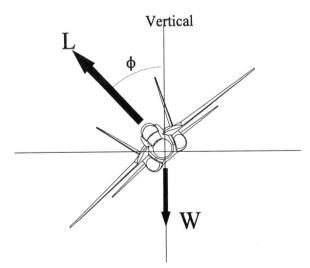

Fig. 13.4   Angle of bank.

create a rolling moment. The aircraft total rolling moment coefficient comprises the four following terms:

$$C_{\ell_{\text{total}}} = \Delta C_{\ell_{(1)}} + \Delta C_{\ell_{(2)}} \cdot \frac{\delta_{\text{LEF}}}{\delta_{\text{LEF}_{\text{max}}}} + \Delta C_{\ell_{(3)}} \cdot \frac{\delta_{\text{TEF}}}{\delta_{\text{TEF}_{\text{max}}}} + \Delta C_{\ell_{\beta,\text{FLEX}}} \cdot \beta \quad (13.6)$$

where

$\Delta C_{\ell(1)}$	= rolling moment coefficient, flaps retracted
$\Delta C_{\ell(2)}$	= LE flap increments referenced to maximum deflection
δ_{LEF}	= LE flap deflection
$\Delta C_{\ell(3)}$	= trailing-edge (TE) increments referenced to maximum deflection
δ_{TEF}	= TE flap deflection
$\Delta C_{\ell\beta,\text{FLEX}}$	= rolling moment flexibility derivative

Each of these elements is affected by the aircraft AoA and flight Mach number. For aircraft equipped with automatic LE and TE flaps, which deflect according to a preset schedule and which depend on Mach number and AoA, this will introduce a rate of change of maximum rolling moment available during the roll recovery.

To make matters even more complicated, the roll rate acceleration is a function of aircraft inertia moments about the rolling axis. Because an aircraft rolls about its velocity vector, at low AoAs, this roll axis is almost parallel to the aircraft longitudinal axis, but at large AoAs, there is a marked difference as noted in the following equations. Equation (13.7) indicates how to convert the body axis moments of inertia into flight-path axis moments of inertia,

$$L_x = I_{xx}\omega_x + I_{xy}\omega_y + I_{xz}\omega_z$$

$$L_y = I_{yx}\omega_x + I_{yy}\omega_y + I_{yz}\omega_z$$

$$L_z = I_{zx}\omega_x + I_{zy}\omega_y + I_{zz}\omega_z$$

$$\begin{Bmatrix} I_{xx_{\text{FP}}} \\ I_{zz_{\text{FP}}} \\ I_{xz_{\text{FP}}} \end{Bmatrix} = \begin{bmatrix} \cos^2(\alpha) & \sin^2(\alpha) & -\sin(2\alpha) \\ \sin^2(\alpha) & \cos^2(\alpha) & \sin(2\alpha) \\ \frac{1}{2}\sin(2\alpha) & -\frac{1}{2}\sin(2\alpha) & \cos(2\alpha) \end{bmatrix} \times \begin{Bmatrix} I_{xx_{\text{B}}} \\ I_{zz_{\text{B}}} \\ I_{xz_{\text{B}}} \end{Bmatrix} \quad (13.7)$$

The need to identify correctly what is the mass distribution of the aircraft (basic empty, stores loading, cargo loading, fuel weight and location) to correctly determine the moment of inertia can take considerable time (relative to the computation time of a GPWS algorithm, typically 100 ms). Also, this approach is prone to errors if the fuel weight and distribution are wrongly calculated or if the weapons loading or cargo loading are wrongly calculated. A simplified approach to the calculation of the time required to recover from a dive is, therefore, required.

A simplified approach will be used to determine the time required for an aircraft in any given situation to recover from a roll. It will be assumed that the aircraft can reach 60-deg/s roll rate in half a second from 0-deg/s roll rate (i.e., 120 deg/s^2) as shown in Fig. 13.5 (this roll performance is typical of a high-performance aircraft

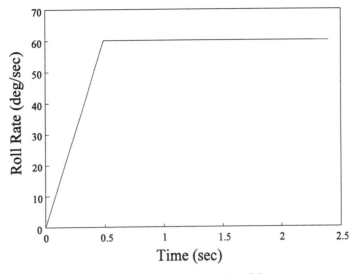

Fig. 13.5   Roll rate as a function of time.

such as a fighter). The roll angle covered during the first half-second, from a 0-deg/s roll rate is

$$\Delta\phi = \left| p\,\Delta t + \tfrac{1}{2}\dot{p}\,\Delta t^2 \right|$$

$$\Delta\phi = \tfrac{1}{2}(120\,\text{deg/s}^2)(0.5)^2 = 15\,\text{deg}$$

This means that an aircraft can recover from a 90-deg bank angle (i.e., $\Delta\phi = 85$ deg) in 1.667 s as will be shown. For the first half-second,

$$P_i = 0 \qquad \dot{p} = 120\,\text{deg/s}^2$$

$$\Delta\phi_1 = \tfrac{1}{2}\dot{p}\,\Delta t^2 = 15\,\text{deg}$$

For the rest of the roll,

$$p = 60\,\text{deg/s} \qquad \dot{p} = 0\,\text{deg/s}^2$$

$$\delta\phi = 70\,\text{deg} \Rightarrow \Delta t = \Delta\phi/p = 1.167\,\text{s}$$

This simplified approach may not be adequate for a lightly loaded fighter that may attain roll rates of in excess of 180 deg/s as well as greater roll rate accelerations that cover that distance in less than 1 s. For the heavily loaded aircraft at slow speed, this may be the maximum it can attain, and safety is of concern here. Also, if the aircraft is already rolling at the end of the pilot reaction time unit and the roll rate is superior to the maximum of 60 deg/s assumed here, then a constant roll rate is assumed from this and it is equal to the last roll rate calculated.

The altitude loss during the roll recovery can be determined using equations from Sec. 13.2. Here one must assume that the aircraft will maintain the dive angle calculated at the end of the pilot reaction time unit (i.e., not a combined roll/dive recovery).

13.4   Dive Recovery Unit

This unit calculates the predicted altitude loss due to the time it takes to recover from a dive (positive climb angle). It will be broken into two parts: First, calculate the loss of altitude to recover from a dive to level flight (flat ground assumption); then, calculate the angle of climb necessary to avoid a rising ground (projection of the rising ground along the instantaneous flight path, not curved).

13.4.1   Recovery to Level Flight

It was determined from Chapter 4 that the radius of curvature of the recovery profile in a dive pullout (Fig. 13.6) is

$$r = \frac{V^2}{g} \frac{1}{[n - \cos(\gamma)]} \tag{13.8}$$

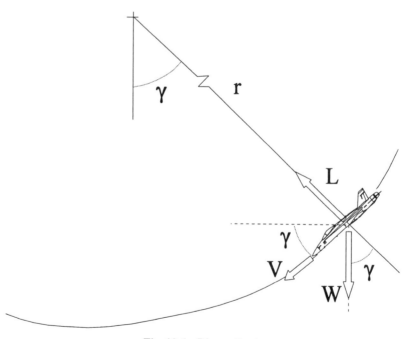

Fig. 13.6   Dive pullout.

Let

$$n_{\text{eff}} = n - \cos(\gamma)$$

$$\frac{dn_{\text{eff}}}{dt} = \frac{dn}{dt} + \sin(\gamma)\frac{d\gamma}{dt} \tag{13.9}$$

$$\frac{d^2 n_{\text{eff}}}{dt^2} = \frac{d^2 n}{dt^2} + \cos(\gamma)\left(\frac{d\gamma}{dt}\right)^2 + \sin(\gamma)\frac{d^2\gamma}{dt^2}$$

where n is the load factor. Thus, n_{eff} is the portion of the total load factor that effectively pulls the aircraft out of the dive. Both n and γ can be provided by the aircraft mission computer. Let us assume that the target load (n_T, the largest load factor during recovery) is 5 g and that the load onset rate (dn/dt) is 2.5 g/s.

To recover, the aircraft has to turn through γ degrees (relative to a flat ground). When the initial dive angle is known,

$$\Delta\gamma = 0 - \gamma_i = -\gamma_i$$

Then,

$$\frac{d\gamma}{dt} = \frac{V}{r} = \frac{g[n - \cos(\gamma)]}{V} = \frac{g n_{\text{eff}}}{V} \tag{13.10}$$

$$\frac{d^2\gamma}{dt^2} = g\left[\left(\frac{dn_{\text{eff}}}{dt}V - n_{\text{eff}}\frac{dV}{dt}\right)\Big/V^2\right] \tag{13.11}$$

As a first approximation of the time required to recover, let

$$\Delta\gamma = \frac{d\gamma}{dt}\Delta t + \frac{1}{2}\frac{d^2\gamma}{dt^2}\Delta t^2 \Rightarrow \frac{1}{2}\ddot{\gamma}\Delta t^2 + \dot{\gamma}\Delta t + \gamma_i = 0$$

$$\Delta t = \frac{-\dot{\gamma} \pm \sqrt{\dot{\gamma}^2 - 4(\frac{1}{2}\ddot{\gamma})(\gamma_i)}}{2(\frac{1}{2}\ddot{\gamma})} \tag{13.12}$$

with the value of the load factor at the end of the recovery being

$$n_f = n_i + \frac{dn}{dt}\Delta t \tag{13.13}$$

Two situations are now possible: the aircraft can recover (flight-path angle of 0 deg) before reaching the target load or the aircraft can reach the target load and then maintain it until it has recovered. If

$$\Delta t \le \left[(n_T - n_i)\Big/\frac{dn}{dt}\right] \tag{13.14}$$

then aircraft recovers before reaching the target load factor. The initial and final flight-path angles are

$$\gamma(0) = \gamma_i \qquad \gamma(\Delta t) = 0$$

At $t = 0$ (for the dive portion), the following initial conditions exist:

$$V_V \approx V \sin(\gamma) \qquad (13.15)$$

$$a_V = \frac{dV}{dt} \sin(\gamma) + g n_{\text{eff}} \cos(\gamma) \equiv \begin{array}{c} \text{acceleration} \\ \text{along the} \\ \text{flight path} \end{array} + \begin{array}{c} \text{acceleration} \\ \text{perpendicular to} \\ \text{the flight path} \end{array} \qquad (13.16)$$

$$\dot{a}_V = \frac{d^2 V}{dt^2} \sin(\gamma) + g \frac{dn_{\text{eff}}}{dt} \cos(\gamma) - g n_{\text{eff}} \sin(\gamma) \frac{d\gamma}{dt} \qquad (13.17)$$

These last three equations are valid for any flight conditions. The loss of altitude (MSL) due to the preceding equations is as per Eq. (13.1), only this time the flight profile is estimated to be as just described, whereas during the pilot response time unit it was assumed that the pilot was continuing whatever maneuver was occurring. Let us break the dive recovery profile into two segments: the segment where the pilot builds up the load factor from the initial value to the target load (n_T) and the segment where the pilot maintains the target load until the aircraft has recovered.

During the segment where the pilot increases the load factor from its initial value at the beginning of the dive recovery up to the target load factor, the values for n_{eff} and (dn_{eff}/dt), as derived from Eq. (13.9), are

$$n_{\text{eff}} = n_i - \cos(\gamma_i)$$

$$\frac{dn_{\text{eff}}}{dt} = \frac{dn}{dt} + \sin(\gamma_i) \left(\frac{d\gamma}{dt} \right)_i \qquad (13.18)$$

where (dn/dt) is the load onset rate and was fixed at 2.5 g/s. Once the target load is reached [i.e., after Δt seconds as calculated by Eq. (13.12)], the conditions during the second segment are

$$n_{\text{eff}} = n_T - \cos(\gamma_2)$$

$$\frac{dn_{\text{eff}}}{dt} = 0 + \sin(\gamma_2) \left(\frac{d\gamma}{dt} \right)_2 \qquad (13.19)$$

It can be seen from Eq. (13.19) that the effective load factor (the one pulling the aircraft out of the dive) decreases as the dive angle decreases and equals ($n_T - 1$) when the aircraft is at 0-deg dive angle. Also, (dn_{eff}/dt) at a dive angle of 0 deg is equal to zero.

13.4.2 Terrain Height Prediction

The terrain height in front of the aircraft can be determined by calculating the horizontal distance traveled by the aircraft during a dive recovery and by knowing the terrain angle at the start of the dive recovery. The simplest way to determine the horizontal distance covered during the dive recovery, although crude, is to calculate the distance covered by the aircraft during the time required to recover from the dive assuming that the aircraft horizontal velocity is the same as its actual flight-path velocity. This will be the worst-case scenario. For our case, with a flat terrain, the aircraft will have recovered (i.e., no risk of hitting the ground) when the dive angle reaches 0 deg.

13.5 Maximum AoA During Recovery

At low velocity, an aircraft will stall (exceed α_{stall}) before reaching its target load (see Fig. 13.3). Exceeding α_{stall} means decreasing lift, leading to lower recovery altitude, and often loss of controllability, which can be deadly in an emergency recovery. Thus, a GPWS algorithm must account for the aircraft specific α_{stall} (thus, $C_{L_{max}}$) while predicting a dive recovery. Using typical aircraft C (Table 13.1) as an example, it can be seen that its α_{stall} is 30 deg. For safety reasons, i.e., to account for the aircraft varying weight during flight, from a maximum gross to almost empty weight, a safety maximum AoA of 25 deg will be used. Now the recovery criteria are either a target load factor of 5 g or a maximum AoA of 25 deg will be reached during the dive recovery, whichever occurs first. This should prevent any stall due to the combination of high-g loading and slow speed. This requirement will translate into a lesser target load at low speed. Assume that, at low speeds, the lift coefficient at 25-deg AoA is approximately equal to 1.5. The target load factor equivalent to such a lift coefficient can be expressed as

$$n_T = \tfrac{1}{2}\rho_{SL}\sigma V^2 S C_{L_{25deg}}/W \qquad (13.20)$$

Table 13.1 Roll angle < 5 deg

Test	γ, deg	IAS,[a] kn	ϕ, deg	Maximum load factor AETE	Maximum load factor GPWS	V_V, ft/s	Altitude loss, ft AETE	Altitude loss, ft GPWS
1	−3	262	2	2.8	2.75	23.13	40	34.9
15	−3	363	3	4.4	3	32.05	60	51.3
16	−3	352	1	4.6	3	31.08	20	49.6
39	−3	456	1	4.9	3.5	40.26	70	67.4
42	−2	451	2	4.3	3	26.55	70	41.1
75	−15	292	0	3.2	—	127.5	290	282.6
79	−15	350	1	4.1	5	152.8	290	356.8
105	−15	463	0	5.5	5	202.2	350	516.3
110	−14	430	1	5.1	5	175	360	427.3
144	−15	553	2	5.1	5	241.5	500	657.9
150	−45	282	0	4.4	—	336.4	1550	1118
155	−43	397	0	5	5	456.8	1690	1754
157	−43	372	0	5	5	428	1630	1589
158	−43	436	0	5.6	—	501.6	2040	2027
179	−60	370	0	4.5	—	540.6	2800	2426

[a]Indicated airspeed.

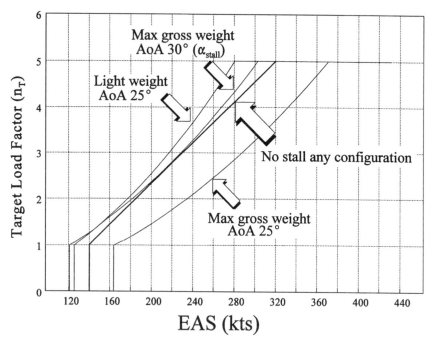

Fig. 13.7 Difference in target load factor depending on aircraft weight, for a maximum recovery AoA of 25 deg.

The term (σV^2) is the equivalent airspeed (EAS) squared. It is readily apparent that the target load factor, if not fixed at 5 g, is highly dependent on the aircraft weight. Figure 13.7 indicates the difference in target load factor with respect to EAS for two extreme weight configurations: one the light fighter at 60% fuel ($\approx$16,000 lb) and the other one at maximum gross weight with external stores ($\approx$28,000 lb).

It is readily apparent that to avoid stall at any configuration (with maximum gross weight being critical) one must assume a maximum load factor when flying at low speeds. In this case, a proposed curve (darkest line in Fig. 13.7) for the variation of the target load factor as a function of equivalent airspeed to prevent stall of the heaviest configuration for typical aircraft C could be (valid for airspeeds above 140 kn)

$$n_T = 5 \qquad \text{for EAS} \geq 320\,\text{kn}$$

$$n_T = 1 + 4\left[\frac{\text{EAS} - 140\,\text{kn}}{320\,\text{kn} - 140\,\text{kn}}\right] \qquad \text{for EAS} \leq 320\,\text{kn} \qquad (13.21)$$

However, playing with the target load factor in such a way may be the cause of some false alarms. This must be investigated during flight test of the particular aircraft/avionics suite combination.

13.6 Comparison Between GPWS Units and Standard Aircraft Performance

The resultant total force vector acting on an aircraft in-flight can be broken down into components parallel to and perpendicular to the flight path as shown in Fig. 13.8. The forces acting along the flight path are

$$T \cos(\alpha_T) - D - W \sin \gamma = (W/g)a \qquad (13.22)$$

The forces perpendicular to the flight path are

$$L + T \sin(\alpha_T) - W \cos(\gamma) = (W/g)(V^2/r) \qquad (13.23)$$

where (V^2/r) is the acceleration perpendicular to the flight path that we will redesignate a_r for now.

The rate of change of acceleration with time is considered here because it incorporates, as it will be shown, such effects as the change of weight over time due to the fuel flow, the rate of change of AoA, etc. It provides one more level of approximation/correction in our estimate of the altitude loss. The rate of change of acceleration along the flight path can be found by differentiating Eq. (13.22) with respect to time. The resulting equation is

$$\frac{\mathrm{d}T}{\mathrm{d}t} \cos(\alpha_T) - T \sin(\alpha_T)\frac{\mathrm{d}\alpha_T}{\mathrm{d}t} - \frac{\mathrm{d}D}{\mathrm{d}t} - \frac{\mathrm{d}W}{\mathrm{d}t} \sin(\gamma)$$

$$- W \cos(\gamma)\frac{\mathrm{d}\gamma}{\mathrm{d}t} = \frac{1}{g}\left[\frac{\mathrm{d}W}{\mathrm{d}t}a + W\dot{a}\right] \qquad (13.24)$$

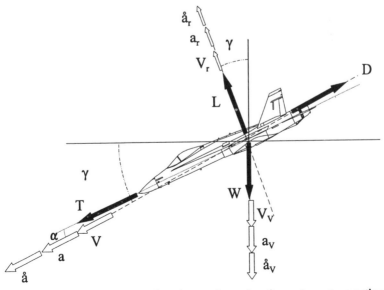

Fig. 13.8 Force, velocity, acceleration, and acceleration rate vectors acting on an aircraft in a dive.

where

$$\frac{\mathrm{d}D}{\mathrm{d}t} = \frac{1}{2}\rho_{\mathrm{SL}}S\left[\frac{\mathrm{d}\sigma}{\mathrm{d}t}V^2 C_D + \sigma 2V \frac{\mathrm{d}V}{\mathrm{d}t}C_D + \sigma V^2 \frac{\mathrm{d}C_D}{\mathrm{d}t}\right] \qquad (13.25)$$

and

$$\frac{\mathrm{d}C_D}{\mathrm{d}t} = \frac{\mathrm{d}C_{D_0}}{\mathrm{d}t} + \frac{\mathrm{d}K}{\mathrm{d}t}C_L^2 + K 2C_L \frac{\mathrm{d}C_L}{\mathrm{d}t} \qquad (13.26)$$

Rewriting Eq. (13.24), we find that the rate of acceleration along the flight path is

$$\dot{a} = \frac{g}{W}\left[\frac{\mathrm{d}T}{\mathrm{d}t}\cos(\alpha_T) - T\sin(\alpha_T)\frac{\mathrm{d}\alpha_T}{\mathrm{d}t} - \frac{\mathrm{d}D}{\mathrm{d}t} - \frac{\mathrm{d}W}{\mathrm{d}t}\sin(\gamma)\right.$$

$$\left. -W\cos(\gamma)\frac{\mathrm{d}\gamma}{\mathrm{d}t} - \frac{\mathrm{d}W}{\mathrm{d}t}\frac{a}{g}\right] \qquad (13.27)$$

The rate of change of acceleration along the flight path is influenced by many parameters that will now be analyzed. We first notice that is inversely proportional to the aircraft weight; thus, the larger the weight is, the smaller the value of $\dot{a}$.

The term $[(\mathrm{d}T/\mathrm{d}t)\cos(\alpha_T)]$ is the rate of change of engine thrust with time. Since thrust T is a function of altitude, velocity, and throttle setting, the variation of thrust with time will depend on the velocity, climb/dive angle, and the rate of change of throttle setting (combined with engine spool up time). This parameter will be scaled by the cosine of the angle between the thrust vector and the flight path (α_T). It can be assumed that the difference between the aircraft AoA and thrust angle is negligible. High-performance aircraft can have large-AoA ranges (0–65 deg) but the main operating ranges are usually limited to below 25 deg (minimum cosine of 0.90631). Thus, the cosine will not have a large effect on the value of this parameter, especially at large airspeeds where α_T is small.

The next term $[-T\sin(\alpha_T)(\mathrm{d}\alpha_T/\mathrm{d}t)]$ is the variation of thrust along the flight path due to the changing angle of thrust (proportional to the change of AoA) with time. This parameter will have a larger influence while maneuvering at low to medium speeds (large AoA and AoA rate of change). Note the negative sign of this parameter, which means that the rate of acceleration will decrease when then AoA increases (less thrust in the flight-path direction).

The parameter representing the rate of change of aircraft drag with time is

$$-\frac{\mathrm{d}D}{\mathrm{d}t} = -\frac{1}{2}\rho_{\mathrm{SL}}S\left[\frac{\mathrm{d}\sigma}{\mathrm{d}t}V^2 C_D + \sigma 2V \frac{\mathrm{d}V}{\mathrm{d}t}C_D + \sigma V^2 \frac{\mathrm{d}C_D}{\mathrm{d}t}\right]$$

$$C_D = C_{D_0} + K C_L^2$$

$$\frac{\mathrm{d}C_D}{\mathrm{d}t} = \frac{\mathrm{d}C_{D_0}}{\mathrm{d}t} + \frac{\mathrm{d}K}{\mathrm{d}t}C_L^2 + K 2C_L \frac{\mathrm{d}C_L}{\mathrm{d}t}$$

As drag increases, the rate of change of acceleration will decrease. The instantaneous variation of drag with time is affected by the aircraft altitude (through σ) and the rate of change of altitude (through $\mathrm{d}\sigma/\mathrm{d}t$). It is also affected by the airspeed (through V) and acceleration along the flight path (through dV/dt). Last, it is affected by the drag coefficient C_D and the rate of change of the drag coefficient (dC_D/dt). This last term includes such effects as the change of minimum drag coefficient (through dC_{D_0}/dt) due to the opening/closing of the speed brake, the flaps, or to Mach number (compressibility). The term (dK/dt) will vary very slightly at low speeds except during the lowering/retracting of the flaps, whereas at high speeds it will be affected by the Mach number. Finally, it can be seen that the rate of acceleration will be affected by the rate of change of aircraft lift coefficient due to the changing airspeed, maneuvering or wind gust.

The next parameter $[-(dW/dt)\sin(\gamma)]$ is somewhat simpler to analyze than the previous one. It represents the effects of the change of aircraft weight with time on the aircraft instantaneous rate of acceleration. Usually, this term corresponds to the aircraft fuel flow (a decrease of weight with time), which means that as fuel is consumed, the rate of change of acceleration increases (with other parameters constant). In a few cases the aircraft weight increases during flight (air-to-air refueling) or there is a very large instantaneous decrease (dropping of external stores). It is also factored by the sine of the flight-path angle; thus, this parameter has minimal impact on $\dot{a}$ except at large γ (dive) and high (dW/dt).

Skipping one parameter in Eq. (13.27), we can notice that the last parameter $[-(dW/dt)a/g]$ is similar to the previous one except that this time it is factored by the relative acceleration (a/g) along the flight path. This term can be large at low-airspeed/low-altitude/high-thrust setting combination.

The parameter

$$-W\cos(\gamma)\frac{d\gamma}{dt} = -W\cos(\gamma)\frac{g[n-\cos(\gamma)]}{V}$$

represents the rate of change of the weight vector parallel to the flight path. As the angle of the flight path decreases ($d\gamma/dt$ negative) from the horizontal, the rate of acceleration will increase because the component of weight along the flight path increases.

From the preceding brief analysis, we can see that the rate of change of acceleration will provide some form of correction to our estimate of the recovery profile, which may take as little as 1.3 s (1 s due to the pilot reaction time) for a 1-deg dive at 250 kn or as much as 8.6 s for a 60-deg dive at 630 kn.

The rate of acceleration perpendicular to the flight path is

$$\dot{a}_r = \frac{g}{W}\left[\frac{dL}{dt} + \frac{dT}{dt}\sin(\alpha_T) + T\cos(\alpha_T)\frac{d\alpha_T}{dt} - \frac{dW_T}{dt}\cos(\gamma)\right.$$

$$\left. + W\sin(\gamma)\frac{d\gamma}{dt} - \frac{dW}{dt}\frac{a_r}{g}\right] \tag{13.28}$$

where

$$\frac{dL}{dt} = \frac{1}{2}\rho_{SL}S\left[\frac{d\sigma}{dt}V^2 C_L + \sigma 2V\frac{dV}{dt}C_L + \sigma V^2\frac{dC_L}{dt}\right] \tag{13.29}$$

There is no need at this time to analyze this equation, but it contains some aircraft performance information that may prove useful to determine the cause of false alarms in the algorithm itself.

13.7 Comparisons

This section presents a comparison of the results obtained with the preceding equations incorporated into an algorithm for a personal computer and the data from a flight test at the Canadian Forces Aerospace Engineering Test Establishment (AETE) in September 1991. The evolution of performance parameters during a CFIT recovery was also monitored using a CF18 simulation within the algorithm (Tables 13.1–13.4 and Figs. 13.9–13.11).

As one can see from Figs. 13.11 and 13.12, the GPWS algorithm constructed using the equations of this chapter tends to overpredict (conservative) the altitude loss for large-roll angles. For low-roll angles the prediction is fairly correct. Thus, our simple roll unit model may somewhat underpredict the roll capacity of the aircraft used at AETE (fighter configuration) but the same roll model might be nonconservative for highly loaded aircraft (although it is unlikely that a heavily loaded aircraft would go into an inverted steep dive relatively close to the ground). This last configuration was not tested for those roll angles in September 1991.

In general, the algorithm created with the equations of this chapter agrees well with the results of the flight test for roll angles less than 45 deg and vertical velocities of less than 500 ft/s (this corresponds to a 45-deg dive at around 420 kn true airspeed). At larger roll angles, this GPWS algorithm tends to overestimate the altitude loss during the recovery, which indicates that the assumption of 60-deg/s roll rate and 120-deg/s^2 roll onset rate may to be slightly low compared to the actual values. These values were selected to ensure that a heavily loaded fighter at slow speed could recover in any normal flight situation. These values could be corrected following a flight test to determine the worst-case scenarios.

Of course, the values obtained are highly dependent on the actual pilot response time, the pilot aggressiveness and aircraft ability to roll out, and the pilot following the prescribed load onset rate and target load factor. To demonstrate this, Table 13.5 indicates recovery values for a 45-deg dive angle at a 45-deg roll angle and a

Table 13.2   Roll angle between 40 and 50 deg

Test	γ, deg	IAS,[a] kn	ϕ, deg	Maximum load factor AETE	Maximum load factor GPWS	V_V, ft/s	Altitude loss, ft AETE	Altitude loss, ft GPWS
6	−3	256	44	3.1	2.75	22.6	70	54.4
23	−3	352	40	3.1	3	31.08	110	75.5
44	−3	460	47	5.4	3.5	40.61	140	106.8
46	−2	447	41	4.8	3	26.31	100	63.1
47	−3	438	41	4.6	3.25	38.67	70	97.2
48	−4	456	44	5.8	3.75	53.66	140	143.9
112	−13	442	43	5.2	5	167.7	490	551.9
153	−45	274	45	3.5	—	326.9	1280	1379
154	−42	230	45	3.2	—	259.6	700	1008
161	−45	400	45	5.1	—	477.2	2230	2341
189	−59	302	45	4.9	—	436.7	2180	2151

[a]Indicated air speed.

Table 13.3   Roll angle = 120 deg

Test	γ, deg	IAS,[a] kn	ϕ, deg	Maximum load factor AETE	Maximum load factor GPWS	V_V, ft/s	Altitude loss, ft AETE	Altitude loss, ft GPWS
30	−12	339	120	4.1	5	118.9	470	517.6
49	−5	448	120	5.2	4	65.87	310	266.2
52	−8	433	120	5	5	101.7	330	432.7
87	−17	362	120	4.8	5	178.6	610	830.4
118	−18	457	120	5.6	5	238.2	710	1173
163	−48	351	120	4.9	—	440	2030	2674
171	−43	452	120	5.4	—	520	2710	3325

[a]Indicated air speed.

Table 13.4   Roll angle = 180 deg

Test	γ, deg	IAS,[a] kn	ϕ, deg	Maximum load factor AETE	Maximum load factor GPWS	V_V, ft/s	Altitude loss, ft AETE	Altitude loss, ft GPWS
34	−3	355	180	4.6	3	31.34	90	149.7
35	−4	347	180	4.4	3.5	40.83	270	198.4
37	−10	341	178	4.8	4.7	99.89	440	522.3
54	−5	446	180	5.3	4	65.57	310	330.9
57	−11	451	180	5.2	5	145.2	650	797.5
95	−17	360	180	5.2	5	177.6	860	1006
98	−22	367	180	5.8	5	231.9	830	1376
121	−18	447	180	5.7	5	233	1020	1380
148	−20	566	180	6.3	5	326.6	1330	2078
176	−44	494	180	5	—	578.9	3390	4616

[a]Indicated air speed.

Table 13.5   Altitude loss prediction units, sensitivity analysis

GPWS algorithm prediction	Altitude loss, ft	Time, s	Extra time, s	Additional altitude loss, ft
Pilot reaction time	695.3	1.00	0.1	69.53
Roll recovery	629.6	0.92	0.1[a]	69.53
Dive recovery	2,442.0	6.68	0.2[b]	215.58
Total loss	3,766.9	8.60	0.4	354.74

[a]Based on a roll onset rate of 100 deg/s^2 and a maximum roll of rate of 52.9 deg/s.
[b]Based on a load onset rate of 2 g/s and a target load factor of 5 g.

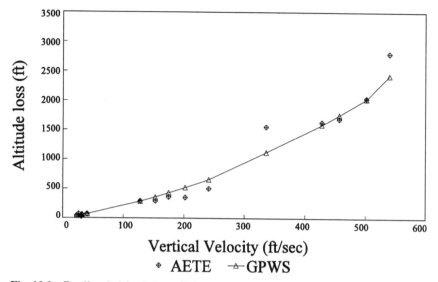

Fig. 13.9   Predicted altitude loss (GPWS) vs actual flight test altitude loss (AETE) for roll angles of less than 5 deg.

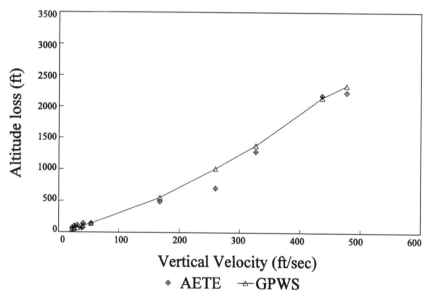

Fig. 13.10   Predicted altitude loss (GPWS) vs actual flight test altitude loss (AETE) for roll angles between 40 and 50 deg.

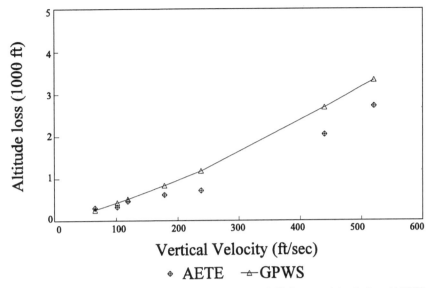

Fig. 13.11   Predicted altitude loss (GPWS) vs actual flight test altitude loss (AETE) for roll angles of 120 deg.

calibrated airspeed of 550 kn corresponding to a vertical velocity of 695.3 ft/s at 4000-ft (ISA).

It is evident that a sluggish reaction from the pilot in the Table 13.5 situation (less than half a second extra time for the entire recovery profile) will negate the effects of getting a GPWS with a minimum recovery altitude of only 100 ft! The extra 0.1 s for the pilot to react takes almost all of the 100-ft buffer away. The opposite is also true in that if pilots react a lot more aggressively than what is programmed into the algorithm they will think they got false alarms as they recovered high above the MRA.

In any case, this GPWS algorithm does provide a means of evaluating the altitude loss for a given set of recovery parameters. These parameters are pilot reaction time, roll onset rate and maximum roll rate, target load factor, and load onset rate. Once these are known, the altitude loss can be calculated and displayed on a graph of altitude loss vs vertical velocity.

13.8   Filtering of Aircraft Data

Many factors can cause the incoming aircraft data to oscillate about a mean value and cause problems for the GPWS algorithm. The algorithm can be particularly very sensitive to fluctuating radar altimeter readings since this, in combination with horizontal velocity, is the parameter used to predict terrain altitude in front of the aircraft.

During testing of this algorithm with a personal computer, it was found that the algorithm is also sensitive to sudden change in aircraft pitch and roll.

It was found in this investigation that using the rate of acceleration and acceleration along the flight path, although providing extra degrees of precision in determining the altitude loss, may cause false alarms during sudden variation of

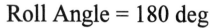

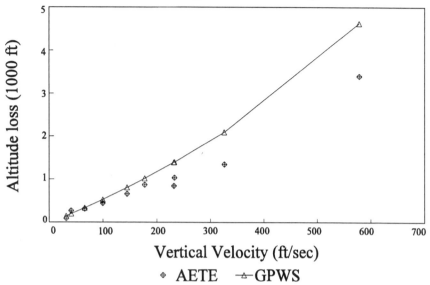

Fig. 13.12   Predicted altitude loss (GPWS) vs actual altitude loss (AETE) for roll angles of 180 deg.

pitch and roll. It is particularly sensitive to pitch because of the effects on the vertical velocity and acceleration, which are used to calculate the predicted altitude loss. The problem comes primarily from the basic assumption that during the pilot reaction time, the pilot will be continuing whatever maneuver was occurring. A sudden pitch down may, for the GPWS algorithm, look like the pilot will reach a large-dive angle at the end of the pilot reaction time of 1 s and may generate a CFIT warning if the calculated flight path detects an impending CFIT.

13.9   Conclusions

This chapter has demonstrated a basic GPWS using aircraft performance equations affecting the dive recovery for a given set of recovery parameters. The theory behind these equations can be found in preceding chapters.

No matter what system is used in an aircraft, the dynamic behavior of the system, i.e., sudden dive or roll close to the MRA, must always be flight tested to determine if any false alarms are detected and if so, what is the sensitivity of the algorithm. Such an algorithm should also be tested while flying over sharp edge cliffs and varying terrain conditions such as flying above a forest of varying tree density and height.

Appendix A: Aircraft Nomenclature

A.1 Introduction

B EFORE we can discuss any flight characteristics of any craft, we need to define the types and categories in which they may belong. We also need to describe them and their many components so as to be aware of the definitions when they are encountered.

Aerospace vehicles include all craft that will operate into the atmosphere and/or the outer space. An example of such an aerospace vehicle is the Space Shuttle, which takeoff like a rocket, maneuvers in space, and glides back to Earth. More recent vehicles in this category include the National Aerospace Plane X-30 (now canceled), which would have been able to takeoff horizontally, climb to orbit, maneuver in space, and fly back to Earth, as well as the X-33 single stage to orbit research vehicle. The aerospace vehicle category can be divided into two subcategories: the space vehicles, which transit through the atmosphere and operate only in space (satellites, orbital stations, etc.), and the atmospheric vehicles or aircraft, which operate only inside the boundary of the atmosphere.

In this book we concentrate on the aircraft side of the tree in Fig. A.1. This subcategory includes two more subcategories: the aerostat, which includes all lighter-than-air machines whose lift is generated from static forces of the atmosphere (balloons, blimps, dirigibles), and the aerodynes, which include all heavier-than-air machines whose lift is generated from aerofoils driven through the atmosphere. This last subcategory is still further split in two to include fixed wings (or airplanes) and rotary wings (or helicopters).

A.2 Aircraft Reference Planes and Axes

To locate components on the aircraft, it is necessary to define coordinate axes and planes. This will also allow us to define angles on the plane. The longitudinal axis goes from nose to tail. The origin is located in front of the nose to account for any equipment that might be installed later (such as a test probe) and the positive side of the axis is toward the tail of the aircraft. The lateral axis goes from wing tip to wing tip with the origin point on the longitudinal axis of the aircraft and the positive side is on the starboard side of the plane. The vertical axis is perpendicular to the other two with the datum point at ground level (while aircraft is on its landing gear) with the positive side pointing up.

The fuselage plane is perpendicular to the longitudinal axis. At a given point (called the fuselage station) along the longitudinal axis, the fuselage plane will show a cross-sectional area of the fuselage. The buttock plane is perpendicular to the lateral axis. A point along the lateral axis is called a wing station. The buttock plane is particularly useful to view the many profiles of the wing along its span. The water plane is perpendicular to the vertical axis and a point along that axis is called a water line (Fig. A.2).

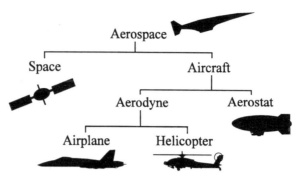

Fig. A.1   Aerospace vehicle subcategories.

A.3   Major Aircraft Components

An aircraft can be divided into five major components. They are: the wing, the fuselage, the propulsion system, the empennage, and the landing gear. The general layout and the way each component is integrated with the others will greatly influence the aircraft's performance.

A.3.1   Wing

The main function of the wing is to generate most of the necessary lift to maintain the airplane in the air. It can also carry external loads (missiles, bombs,

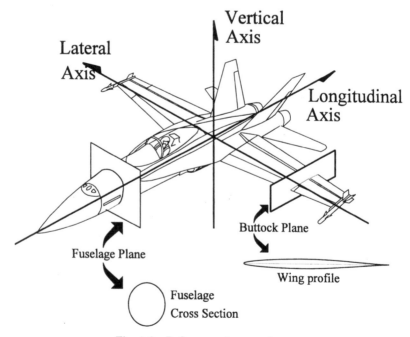

Fig. A.2   Reference planes and axes.

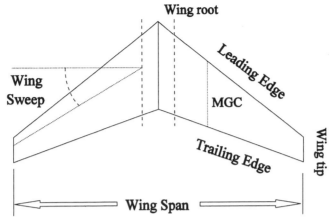

Fig. A.3   Description of a wing.

external fuel tank, etc.), support engines, contain internal fuel tanks, and house flight controls and landing gear.

A wing can be describe by its physical aspect according to the following parameters: surface, planform, wing span, mean geometric chord (MGC), MAC, taper ratio, AR, sweep angle, aerofoils, geometric twist, aerodynamic twist, flight controls, etc. (Fig. A.3).

The wing span b of the wing is the distance between both wing tips, perpendicular to the longitudinal axis of the aircraft. This distance is generally fixed, but some planes have variable geometry or folding wings that alter the wing span. These two topics will be discussed later.

The surface S of the wing is the area it covers when projected onto a water plane. We can distinguish three types of wing surfaces (Fig. A.4): The exposed

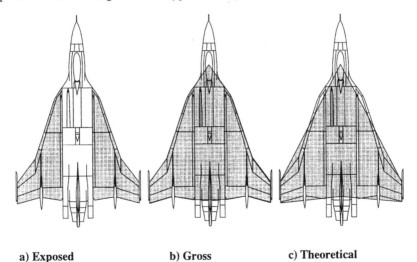

a) Exposed   b) Gross   c) Theoretical

Fig. A.4   Wing surface definition.

wing surface is the surface that you can actually touch (it is completely outside of the fuselage). The gross wing surface includes the exposed wing surface plus the projection of the wing within the fuselage; it is usually the reference value in aerodynamics. Last, when a wing has no straight LE and TE, a theoretical wing surface is then defined; it has the same size as the gross wing surface. It then becomes easier to define taper ratio and wing sweep. The planform is the actual shape of the wing as projected on the water plane. Some types of planform are: delta (Mirage 2000), rectangular (Piper Cherokee), swept wing (Super Étandard), tapered (CF-18), etc.

The chord is the distance from the LE to the TE of the wing, parallel to the longitudinal axis. This distance will generally decrease from root to tip but some aircraft have rectangular wings (constant chord) and the Republic XF-91 of the 1950s even had an increasing chord from root to tip. The taper ratio λ of a wing is the ratio of the wing tip chord c_t to the wing root chord c_r. A taper ratio of 0.5 means that the wing tip chord is half the size of the wing root chord

$$\lambda = c_t/c_r \tag{A.1}$$

It is often desirable to have a constant value instead of one that varies. The MGC is the chord that the wing would have if it were rectangular with the same surface and span

$$MGC = S/b \tag{A.2}$$

Another imaginary value, which is useful, is the MAC or $\bar{c}$. It is a theoretical chord for a rectangular wing, which has the same force vectors as the actual wing. For many planform/wing designs MAC $\approx$ MGC and the easier definition of MGC is usually used as first estimation. The MAC can be calculated in the following way:

$$MAC = \frac{2}{S} \int_0^{b/2} c(y)^2 dy \tag{A.3}$$

$$x_{MAC,i} = \frac{2}{C_L S} \int_0^{b/2} C_l c x_i dy \quad \text{where } x_i \equiv (x, y, z) \tag{A.4}$$

and where C_l is the local lift coefficient, c is the local chord and x_i is the position of the local aerodynamic center (generally, $c/4$ for subsonic flow and $c/2$ for supersonic flow).

An important aerodynamic parameter is the AR or $\mathcal{R}$. It is the ratio of the wing span to the MGC, or from the definition of MGC, the ratio of the span square to the surface. An AR of 8 means that the wingspan is 8 times the length of the MGC. Some typical values of AR are fighters 3, transport aircraft 8, gliders 15.

$$AR = \frac{b}{MGC} = \frac{b^2}{S} \tag{A.5}$$

The sweep angle Λ is the angle between the lateral axis and a reference line along the span. The reference line follows a constant relative chord position along

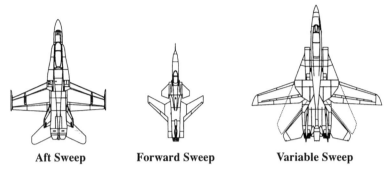

| Aft Sweep | Forward Sweep | Variable Sweep |

Fig. A.5   Different quarter-chord sweeps.

the span. This point is usually the quarter-chord position or the LE of the wing. A positive sweep angle is one where the wing tip reference point is behind the root chord one. The negative swept wing (also called forward swept wing) is just the opposite. The Grumman X-29 is one example of a forward swept wing aircraft. A variable sweep angle wing aircraft has the possibility of symmetrically modifying the wing sweep in-flight, to increase the efficiency of the wing for a wide range of velocities, as well as on the ground, to reduce the hangar space required to park it. An oblique wing aircraft has one wing swept forward and one wing swept back (Fig. A.5).

The geometric twist angle ϵ of a wing is the angle between the root chord line and the wing tip chord line (Fig. A.6). If the LE of the wing tip points more downward than the LE of the wing root, ϵ is positive. We also call a positive twist washout. A washin is the opposite.

The aerodynamic twist is the change of aerofoil camber and position of maximum camber along the span of the wing to change the lift curve. This changes the zero lift AoA along the span, simulating a geometric twist without physically twisting the wing. A decreasing airfoil camber from root to tip is called a washout.

The dihedral angle Γ is the angle made between a line joining the wing root to the wing tip and the water plane (Fig. A.7). The angle is positive if the wing tip is above the wing root. A negative dihedral angle is also called anhedral angle. Some aircraft may not have the entire wing at the same dihedral angle. The McDonnell Douglas F-4 is an example; the inner wing is at zero dihedral while the outer wing is at 12-deg dihedral angle.

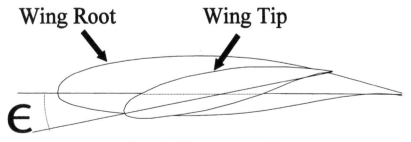

Fig. A.6   Wing geometric twist.

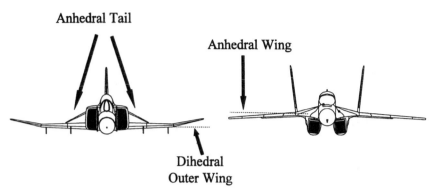

Fig. A.7   Wing dihedral.

Some aircraft have the ability to fold their wings while on the ground (Fig. A.8). This greatly eases aircraft ground maneuvering in constricted areas and reduces parking space requirements. This feature is mainly used by carrier-borne aircraft.

LEXs or LE root extensions are mainly used on fighter type aircraft. They produce a vortex at high AoA that delays stall and helps bring some order to an otherwise chaotic flow (Fig. A.9).

A.3.2   Fuselage

The fuselage is the body of the aircraft. It is designed to accommodate the crew, the cargo and/or the passengers, the avionics, and some or most of the fuel. It might carry the engines and house the landing gear. As well, for combat aircrafts, the fuselage will most likely contain the gun and its ammunition, some weapons, and their associated launchers. The cockpit must be positioned in such a way as to offer the maximum visibility forward (for landing, takeoff, and flight) and all around in the case of fighters. In most of today's modern jet aircraft, the fuselage will also house the auxiliary power unit, which can enable the aircraft to start its engines or operate without the need for aircraft maintenance support equipment, also known as ground support equipment. All of this must fit inside or on the most aerodynamically efficient airframe possible to reduce as much as possible the drag of the fuselage (Fig. A.10).

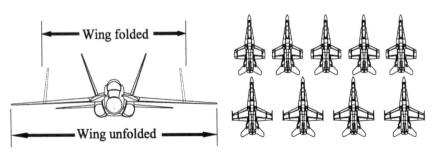

Fig. A.8   Folding wing.

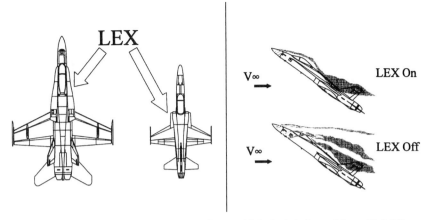

Fig. A.9   Effect of LEX on forebody flow at high AoA (adapted from Ref. 22).

A.3.3   Empennage

The empennage is the traditional grouping of the vertical and horizontal tails. Its main functions are to provide stability at all flight regimes, as well as control during maneuvering. The horizontal tail is traditionally made up of a nonmoving part, called the horizontal stabilator, and of a moving part, called the elevator. The vertical tail is traditionally made up of a nonmoving part, called the vertical stabilator, and of a moving part, called the rudder.

The word traditional was used many times because it is how most people have come to see an aircraft layout. The general population is exposed to airline type transporters and small commuter planes where this layout is most common. But other layouts, such as canards or twin vertical tails, which achieve the same functions of providing stability and control are being used more often now than in the past and may be a common feature on future transport aircraft.

A canard is the generic term given to an horizontal tail that is placed ahead of the wing. This layout permits the reduction of trim drag of the aircraft by providing an upward force for trim instead of the normal downward force for aft-located horizontal tails. But the location of the canards will have a large impact on the wing aerodynamic characteristics by changing the direction of the airflow coming to the wing. Some aircraft, such as the Piaggio Avanti, use a three-surface configuration

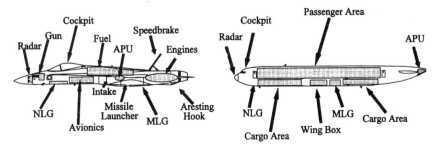

Fig. A.10   Fuselage of a fighter and of a passenger jet.

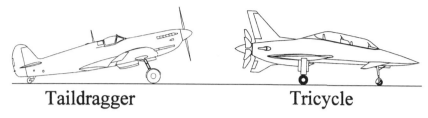

Taildragger   Tricycle

Fig. A.11   Two landing gear arrangements.

(canard for trim, wing, and horizontal tail for control), whereas other aircraft were designed without any empennage at all (flying wings such as the B-2).

A.3.4 Landing Gear

The landing gear is used to provide support to the aircraft while on the ground. It must be strong enough to absorb the normal impact of the aircraft during landing. The landing gear of most airliners will be designed to absorb a sink rate on touchdown of approximately 10 ft/s (a bad landing for most passenger would be of the order of 5 ft/s). Usually, an aircraft must flare before touching down, thus decreasing the rate of descent, but Navy aircraft are forced down from a few feet above the ground/deck by arresting cables, which means that the gear and surrounding aircraft structure must be made stronger to withstand the impact (landing gear designed to withstand sink rates of the order of 20 ft/s).

A tricycle landing gear arrangement is one where there are two main landing gear close to, but behind, the c.g. plus a nose landing gear. The main landing gear are designed to absorb the impact on landing and to support most of the weight of the aircraft while on the ground. The nose landing gear is designed to provide a balancing vertical force to the aircraft and, most of the time, to provide some steering capability for ground maneuvering (Fig. A.11).

A taildragger arrangement (or conventional arrangement) is one where there are two main gear ahead of the c.g. with a tail wheel for the balancing moment. This used to be the most widely used landing gear arrangement up to the end of World War II, after which the tricycle arrangement became dominant.

There are several other types of arrangements of landing gear in use, although not as widely spread as the two described. The selection of a particular arrangement will be dictated by the mission requirements of the aircraft, the required ground handling capability, and the flotation requirements (i.e., maximum load permitted per tire for a given runway surface), as well as several other factors.

A.3.5 Engine

This section was covered extensively in Chapter 2. Needless to say, an aircraft powerplant will be selected to provide the optimum performance to meet the mission requirements.

A.4 Profiles

A profile is a wing section parallel to the vertical plane. The aerodynamic characteristics of a wing are strongly affected by the shape of the profiles from which it is built and how these profiles vary from wing root to wing tip, both in shape and orientation (twist).

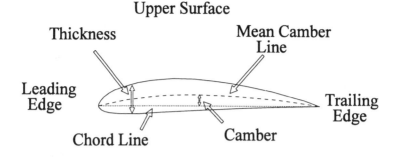

Fig. A.12 Profile description.

A.4.1 Profile Description

Figure A.12 shows the main parameters used to described a wing profile. The mean camber line is the line that is located midway between the upper and lower surface of the profile. The LE is the forward extremity (facing the incoming flow) of the mean camber line, while the TE is the aft extremity of the mean camber line. The chord line is the straight line joining the LE and TE. The maximum camber of the profile is located at a position along the chord line where the distance between the chord line and the mean camber line is maximum.

A.4.2 Profile Types

Up to the mid-1920s, little progress had been made on understanding wing profile shape effects on aircraft aerodynamics and performance. Up to the end of World War I, wing profiles were very thin and highly cambered. Tests performed at Göttingen in the mid-1920s provided significant information for the development of more modern wing profiles. But the work done by the National Advisory Committee for Aeronautics (NACA) in the U.S. proved outstanding in terms of data collected and classification of the effects of the various profile parameters on its aerodynamic behavior. The NACA series airfoils are the most widely used airfoil sections in the world, either in their basic form or in a modified form.

NACA series. In 1929, NACA initiated the development of a series of wing sections to systematically investigate the effects of changes in shape of the mean camber line and thickness distribution on the aerodynamic performance of the profiles. It was assumed by NACA that the thickness distribution was the least important parameter of the wing profile (parameters: shape of the mean line, thickness, thickness distribution). NACA based the average thickness distributions on two well-known wing sections for their early test: the Clark Y (U.S.) and the Göttingen 398 (Germany).[23] NACA provided multiple series of well-defined wing profiles with aerodynamic data (for lift, drag, and moment coefficients) up to high Reynolds numbers.

Four-digit series profiles are based on a mean line defined by two second-degree parabolas that are tangent at the point of maximum camber. The code used to define these profiles is defined in Fig. A.13. Symmetrical profiles start with two zeros,

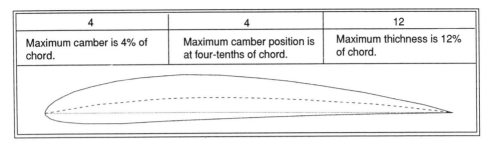

4	4	12
Maximum camber is 4% of chord.	Maximum camber position is at four-tenths of chord.	Maximum thichness is 12% of chord.

Fig. A.13   NACA 4412.

2	30	12
Maximum camber in terms of the relative magnitude of the design C_L. The design C_L in tenths is 1.5 times the first integer. Here $C_{L,des} = 0.3$	Twice the distance of the maximum camber in percent of chord from the LE.	Maximum thichness is 12% of chord.

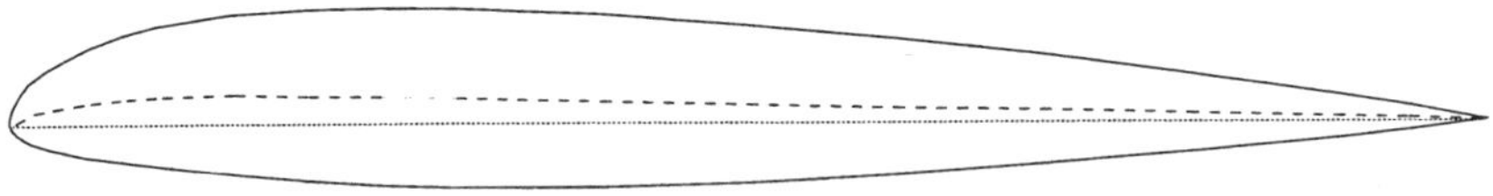

Fig. A.14   NACA 23012.

6	5	1	0	12
Laminar flow aerofoil.	Chordwise position of minimum pressure in tenths.	Range of C_L in tenths above and below the design C_L.	Design C_L in tenths.	Maximum thichness is 12% of chord.

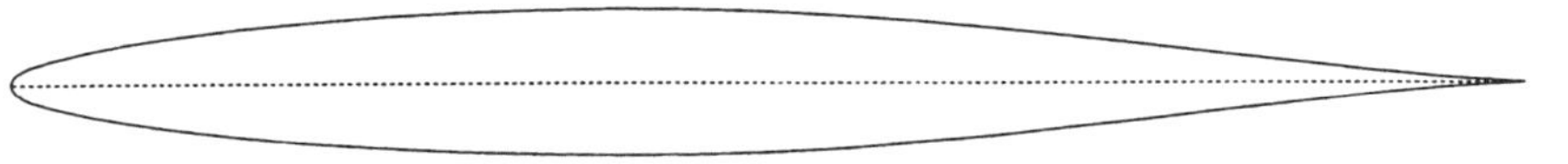

Fig. A.15   NACA 65_1-012.

for example, a NACA 0012 is a symmetrical wing profile with 12% maximum thickness.

Five-digit series have the same thickness distribution as the four-digit series but are based on a redefined mean line that is more suitable for extreme forward position of the maximum camber. The code used to define such profiles is shown in Fig. A.14. Note that there are no symmetrical profiles in the five-digit series.

Six-digit series were the first successful attempts to design efficient wing profiles with extensive laminar BL (to reduce drag). The code used to describe such profiles is shown in the Fig. A.15. These three series are the most commonly used and most well-known series.

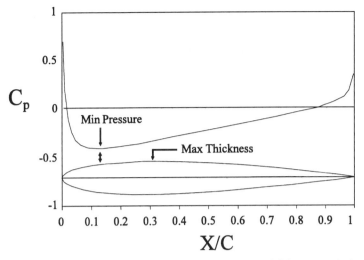

Fig. A.16   Pressure distribution over a NACA 0012 airfoil at zero AoA.

Comparison between four-digit series and six-digit series.   The six-digit series NACA airfoil have larger regions of laminar flow compared to the four-digit series. This is achieved by having the most negative pressure on the profile occur closer to the TE than on the four- or five-digit series and by reducing the absolute value of the most negative pressure (see Figs. A.16 and A.17). This reduces skin-friction drag and improves the high-speed characteristics of the profiles.

Supercritical.   Aircraft designed to operate at high speed require profiles that will have relatively low drag even in the transonic regime. Supercritical wing profiles are designed to place the drag-divergence Mach number as close to 1.0

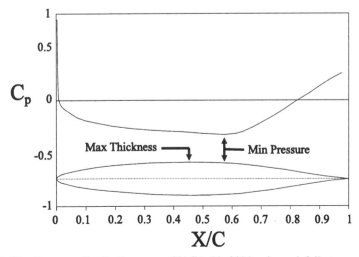

Fig. A.17   Pressure distribution over a NACA 66_1-012 laminar airfoil at zero AoA.

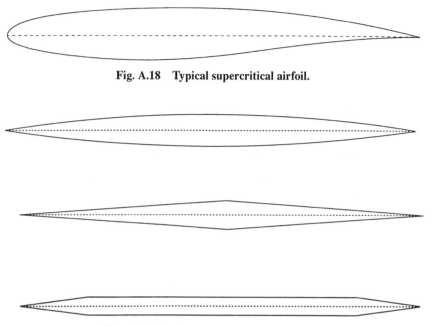

Fig. A.18   Typical supercritical airfoil.

Fig. A.19   Typical supersonic wing profiles.

as possible. This is achieved by having a nearly flat upper surface and zero (or near-zero) camber in the forward part of the profile. Any required camber would be located in the aft portion of the profile, close to the TE (Fig. A.18).

Variable camber.   NASA investigated the aerodynamic characteristics of a smooth surface, variable camber wing. The basic principle is that the camber could be tailored to provide roll control and increased lift at low speeds (but without the drag rise associated with conventional slats and flaps). Also, the control surfaces can be deflected downwards for optimum lift during transonic maneuvers and constantly seek the most efficient setting for cruise at constant power and altitude. This wing was installed on a modified F-111 and the camber was controlled by computer. A less costly alternative is computer controlled LE and TE flaps that deflect at preset conditions of Mach number and AoAs to constantly provide the best drag polar achievable by the aircraft.

Supersonic design.   The most aerodynamically efficient shape to provide lift and minimize drag at supersonic speeds is a flat plate with near zero thickness. Obviously, this shape would not have any structural strength and would have disastrous aerodynamic characterisitcs at subsonic speeds. Thus, other shapes are used such as the ones shown in Fig. A.19.

Appendix B: Atmosphere

B.1 Introduction

A GOOD knowledge of the atmosphere is very important for flight of atmospheric vehicles. The flow of air over the surface of the aircraft generates the
necessary lift to allow it to fly. Air is needed by the aircraft's engines to generate
thrust. The aircraft is shaped in such a way as to maximize lift and to minimize drag. Many aircraft instruments also need the atmosphere to work (airspeed,
altitude, etc.).

The atmosphere is a mixture of gases, which can be considered for aerodynamic
purposes as a single gas called air. This assumption is valid for flows at speeds
lower than about Mach 5; above these speeds dissociation of air particles begins.
The air consists, by weight, of 76% nitrogen, 23% oxygen, and 1% of other gases
(such as neon, freon, helium, water vapor, etc.). By volume, the composition of air
is 78% nitrogen, 21% oxygen, and 1% other gases. Although water is not a major
element of the atmosphere, it is the small amount of water in the air that creates
all of the weather (clouds, rain, storms, etc.). The atmosphere will be considered
as being dry air to facilitate our calculations, but the presence of water (humidity)
will decrease the actual density of air for the same temperature and pressure.

B.2 Properties

Pressure is stress and its units are, therefore, force per unit area. Also, pressure
is isotropic (its value being the same in every direction). Static pressure (p or p_s)
can be viewed as the weight per unit area of a column of fluid above the point of
measurement. It is the potential energy of the fluid. In the case of a fluid in motion,
an additional pressure, called impact pressure q_c, will arise due to the velocity
of the fluid. It is the kinetic energy of the fluid in movement. Total pressure (p_0
or p_T) is the summation of the static and impact pressures. Most instruments
measure the gauge pressure, which is a relative pressure in that the pressure
indicated by the instrument is the difference between the measured pressure and a
reference pressure. Relative pressure can be negative or positive, but static pressure
is always positive (vacuum or zero pressure being the lower limit). Common units
of pressure are newton per square meter (N/m^2) or pascal (Pa), pounds per square
inch ($lb/in.^2$), inches of mercury (in.Hg), and atmosphere (atm), and

$$p_0 = p_s + q_c \qquad \text{(B.1)}$$

$$1 \text{ atm} = 101,325 \text{ N/m}^2 = 101,325 \text{ Pa} = 14.7 \text{ lb/in.}^2 = 29.92 \text{ in.Hg}$$

Density ρ is the mass of a substance per unit volume. Units are usually kilograms
per cubic meter (kg/m^3) or pounds mass per cubic foot (lbm/ft^3) or slug per cubic
foot ($slug/ft^3$).

Temperature T is a measure of the average kinetic energy of the particles in the
gas. This means that the higher the temperature of the gas, the higher the speeds
of the molecules in it. Absolute temperature is always positive and is given in

Kelvin (K) for SI units and in degree Rankine (°R) for the English units. Relative temperature can be positive or negative and is given in degree Celsius (°C) for SI units and degree Fahrenheit (°F) in English units:

$$0°C = 273.16 \text{ K}$$

$$0°F = 459.69°R$$

$$1 \text{ K} = 1.8°R$$

Absolute (or dynamic) viscosity μ is a measure of the internal resistance exhibited as one layer of a fluid is moved in relation to another layer. It is one of the most important properties in fluid dynamics because all real fluids have a nonzero value of viscosity. The units are kilograms per meter second [kg/(ms)] or pounds per foot second [lb/(fts)]. We can also define the kinematic viscosity ν as the ratio of the absolute viscosity to density of the fluid. Its units are meters squared per second (m²/s), feet squared per second (ft²/s) or the equivalent in other units and

$$\nu = \mu/\rho \tag{B.2}$$

The air, under normal atmospheric conditions is considered a perfect gas (where the intermolecular forces are negligible). The relation between the pressure, the density, and the temperature is then

$$p = \rho RT \tag{B.3}$$

where R is the specific gas constant. For air, R is

$$R = 287 \text{ J/kg-K} = 1716 \text{ ft-lbf/slug-°R} = 53.35 \text{ ft-lbf/lbm-°R}$$

Air can be considered a perfect gas at temperature up to 2500 K (2327°C). At temperatures higher than 2500 K, oxygen begins to dissociate.

B.3   Regions to 100,000 Ft

The atmosphere is divided into many distinctive layers. Those of interest are below 100,000 ft above the MSL because most aircraft have a normal operating ceiling of under 60,000 ft with a few exceptions such as the SR-71.

The troposphere is the lowest layer. This is where most atmospheric flights occur. It is characterized by a negative temperature gradient (or temperature lapse rate). This means that the temperature decreases as the altitude increases ($dT/dh < 0$). The height of the troposphere will vary according to the geographical location from approximately 28,000 ft at the poles to about 55,000 ft at the equator. It also varies between summer and winter.

The tropopause is the upper limit of the troposphere. At this point the temperature gradient gradually changes to zero ($dT/dh = 0$). And the temperature gradient remains zero throughout the stratosphere, which is the second layer. In this layer the water vapor and air currents are almost nonexistent. The mesosphere is the next layer. It used to be considered part of the stratosphere, but since the temperature in the mesosphere changes with altitude (which contradicts the basic definition of the stratosphere) some authors prefer to separate them in that way. The temperature increase encountered in the mesosphere is due to the presence of a layer of ozone, which absorbs more of the sun's radiation. The top of the mesosphere is the mesopause where the temperature drops again.

The jet stream is made of horizontal high-velocity winds in the altitude range of 30,000–40,000 ft (around the tropopause) that blow west to east at velocities of about 100 mph. This can obviously affect jet aircraft transit time between North America and Europe.

The two upper layers of the atmosphere are called the thermosphere and the exosphere. The ionosphere is part of the thermosphere, and it is a layer of high-free-electron density, which affects radio communication. In this layer, low to high frequency waves are reflected back to Earth.

B.4 Standard Atmosphere

Because the density, the pressure, and the temperature of the air depend on many circumstances of date, position, humidity, and so on, it is usual in aerodynamics to postulate certain standard values for these fundamental quantities. This is necessary to make the comparison between the performance of different aircraft possible.

The properties of air change with altitude. As the air gets thinner, the pressure will decrease and so will the density. The temperature will vary from layer to layer. Figure B.1 gives an idea of how the properties' ratios (defined in the next section) vary with the altitude. Note that the troposphere extends from 0 to 11 km and the stratosphere from 11 to 25 km for the standard atmosphere.

In the standard atmosphere it is assumed that the atmosphere is a perfect gas, the air is dry, the gravitational acceleration decreases with altitude (for altitudes below the stratopause the gravitational acceleration can be considered constant

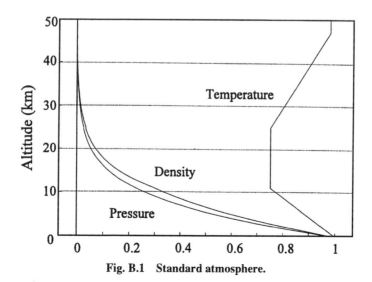

Fig. B.1 Standard atmosphere.

and equal to the sea level value, $g = g_{SL}$), and hydrostatic equilibrium exists ($dp = -\rho g dh$). The values of temperature, pressure, density, and gravitational acceleration at MSL, in the standard atmosphere, are

$$T_{SL} = 288.16 \text{ K} = 518.69°\text{R} = 15°\text{C} = 59°\text{F}$$

$$p_{SL} = 101,325 \text{ N/m}^2 = 2116.2 \text{ lb/ft}^2 = 14.7 \text{ psi}$$

$$\rho_{SL} = 1.225 \text{ kg/m}^3 = 0.002377 \text{ slug/ft}^3 = 0.0765 \text{ lbm/ft}^3$$

$$g_{SL} = 9.81 \text{ m/s}^2 = 32.17 \text{ ft/s}^2$$

The standard atmosphere concept came from the need for scientists to compare performance between aircraft and to estimate the performance of any aircraft at any altitudes. A first formula for the decrease of the temperature with height was proposed by A. Toussaint in 1920,

$$T = 15 - 0.0065h \qquad (B.4)$$

where T is in degrees Celsius and h in meters. Then, in 1925, a NACA report (TR 218) entitled "Standard Atmosphere" was released. It used metric and English units. It gave the properties of the air up to an altitude of 20 km. It was based on observations made over many years, thus giving an average value of the properties. More tables were made later on and most are in agreement for the temperature under about 30 km. We will use the ICAO standard atmosphere, which is valid for properties at about 40°N latitude. Other standard atmospheres include the ARDC 1959 model atmosphere and the U.S. standard atmosphere—1962. All three of these standard atmospheres are basically the same up to an altitude of approximately 66,000 ft.

B.5   Temperature, Pressure, and Density Altitudes

Three new types of altitudes are defined: the pressure (h_p), the temperature (h_T), and the density (h_ρ) altitudes. They are the altitudes in a standard atmosphere corresponding to a measured property of the air. To evaluate these altitudes, the variations of the properties with (true) altitude need to be known. For the troposphere, the temperature is considered to decrease linearly with altitude

$$T_h = T_{SL} + a_{LR}h \qquad (B.5)$$

where $a_{LR} = -0.0065 \text{ K/m} = -0.0065°\text{C/m} = -0.003566°\text{F/ft} \equiv$ temperature lapse rate.

The pressure is considered to decrease with altitude in the following way:

$$p_h = p_{SL}\left(\frac{T_h}{T_{SL}}\right)^{(-g/a_{LR}R)} \qquad (B.6)$$

where $g = 9.81 \text{ m/s}^2 = 32 \text{ ft/s}^2 \equiv$ gravitational acceleration and so

$$p_h = p_{SL}\left(1 + \frac{a_{LR}}{T_{SL}}h\right)^{(-g/a_{LR}R)} \qquad (B.7)$$

The density will vary both with the pressure and the temperature, and using Eq. (B.3) gives

$$\rho_h = \rho_{SL} \left(\frac{T_h}{T_{SL}} \right)^{(-1-g/a_{LR}R)} \tag{B.8}$$

It should be noted that the value of the exponent is the same in both sets of units. These values are

$$-g/(a_{LR}R) \approx 5.2621$$

and

$$-[(g/\{a_{LR}R\}) + 1] \approx 4.2621$$

A temperature ratio θ is defined as the ratio of the temperature at altitude over the temperature at MSL. The pressure ratio (δ) is the pressure at altitude over the pressure at MSL. And the density ratio (σ) is the ratio of the density at altitude over the density at MSL

$$\theta = \frac{T_h}{T_{SL}} \qquad \delta = \frac{p_h}{p_{SL}} \qquad \sigma = \frac{\rho_h}{\rho_{SL}} \tag{B.9}$$

Lift, being a function of the density, will decrease as altitude increases if the velocity and the AoA (lift coefficient) are kept constant. The same is applied for drag. Under these conditions, the lift at a certain altitude will be equal to the lift at sea level times the density ratio

$$L_h = \sigma L_{SL} \tag{B.10}$$

The property ratios can also be evaluated as a function of altitude

$$\theta = 1 - (2.2557 \times 10^{-5})h = 1 - (6.875 \times 10^{-6})h$$
$$h \rightarrow \text{meters} \qquad h \rightarrow \text{feet} \tag{B.11}$$
$$\delta = \theta^{5.2621} \qquad \sigma = \theta^{4.2621}$$

Now for the stratosphere, knowing that the temperature is constant, we get the following equations (for the first equation of both δ and σ, h is in meters and for the second, h is in feet):

$$\theta = 0.7519$$
$$\delta = 0.223 \exp[-(0.0001578)(h - 11000)]$$
$$= 0.223 \exp[-(0.00004811)(h - 36089)] \tag{B.12}$$
$$\sigma = 0.297 \exp[-(0.0001578)(h - 11000)]$$
$$= 0.297 \exp[-(0.00004811)(h - 36089)]$$

Given an actual atmospheric property, it is a simple matter to determine the corresponding temperature, pressure, or density altitude.

Table B.1 was created using these equations and is provided for quick reference to the atmospheric properties variation with altitude.

Table B.1   Standard atmosphere

Altitude, ft	Altitude, m	θ	δ	σ	Speed of sound, ft/s
0	0	1.0000	1.0000	1.0000	1,116
1,000	305	0.9931	0.9643	0.9710	1,112
2,000	610	0.9862	0.9297	0.9427	1,109
3,000	914	0.9794	0.8961	0.9150	1,105
4,000	1,219	0.9725	0.8635	0.8880	1,101
5,000	1,524	0.9656	0.8319	0.8615	1,097
6,000	1,829	0.9587	0.8012	0.8357	1,093
7,000	2,134	0.9518	0.7714	0.8105	1,089
8,000	2,438	0.9449	0.7425	0.7858	1,085
9,000	2,743	0.9381	0.7145	0.7617	1,081
10,000	3,048	0.9312	0.6874	0.7382	1,077
11,000	3,353	0.9243	0.6611	0.7153	1,073
12,000	3,658	0.9174	0.6357	0.6929	1,069
13,000	3,962	0.9105	0.6110	0.6710	1,065
14,000	4,267	0.9036	0.5871	0.6497	1,061
15,000	4,572	0.8969	0.5640	0.6288	1,057
16,000	4,877	0.8899	0.5416	0.6086	1,053
17,000	5,182	0.8831	0.5199	0.5888	1,049
18,000	5,486	0.8762	0.4990	0.5695	1,045
19,000	5,791	0.8694	0.4787	0.5507	1,041
20,000	6,096	0.8625	0.4591	0.5324	1,037
21,000	6,401	0.8556	0.4402	0.5145	1,032
22,000	6,706	0.8487	0.4219	0.4971	1,028
23,000	7,010	0.8419	0.4042	0.4802	1,024
24,000	7,315	0.8348	0.3871	0.4637	1,020
25,000	7,620	0.8279	0.3706	0.4477	1,016
26,000	7,925	0.8210	0.3547	0.4321	1,011
27,000	8,230	0.8142	0.3394	0.4168	1,007
28,000	8,534	0.8073	0.3246	0.4021	1,003
29,000	8,839	0.8004	0.3103	0.3877	999
30,000	9,144	0.7935	0.2965	0.3737	994
31,000	9,449	0.7866	0.2832	0.3601	990
32,000	9,754	0.7797	0.2705	0.3469	986
33,000	10,058	0.7729	0.2581	0.3340	981
34,000	10,363	0.7660	0.2463	0.3215	977
35,000	10.668	0.7591	0.2349	0.3094	973
36,089	11,000	0.7519	0.2229	0.2966	968
40,000	12,192	0.7519	0.1856	0.2469	968
45,000	13,716	0.7519	0.1460	0.1942	968

Table B.1 Standard atmosphere (continued)

Altitude, ft	Altitude, m	θ	δ	σ	Speed of sound, ft/s
50,000	15,240	0.7519	0.1148	0.1527	968
55,000	16,764	0.7519	0.0903	0.1201	968
60,000	18,288	0.7519	0.0710	0.0944	968
65,000	19,812	0.7519	0.0558	0.0743	968
70,000	21,336	0.7519	0.0439	0.0584	968
75,000	22,860	0.7519	0.0345	0.0459	968
80,000	24,384	0.7519	0.0271	0.0361	968

B.6 Airspeed Measurement

The airspeed is an important parameter of aircraft performance. A pilot must know the airspeed to takeoff and to land safely. There are also airspeeds for the steepest climb, for the fastest climb, for the longest flight and glide descent, etc. Aircraft designers and evaluators must know the airspeed in a wind tunnel to evaluate a model's performance.

B.6.1 Speed of Sound and Mach Number

The speed of sound is the speed at which disturbances travel in the medium they were created in. The speed of sound for air can be determined as follows (isentropic flow):

$$a = \sqrt{\gamma p / \rho} = \sqrt{\gamma R T} \qquad (B.13)$$

or

$$a = \sqrt{\gamma R \theta T_{SL}} \qquad (B.14)$$

where γ is the ratio of specific heats and is equal to 1.4 for air. The speed of sound is a reference value for atmospheric flights. For low-aircraft airspeeds with respect to the speed of sound, one can neglect the effects of compressibility on aircraft drag, lift, and thrust. As the aircraft airspeed approaches the speed of sound, compressibility effects are more apparent, and small shock waves may develop on some parts of the aircraft. When the aircraft airspeed reaches the speed of sound ($V = a$), any disturbances created by the aircraft will pile up in front of the source to form a normal shock wave. This shock wave was impossible to penetrate for the first aircraft to encounter it in the late 1940s and was coined the sound barrier. It was first broken by Charles E. Yeager on Oct. 14, 1947 while he was flying the Bell XS-1. When the airspeed exceeds the speed of sound ($V > a$), the entire aircraft flies within a cone of disturbance as shown in Fig. B.2.

The Mach number, first used in 1929 in honor of Ernst Mach (1838–1916) for his contribution to the study of supersonic flow, is the ratio of the aircraft velocity to the speed of sound,

$$M = V/a \qquad (B.15)$$

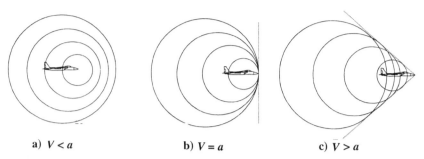

a) $V < a$ b) $V = a$ c) $\overline{V} > a$

Fig. B.2   Disturbances (circles) generated by nose of the aircraft at regular intervals.

The Mach number, in addition to the Reynolds number, is a governing parameter in aircraft aerodynamics. Usually, the flow is considered incompressible for Mach numbers lower than 0.3. For Mach numbers greater than 0.3, each case must be investigated individually to determine the effects of compressibility. The cone of the oblique shock wave (Fig. B.2c) has the following angle, with respect to its axis of symmetry:

$$\mu = \arcsin(1/M) \tag{B.16}$$

Thus, at Mach 1, the angle is 90 deg (normal shock wave) and at Mach numbers greater than 1, the angle is less than 90 deg (at Mach 2, $\mu = 30$ deg).

B.6.2 Airspeed Measurement Systems

Venturi.   A venturi consists of a convergent–divergent duct through which the air is allowed to flow (see Fig. B.3). For an incompressible flow ($M < 0.3$, $\rho =$ const), Bernoulli's equation can be used to determine the freestream airspeed

$$p_2 - p_1 = \tfrac{1}{2}\rho\left(V_1^2 - V_2^2\right) \tag{B.17}$$

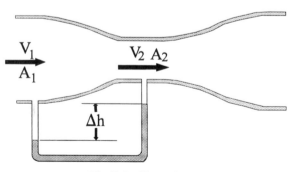

Fig. B.3   Venturi tube.

Neither V_1 nor V_2 are known, but the areas A_1 and A_2 are by design. From the continuity equation for an incompressible flow ($VA = $ const),

$$V_2 = V_1(A_1/A_2) \tag{B.18}$$

and substituting into Eq. (B.17)

$$2\frac{(p_1 - p_2)}{\rho} = V_1^2\left(\frac{A_1^2}{A_2^2} - 1\right) \tag{B.19}$$

which finally leads to

$$V_1 = \sqrt{\left[2(p_1 - p_2)\Big/\rho\left(\frac{A_1^2}{A_2^2} - 1\right)\right]} \tag{B.20}$$

The value of $(p_1 - p_2)$ is obtained from a differential pressure measuring instrument such as a U-tube manometer (Fig. B.3) commonly used with small wind tunnels. In this case the differential pressure is equal to the weight of the column of liquid of height Δh. If the density of that liquid is ρ_ℓ, then the differential pressure is

$$(p_1 - p_2) = \rho_\ell g \Delta h \tag{B.21}$$

A liquid-filled manometer cannot be used in an aircraft and is replaced by a differential pressure capsule. A venturi has some disadvantages: as the density at the throat (ρ_2) decreases with increasing velocity V_2, the danger of icing increases; this will reduce the throat area (A_2) and the tube loses its calibration. The tube also loses its calibration if the Mach number at the throat is greater than 0.3 because the incompressible flow hypothesis is no longer valid. A venturi is very sensitive to off-angle flow (position errors) and has a relatively high-drag coefficient. It was used on older design, low-airspeed aircraft.

Pitot tube.   The instrument that is now widely used on aircraft to measure the total pressure is the pitot tube (Fig. B.4), named after Henri Pitot who first used it to measure the flow velocity of the Seine River in Paris in 1732. Unlike the venturi, the velocity of the fluid inside the pitot tube is zero. The fluid is slowed down isentropically from the freestream velocity to zero at the mouth of the instrument. By assuming incompressible flow and using Bernoulli's equation, the freestream velocity can be found

$$p_2 - p_1 = \tfrac{1}{2}\rho V_1^2 \tag{B.22}$$

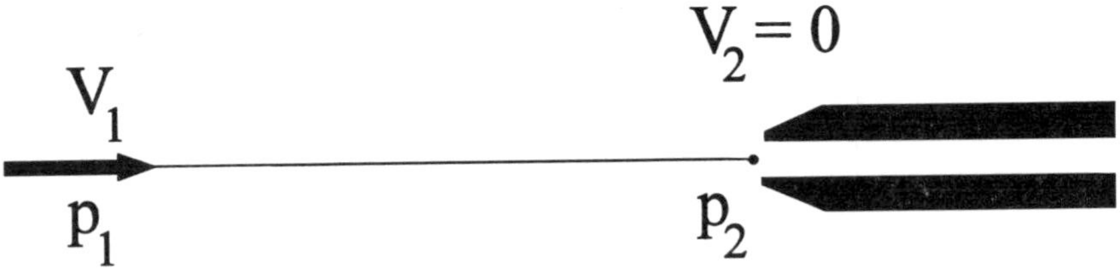

Fig. B.4   Pitot tube.

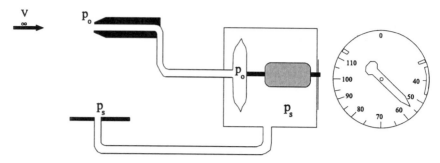

Fig. B.5 Pitot tube for total pressure and static pressure port.

The value of $(p_2 - p_1)$ is found with the help of a differential capsule. The movement of the capsule wall is translated into a velocity with the help of a mechanical device (Fig. B.5). This device can be calibrated so as to assume that the airflow is incompressible ($\frac{1}{2}\rho V^2$), the velocity is then

$$V = \sqrt{2(p_0 - p_s)/\rho} \qquad (B.23)$$

If the flow can not be considered incompressible ($M > 0.3$), but it is still slowed down isentropically (no shock wave), then Euler's equation must be used to account for the varying air density. The equation must be integrated from a point far upstream to the mouth of the pitot tube,

$$\int_1^2 \frac{dp}{\rho} = -\int_1^2 V\, dV \qquad (B.24)$$

At point 2 the velocity is zero and the pressure is equal to the total pressure, whereas at point 1 the pressure is equal to the static pressure and the velocity is the freestream velocity. Because the flow is isentropic the following equation holds:

$$p/\rho^\gamma = \text{const} \qquad (B.25)$$

Integrating Eq. (B.24) leads to

$$\frac{\gamma}{\gamma-1}\left[\frac{p_2}{\rho_2} - \frac{p_1}{\rho_1}\right] = \frac{V_1^2}{2}$$

$$\frac{\gamma}{\gamma-1}\left[\rho_1\frac{p_2}{\rho_2} - p_1\right] = \rho_1\frac{V_1^2}{2} \qquad (B.26)$$

$$\frac{1}{\gamma-1}\left[\left(\frac{p_2}{p_1}\right)^{\frac{\gamma-1}{\gamma}} - 1\right] = \frac{\rho_1}{\gamma p_1}\frac{V_1^2}{2} = \frac{1}{2}\frac{V_1^2}{a_1^2} = \frac{1}{2}M_1^2$$

Finally, the velocity of the airstream is (replacing p_2 by p_0 and the values at point 1 by their freestream values)

$$V_\infty = \sqrt{\frac{2a_\infty^2}{\gamma - 1}\left[\left(\frac{p_0}{p_s}\right)^{\frac{\gamma-1}{\gamma}} - 1\right]} \tag{B.27}$$

or the impact pressure

$$q_c = p_s\left[\left(1 + \frac{(\gamma - 1)}{2}\frac{\rho_\infty}{\gamma p_s}V_\infty^2\right)^{\frac{\gamma}{\gamma-1}} - 1\right]$$

$$q_c = p_s\left[\left(1 + \frac{(\gamma - 1)}{2}M_\infty^2\right)^{\frac{\gamma}{\gamma-1}} - 1\right] \tag{B.28}$$

Impact pressure is somewhat different from the dynamic pressure. Dynamic pressure q represent the grouping $\frac{1}{2}\rho V^2$ for any flow from low subsonic to hypersonic. Because of its simplicity, it is often used to replace the impact pressure in the calculation of velocity at low speed (flow assumed incompressible). By comparing Bernoulli's equation with the equation for total pressure we can see the similarity of both equations,

$$p + \frac{1}{2}\rho V^2 = \text{const}$$

$$p + q_c = p_0 \tag{B.29}$$

But it must be remembered that the first equation (Bernoulli's) is valid for incompressible flows only while the impact pressure equation is valid for all isentropic flows. Figure B.6 shows that the difference between the impact pressure and the dynamic pressure is less than 2.3% for Mach numbers smaller than 0.3.

When the flow becomes supersonic ($M > 1$) a shock wave forms in front of the pitot's mouth (Fig. B.7). The flow across a shock wave is not isentropic but the flow in front and behind it are. To solve this problem, the flow is separated into three regions: the flow in front of the shock wave (undisturbed by the pitot tube), the flow through the shock wave in front of the pitot tube (normal shock wave immediately in front of the pitot's mouth), and the isentropic deceleration of the flow between the shock wave and the pitot's mouth. The ratio of the total pressure measured by the pitot tube (p_{0_2}) to the atmospheric static pressure (p_1 or p_s) is then

$$\frac{p_{0_2}}{p_1} = \left[\frac{\gamma + 1}{2}M_\infty^2\right]^{\frac{\gamma}{\gamma-1}}\left[\frac{\gamma + 1}{2\gamma M_\infty^2 - \gamma + 1}\right]^{\frac{1}{\gamma-1}} \tag{B.30}$$

To get the correct value of impact pressure, the pitot tube and static pressure port must be located on the aircraft where the local flow conditions are closest to the freestream values and for all attitudes the aircraft can be called upon to fly. The pitot tube must be clear of the BL and of separated and/or vortex flows. Figure B.8 shows such a location on a CF-18 forward fuselage. At very low speed, the fuselage

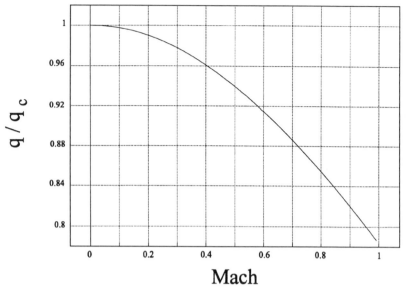

Fig. B.6   Ratio of the dynamic pressure to the impact pressure.

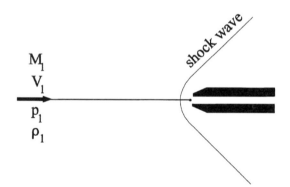

Fig. B.7   Pitot tube with shock wave.

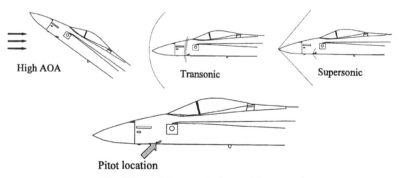

Fig. B.8   Airflow variation at Pitot location.

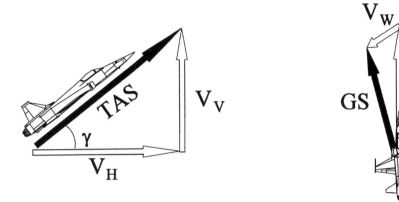

Fig. B.9   Comparison of TAS and GS.

can be angled at AoAs as high as 55–60 deg. As the speed increases, the AoA will go down. The flow will go from low subsonic (assumed incompressible) to high subsonic (compressible). In the transonic regime, local supersonic flow arises and shock waves will travel along the fuselage, which may affect the reading of the local total pressure. In the purely supersonic regime, a shock wave will form in front of the pitot tube.

The static pressure port must be located so as to be always perpendicular to the local flow. This way it will not register impact pressure and so when the measured static pressure is combined with the reading of total pressure from the pitot tube, the result will be a better estimate of the impact pressure.

B.6.3 Airspeed Types

The speed of the aircraft relative to the air is called the true airspeed (TAS) and the speed of the aircraft relative to the ground is the ground speed (GS). The GS is a two-dimensional value that is used for navigational purposes. It differs from the TAS, which is a three-dimensional value (i.e., along the flight path of the aircraft), depending on the climb (or descent) angle (γ, not to be confused with the ratio of specific heats), as well as on the wind speed and direction (wind velocity V_W) (Fig. B.9).

Wind velocity can help increase the range of the aircraft if it is blowing from the rear quadrants, with the maximum effects for tail winds aligned with the horizontal speed vector. On the other hand, it can also sharply decrease the range if the wind comes from any of the two front quadrants (further discussion about Range in Chapter 3). The GS will equal the TAS for the level flight zero wind condition only.

The airspeed indicator ASI in the cockpit does not, usually, indicate the TAS. The airspeed on the ASI is the IAS. To get the TAS one must first correct for position errors (dependent on the flight conditions and the location on the aircraft of the pitot tube and the static pressure port) and for instrument error. Once these corrections are made, one gets the calibrated airspeed (CAS),

$$CAS = IAS + \Delta V_{pos} + \Delta V_{inst} \tag{B.31}$$

Now, there are many types of gear trains (or translators) that can be used in an ASI to convert pressure measurement into CAS. These can be designed for low speed, high speed, or air data computer for all speeds. For the low-speed ASI, the flow is considered incompressible, then Eq. (B.23) is the TAS as long as the density of the air is known. But for a low-speed ASI the translator is usually of simple design, just a few gears, and the density is replaced by the sea level standard atmosphere value. Thus,

$$CAS = \sqrt{2(p_0 - p_s)/\rho_{SL}} \tag{B.32}$$

If a low-speed ASI is used at high speeds ($M > 0.3$) the incompressible flow assumption is no longer valid and some corrections must be made to take into account the effects of compressibility; thus, one gets an EAS,

$$EAS = CAS + \Delta V_c \tag{B.33}$$

The value of ΔV_c will always be smaller than or equal to zero because the impact pressure is underestimated by using Eq. (B.22) but under $M = 0.3$ the error is less than 2.3% (see Figs. B.6 and B.10). For Mach numbers below 0.3 the EAS can be said to be equal to the CAS. The TAS is finally obtained by correcting for the altitude (the ASI uses the sea level density value),

$$TAS = EAS\sqrt{\frac{\rho_{SL}}{\rho}} = \frac{EAS}{\sqrt{\sigma}} \tag{B.34}$$

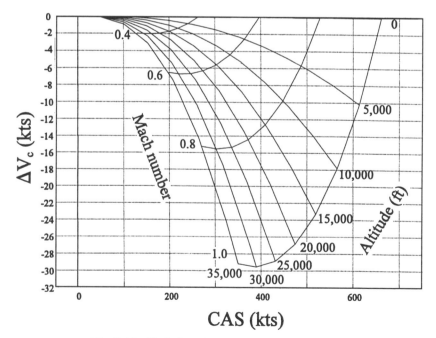

Fig. B.10   Variation of ΔV_c with CAS and altitude.

When a high-speed ASI is used to calculate the velocity of the aircraft, then the approach is slightly different. For a compressible flow, the TAS is equal to Eq. (B.27), or using a differential pressure capsule (difference between the total pressure measured by the pitot tube and the static pressure measured by the static pressure port), one gets

$$\text{TAS} = \sqrt{\left(\frac{2\gamma}{\gamma - 1}\right)\left(\frac{p}{\rho}\right)\left[\left(1 + \frac{p_0 - p}{p}\right)^{\frac{(\gamma-1)}{\gamma}} - 1\right]} \qquad (B.35)$$

Now, for the conversion of the TAS to the EAS, Eq. (B.34) is still valid. Thus, the following equation for EAS is obtained (note the sea level value of the air density within the equation):

$$\text{EAS} = \sqrt{\left(\frac{2\gamma}{\gamma - 1}\right)\left(\frac{p}{\rho_{SL}}\right)\left[\left(1 + \frac{p_0 - p}{p}\right)^{\frac{(\gamma-1)}{\gamma}} - 1\right]} \qquad (B.36)$$

Again here, it is assumed that the values of the pressure and air density are known to calculate the TAS. To get the CAS, these values must be replaced in the preceding TAS equation by their sea level standard values,

$$\text{CAS} = \sqrt{\left(\frac{2\gamma}{\gamma - 1}\right)\left(\frac{p_{SL}}{\rho_{SL}}\right)\left[\left(1 + \frac{p_0 - p}{p_{SL}}\right)^{\frac{(\gamma-1)}{\gamma}} - 1\right]} \qquad (B.37)$$

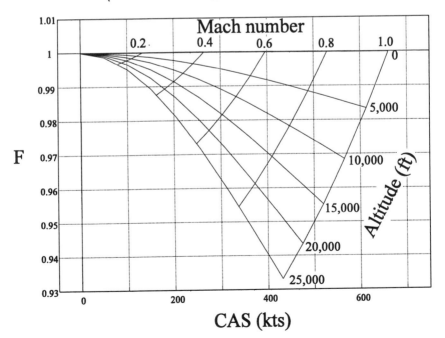

Fig. B.11 Conversion factor F at different CAS and altitude values.

Note here that the differential pressure uses the actual total and static pressure values because they are readily available from the pitot tube and static pressure port. To convert the CAS to the EAS, a conversion factor F is used,

$$EAS = F\,CAS \tag{B.38}$$

where

$$F =$$

$$\sqrt{\left\{\left(\frac{p}{\rho_{SL}}\right)\left[\left(1 + \frac{p_0 - p}{p}\right)^{\frac{(\gamma-1)}{\gamma}} - 1\right] \Big/ \left(\frac{p_{SL}}{\rho_{SL}}\right)\left[\left(1 + \frac{p_0 - p}{p_{SL}}\right)^{\frac{(\gamma-1)}{\gamma}} - 1\right]\right\}} \tag{B.39}$$

This factor is a function of pressure, thus altitude, and of velocity (through the pressure differential, i.e., the impact pressure). Figure B.11 is obtained when F is plotted against the CAS.

Problems

B.1.   Produce a small computer program to generate the standard atmosphere values of temperature, pressure, and density ratios, as well as the speed of sound for altitudes varying from 0 to 45,000 ft, printing the results every 5,000 ft. Present the results in a table format.

B.2.   An aircraft is flying at 27,000 ft in a standard atmosphere. Its TAS is 325 kn. What is the EAS and CAS? What is the aircraft Mach number? If the velocity correction for position (ΔV_{pos}) is +3.0 kn and the correction for instrument error (ΔV_{inst}) is −1.0 kn at that speed, what is the IAS of the aircraft?

B.3.   For a standard atmosphere altitude of 12,300 ft and a Mach number of 0.55, calculate the following values: θ, TAS, EAS, F, CAS, ΔV_c, q_c, and p_∞ (static pressure at that altitude).

B.4.   An altimeter is calibrated to the standard atmosphere value and is measuring a pressure altitude of 7450 ft. The outside temperature is 27°F (−3°C). The aircraft is using a low-speed ASI, which is reading 135-kn IAS ($\Delta V_{pos} = −5$ kn, $\Delta V_{inst} = 0$ kn). What is the value of the air density? Find the following values: Mach number, TAS, EAS, CAS, ΔV_c, temperature altitude h_T, and density altitude h_ρ.

Appendix C: Aircraft Performance Summary

Table C1 summarizes the various design parameters as they affect various aspects of an aircraft's performance.

Table C.1   Performance summary

	Jet aircraft	Prop aircraft
Velocity	High T/W	High P_a/W
	High L/D	High L/D
	High T/S	Low C_{D_0}
	High altitude	Low altitude (unsupercharged)
	Low C_{D_0}	h_{crit} (supercharged)
Maximum range	High wing loading (W/S)	High L/D
	High L/D	High η_p
	High altitude	Low PSFC
	High fuel weight fraction	High fuel weight fraction
	Low TSFC	
Maximum endurance	High L/D	High L/D
	Low TSFC	High η_p
	High fuel weight fraction	Low PSFC
		High fuel weight fraction
		Low W/S
		Low altitude (unsupercharged)
		h_{crit} (supercharged)
Steepest climb	High T/W	High P_a/W
	High L/D	High L/D
		Low velocity
Fastest climb	High T/W	High P_a/W
	High thrust loading (T/S)	High L/D
	High L/D	
Fastest turn	High T/W	High P_a/W
	High L/D	High L/D
	Low W/S	Low W/S
	Low altitude	Low altitude
Tightest turn	High T/W	High P_a/W
	Low W/S	Low W/S
	Low altitude	Low altitude
	High L/D	High L/D
Longest glide	High L/D	High L/D

Table C.1 Performance summary (continued)

	Jet aircraft	Prop aircraft
Minimum sinking rate	Low W/S	Low W/S
	Low altitude	Low altitude
	High $C_L^{(3/2)}/C_D$	High $C_L^{(3/2)}/C_D$
Minimum stall velocity	Low altitude	Low altitude
	High $C_{L_{max}}$	High $C_{L_{max}}$
	Low W/S	Low W/S
Minimum takeoff distance	Low W/S	Low W/S
	Low altitude	Low altitude
	High T/W	High P_a/W
	High $C_{L,TO}$	High $C_{L,TO}$
Minimum landing distance	Low W/S	Low W/S
	Low altitude	Low altitude
	High $C_{L,L}$	High $C_{L,L}$
	High μ	High μ

Appendix D: Development of Equations

D.1 Range Equations: Constant Velocity and Altitude

From Eq. (3.26),

$$X = \frac{-1}{SFC_T} \int_1^2 \frac{V}{D} dW \tag{D.1}$$

Because the velocity and altitude are constant, Eq. (D.1) becomes

$$[X]_{V,h} = \frac{-V}{SFC_T} \int_{W_1}^{W_2} \frac{1}{D} dW = \frac{-V}{SFC_T} \int_{W_1}^{W_2} \frac{E}{W} dW \tag{D.2}$$

The aerodynamic efficiency can be written as

$$E = \frac{C_L}{C_D} = \frac{C_L}{C_{D_0} + KC_L^2} = \frac{2W\rho V^2 S}{C_{D_0}(\rho V^2 S)^2 + 4KW^2} \tag{D.3}$$

Therefore, Eq. (D.2) becomes

$$[X]_{V,h} = \frac{-V}{SFC_T} \left[2\rho V^2 S \int_{W_1}^{W_2} \frac{dW}{(C_{D_0}\rho V^2 S^2 + 4KW^2)} \right] \tag{D.4}$$

Let

$$a = C_{D_0}(\rho V^2 S)^2$$
$$b = 4K$$

and so

$$\sqrt{ab} = 2\sqrt{C_{D_0}K}\,\rho V^2 S$$

$$\sqrt{\frac{b}{a}} = \frac{2}{\rho V^2 S}\sqrt{\frac{K}{C_{D_0}}}$$

Inserting into Eq. (D.4) gives

$$[X]_{V,h} = \frac{-V}{SFC_T} \left[2\rho V^2 S \int_{W_1}^{W_2} \frac{dW}{a + bW^2} \right] \tag{D.5}$$

$$[X]_{V,h} = \frac{-V}{SFC_T} \left[2\rho V^2 S \left\{ \frac{1}{\sqrt{ab}} \arctan\left(\sqrt{\frac{b}{a}}W\right)_{W_1}^{W_2} \right\} \right] \tag{D.6}$$

$$[X]_{V,h} = \frac{-V}{SFC_T}\left[2E_m\left\{\arctan\left(\frac{2}{\rho V^2 S}\sqrt{\frac{K}{C_{D_0}}}W\right)\right\}_{W_1}^{W_2}\right] \qquad (D.7)$$

Finally,

$$[X]_{V,h} = \frac{2E_m V}{SFC_T}\left[\arctan\left(\frac{2}{\rho V^2 S}\sqrt{\frac{K}{C_{D_0}}}W\right)\right]_{W_2}^{W_1} \qquad (D.8)$$

Note that the minus sign was dropped when the integration limits were inverted. Knowing that

$$C_{L_{Em}} = \sqrt{C_{D_0}/K} \qquad \text{and} \qquad C_L = 2W/\rho V^2 S$$

Eq. (D.8) can be written as follows:

$$[X]_{V,h} = \frac{2E_m V}{SFC_T}\left[\arctan\left(\frac{C_L}{C_{L_{Em}}}\right)\right]_{C_{L_2}}^{C_{L_1}} \qquad (D.9)$$

It is more practical to write the range equation in term of fuel–weight fraction ζ, and so further reduction is required. Let

$$A = \arctan\left(\frac{C_{L_1}}{C_{L_{Em}}}\right)$$

$$B = \arctan\left(\frac{C_{L_2}}{C_{L_{Em}}}\right) = \arctan\left(\frac{C_{L_1}(1-\zeta)}{C_{L_{Em}}}\right)$$

Using the trigonometric solution to

$$\tan(A - B) = \frac{\tan(A) - \tan(B)}{1 + \tan(A)\tan(B)}$$

$$\tan(A - B) = \frac{\zeta C_{L_1} C_{L_{Em}}}{C_{L_{Em}}^2 + C_{L_1}^2(1-\zeta)} = \frac{\zeta C_{L_1}\sqrt{C_{D_0}/K}}{(C_{D_0}/K) + C_{L_1}^2(1-\zeta)} \qquad (D.10)$$

Therefore,

$$\tan(A - B) = \frac{K\zeta C_{L_1}\sqrt{C_{D_0}/K}}{(C_{D_0} + KC_{L_1}^2) - K\zeta C_{L_1}^2} = \frac{\zeta E_1}{2E_m(1 - KC_{L_1}E_1\zeta)} \qquad (D.11)$$

which, once returned to Eq. (D.9), gives

$$[X]_{V,h} = \frac{2E_m V}{SFC_T} \arctan[\tan(A - B)] = \frac{2E_m V}{SFC_T} \arctan\left[\frac{\zeta E_1}{2E_m(1 - KC_{L_1}E_1\zeta)}\right]$$
(D.12)

This is Eq. (3.28).

D.2 Range Equations: Constant Altitude and Lift Coefficient

For constant altitude and lift coefficient, Eq. (D.1) becomes

$$[X]_{h,C_L} = \frac{-E}{SFC_T} \int_{W_1}^{W_2} \frac{V}{W} dW$$
(D.13)

Because the lift is equal to the weight (steady-state level flight),

$$W = \tfrac{1}{2}\rho V^2 S C_L$$

$$dW = \tfrac{1}{2}\rho S C_L (2V\,dV)$$
(D.14)

Equation (D.13) reduces to

$$[X]_{h,C_L} = \frac{-2E}{SFC_T} \int_{V_1}^{V_2} dV = \frac{-2E}{SFC_T}(V_2 - V_1)$$
(D.15)

Equation (3.35) states that

$$V_2 = V_1\sqrt{1 - \zeta}$$

and so the range equation for constant altitude and velocity is

$$[X]_{h,C_L} = \frac{2E}{SFC_T} V_1(1 - \sqrt{1 - \zeta})$$
(D.16)

which is Eq. (3.32).

D.3 Fastest Turn Velocity

To determine the fastest turn velocity, the following equation was solved:

$$n^2 - 1 - n\rho V^2 \frac{\partial n}{\partial q} = 0$$
(D.17)

since

$$q = \tfrac{1}{2}\rho V^2$$
(D.18)

Eq. (D.17) can be rewritten as

$$n^2 - 1 - 2nq\frac{\partial n}{\partial q} = 0 \qquad \text{(D.19)}$$

where n was defined in Eq. (5.17), rewritten here,

$$n = \frac{q}{(W/S)}\sqrt{\frac{1}{K}\left[\frac{(T/S)}{q} - C_{D_0}\right]} \qquad \text{(D.20)}$$

The derivative of n with respect to q is

$$\frac{\partial n}{\partial q} = \frac{\partial}{\partial q}\left\{\frac{q}{(W/S)}\sqrt{\frac{1}{K}\left[\frac{(T/S)}{q} - C_{D_0}\right]}\right\} \qquad \text{(D.21)}$$

$$\therefore \frac{\partial n}{\partial q} = \frac{1}{(W/S)\sqrt{K}}\left\{\sqrt{\frac{(T/S)}{q} - C_{D_0}} - \frac{(T/S)}{2q}\left[1\bigg/\sqrt{\frac{(T/S)}{q} - C_{D_0}}\right]\right\}$$

Equation (D.19) can now be solved,

$$\left\{\frac{q^2}{(W/S)^2 K}\left[\frac{(T/S)}{q} - C_{D_0}\right]\right\} - 1 - 2\left\{\frac{q}{(W/S)\sqrt{K}}\sqrt{\frac{(T/S)}{q} - C_{D_0}}\right\}q\frac{\partial n}{\partial q} = 0$$

$$\frac{q^2}{(W/S)^2 K}\left[\frac{(T/S)}{q} - C_{D_0}\right] - 1 - \frac{2q^2}{(W/S)^2 K}\left[\left(\frac{(T/S)}{q} - C_{D_0}\right) - \frac{(T/S)}{2q}\right] = 0$$

$$\left[C_{D_0}q^2/(W/S)^2 K\right] - 1 = 0$$

$$q^2 = (W/S)^2(K/C_{D_0})$$

Therefore, the fastest turn dynamic pressure is

$$q_{\text{FT}} = (W/S)\sqrt{K/C_{D_0}} \qquad \text{(D.22)}$$

which is Eq. (5.19). The velocity is obtained by combining Eqs. (D.22) and (D.18) to give Eq. (5.20).

D.4 Tightest Turn Velocity

The tightest turn velocity is obtained in much the same way as the fastest turn velocity. Starting with Eq. (5.23), rewritten here,

$$\left(n^2 - 1 - qn\frac{\partial n}{\partial q}\right)_{\text{TT}} = 0 \qquad \text{(D.23)}$$

and using Eqs. (D.20) and (D.21) gives

$$\left\{ \frac{q^2}{(W/S)^2 K}\left[\frac{(T/S)}{q} - C_{D_0} \right] \right\} - 1 - \left\{ \frac{q}{(W/S)\sqrt{K}}\sqrt{\frac{(T/S)}{q} - C_{D_0}} \right\} q\frac{\partial n}{\partial q} = 0$$

$$\frac{q^2}{(W/S)^2 K}\left[\frac{(T/S)}{q} - C_{D_0} \right] - 1 - \frac{q^2}{(W/S)^2 K}\left[\left(\frac{(T/S)}{q} - C_{D_0} \right) - \frac{(T/S)}{2q} \right] = 0$$

Therefore,

$$q_{TT} = \frac{2(W/S)K}{(T/W)} \tag{D.24}$$

which is Eq. (5.24). The velocity is obtained by combining Eqs. (D.24) and (D.18) to give Eq. (5.25).

D.5 Approximations for the Takeoff Length (Jet-Powered Aircraft)

Several approximations can be used to solve the takeoff distance Eq. (7.7). The most simple approximation is to assume that there is no friction force and no drag during the ground roll part of the takeoff segment. This first approximation resulted in the second line of Eq. (7.7).

A second approximation is to neglect the drag and lift forces during the ground run (this again neglects the effects of rotation on the sharp increase in drag as seen in Fig. 7.5), but it does account for part of the dissipative forces by assuming that they are equal to the rolling friction force. Looking at Fig. 7.5, it can be seen that the total dissipative force will remain approximately constant up to the point of rotation. This approximation does account for a reduction in excess thrust, thus a reduction in acceleration on the ground. It will yield a somewhat longer ground run distance. Equation (7.4) reduces to

$$dX = \frac{WV\,dV}{g(T - \mu_r W)} \tag{D.25}$$

which can be solved to give

$$X_{GR} = V_{LO}^2 \left/ 2g\left(\frac{T}{W} - \mu_r \right) \right. \tag{D.26}$$

Now, if the only approximation made for the takeoff ground run is that the lift and drag coefficient remain constant (remembering that the thrust is assumed constant for all basic aircraft performance analysis), thus neglecting the effects of rotation on the takeoff acceleration, Eq. (7.10) can be solved in the following way:

$$dX = \frac{WV\,dV}{g\left[(T - \mu_r W) - \frac{1}{2}\rho_{SL}\sigma S(C_{D_T} - \mu_r C_{L_T})V^2 \right]} \tag{D.27}$$

where C_{D_T} and C_{L_T} are the drag and lift coefficient during the takeoff ground run.
Let

$$c = T - \mu_r W$$
$$e = \tfrac{1}{2}\rho_{SL}\sigma S(C_{D_T} - \mu_r C_{L_T}) \tag{D.28}$$

then

$$X_{GR} = \frac{W}{g}\int_0^{V_{LO}} \frac{V\,dV}{(c - eV^2)} \tag{D.29}$$

Now, let

$$f^2 = c/e \tag{D.30}$$

Equation (D.29) may then be rewritten and solved,

$$X_{GR} = \frac{W}{ge}\int_0^{V_{LO}} \frac{V\,dV}{f^2 - V^2} = \frac{W}{ge}\left[-\frac{1}{2}\ell_n(f^2 - V^2)\right]_0^{V_{LO}}$$

$$X_{GR} = \frac{W}{2ge}\left[\ell_n(f^2) - \ell_n\left(f^2 - V_{LO}^2\right)\right] \tag{D.31}$$

Substituting the values of e and f back into Eq. (D.31) gives the following takeoff
ground run distance equation:

$$X_{GR} = \frac{W}{g\rho_{SL}\sigma S(C_{D_T} - \mu_r C_{L_T})}\,\ell_n\left(\frac{T - \mu_r W}{T - \mu_r W - \tfrac{1}{2}\rho_{SL}\sigma S(C_{D_T} - \mu_r C_{L_T})V_{LO}^2}\right) \tag{D.32}$$

This equation can be simplified by introducing a new variable (Ω), which is equal
to

$$\Omega = \sqrt{\frac{W}{\tfrac{1}{2}\rho_{SL}\sigma S(C_{D_T} - \mu_r C_{L_T})}} \tag{D.33}$$

Equation (D.32) then takes the form

$$X_{GR} = \frac{\Omega^2}{2g}\,\ell_n\left(\frac{(T/W) - \mu_r}{(T/W) - \mu_r - (V_{LO}/\Omega)^2}\right) \tag{D.34}$$

where Ω has units of velocity. Writing V_{LO} in full gives the final form for the
takeoff ground run distance,

$$X_{GR} = \frac{\Omega^2}{2g}\,\ell_n\left\{\left[\left(\frac{T}{W}\right) - \mu_r\right]\Big/\left[\left(\frac{T}{W}\right) - \mu_r - \left(\frac{(1.2)^2(C_{D_T} - \mu_r C_{L_T})}{C_{L_{maxTO}}}\right)\right]\right\} \tag{D.35}$$

$$X_{GR} = \frac{\Omega^2}{2g}\,\ell_n\left\{\left[\left(\frac{T}{W}\right) - \mu_r\right]\Big/\left[\left(\frac{T}{W}\right) - \mu_r - \left(\frac{C_{D_T} - \mu_r C_{L_T}}{C_{L_{LO}}}\right)\right]\right\}$$

Appendix E: Aircraft Comparisons

Table E.1   Transport aircraft

	Airbus		Boeing	
	A320-200	A340-300	B737-400	B747-400
b, ft	111.25	197.83	94.75	211
S, ft^2	1,317.5	3,908.4	1,135	5,500
AR	9.4	10.0	7.9	8.1
W_{OE}, lb	91,073	277,917	73,710	398,700
W_{PLmax}, lb	42,307	105,687	35,000[b]	144,000
$W_{fuel,max}$, lb	42,238	231,811	34,520	386,674
$W_{TO,max}$, lb	162,040	558,870	138,500	800,000
$W_{Landing,max}$, lb	142,195	410,060	121,000	574,000
BFL, ft	7,677	9,500[b]	7,600	11,400
Landing, ft	4,823	6,500[b]	5,650	7,000
Range,[a] n mile	2,870	6,750	2,700	7,340
T, lb	2 × 25,000	4 × 31,200	2 × 23,500	4 × 56,750
Bypass	6	6.6	5.0	4.85
TSFC, lb/h/lb	0.596	0.567	0.661	0.537
Passengers	150	295	146	400

[a]With number of passenger listed plus the associated luggage and international fuel reserve.
[b]Estimated value; others are from Jane's *All The World's Aircraft*.

Table E.2   Fighter aircraft

	CF-18A	F-15C	F-16C	MiG-29	Su-27
W_{combat},[a] lb	32,000	40,000	23,000	30,000	42,000
$T/W_{SL,max}$	1.0	1.2	1.26	1.22	1.31
W/S	80.0	65.8	76.7	79.2	65.0
C_{D_0}[b]	0.025	0.023	0.018	0.023	0.025
AR	3.5	3.0	3.0	3.67	3.6
K[b]	0.1137	0.1326	0.1326	0.1084	0.1105
n_{max}	7.5	9.0	9.0	9.5	9.0
C_{Lmax}	1.9	1.75	1.65	1.75	1.8
α_T at C_{Lmax},[b] deg	35	30	25	30	30

[a]Combat weight is aircraft loaded with approximately 50% internal fuel and air-to-air missiles only.
[b]Estimated values.

Table E.3   Regional aircraft (turboprop and turbofan)[a]

| Aircraft | | Engine | | | Weight, kg | | Maximum | 400-n mile trip | | | |
| Type | No. seats | No. | Propulsion[c] | W_{OE} | W_{TO}, maximum | W_f, maximum | range,[b] n mile | Minimum time | | Minimum fuel | |
								Fuel, kg	Time, h	Fuel, kg	Time, h
ATR 42-320	48	2	2,100 SHP	10,402	16700	4500	535	898	1.4	789	1.51
ATR 72-210	66	2	2,750 SHP	12,582	21500	5000	629	1220	1.36	991	1.5
Avro RJ70	70	4	6,130 lb	23,922	38102	9367	1114	2451	1.09	2334	1.15
Avro RJ85	85	4	7,000 lb	24,501	42184	9367	1367	2578	1.09	2474	1.15
Avro RJ100	100	4	7,000 lb	25,477	44225	9367	1275	2680	1.1	2609	1.14
Avro RJ115	116	4	7,000 lb	25,777	46040	10305	1283	2680	1.1	2607	1.14
Canadair RJ	50	2	8,729 lb	13,913	23133	6489	1231	1337	1.03	1176	1.1
Dash 8-100B	37	2	2,150 SHP	10,271	16466	2576	560	968	1.48	808	2.08
Dash 8-300B	50	2	2,500 SHP	11,657	19505	2576	550	1158	1.42	919	1.57
Do 328-100	30	2	2,180 SHP	8,945	13640	3410	645	821	1.23	719	1.36
EMB-120 Brasilia	30	2	1,800 SHP	7,481	11500	2656	550	686	1.33	648	1.4
Jetstream J41	29	2	1,650 SHP	6,562	10886	2639	627	712	1.34	652	1.43
Jetstream J61	68	2	2,050 SHP	14,578	23680	5091	660	1148	1.41	1054	1.54
Saab 340B	35	2	1,870 SHP	8,329	13155	2576	606	721	1.35	656	1.45
Saab 2000	50	2	4,152 SHP	13,614	22000	4148	940	1236	1.15	1061	1.26

[a]Source: Air International, February 1995.
[b]Maximum range to ERA91/01 standard fuel for 45 min hold plus 100 n mile to alternate.
[c]Turboprop aircraft propulsion is rated in terms of SHP and jets are rated in terms of pounds of thrust.

References

[1]Poisson-Quinton, Ph., "Energy Conservation Aircraft Design and Operational Procedures," AGARD-LS-96, Paper 9, Oct. 1978.

[2]Poisson-Quinton, Ph., "Parasitic and Interference Drag Prediction and Reduction," AGARD-R-723, Paper 6, Aug. 1985.

[3]Dommash, D. O., Sherby, S. S., and Connolly, T. F., *Airplane Aerodynamics,* 4th Ed., Pitman Publishing, Toronto, 1967.

[4]Johnson, C. L., "Some Development Aspects of the YF-12A Interceptor Aircraft," AIAA Paper 69-757, 1969.

[5]Vidal, G., and Deschamps, J., "Étude de l'effet de sol au CEAT, exploitation des résultats," AGARD-CP-465, Paper 20, 1989.

[6]Paulson, J. P., Jr., Kemmerly, G. T., and Gilbert, W. P., "Dynamic Ground Effects," AGARD-CP-465, Paper 21, 1989.

[7]Flaig, A., and Hilbig, R., "High-Lift Design for Large Civil Aircraft," AGARD-CP-515, Paper 31, 1993.

[8]Wedekind, G., and Mangold, P., "Aerodynamic Interferences of In-Flight Thrust Reversers in Ground Effect," AGARD-CP-465, Paper 19, 1989.

[9]Landry, R. F., "G-Induced Loss of Consciousness (GLC)," AGARD-CP-377, Paper B2, March 1985.

[10]Organ, R., Page, R., Watson, D., and Wilkinson, L., *Avro Arrow,* Boston Mills Press, Boston, MA, 1980.

[11]Raymer, D. P., *Aircraft Design: A Conceptual Approach,* AIAA Education Series, AIAA, Washington, DC, 1989.

[12]Roskam, J., *Airplane Design—Part I to VIII,* Roskam Aviation and Engineering, Ottawa, KS, 1991.

[13]Hefner, J. N., and Bushnell, D. M., "An Overview of Concepts for Aircraft Drag Reduction," AGARD-R-654, Paper 1, April 1977.

[14]Bradley, R. G., "Practical Aerodynamic Problems—Military Aircraft," *Transonic Aerodynamics,* Vol. 81, Progress in Astronautics and Aeronautics, edited by David Nixon, AIAA, New York, 1982, Chap. 3.

[15]Reinmann, J. J., "Icing: Accretion, Detection, Protection," AGARD-LS-197, Paper 4, Nov. 1994.

[16]Asselin, M., "Analyse du Mouvement des Gouttelettes d'eau et Formation de la Glace en Aéronautique," Mémoire de maîtrise Es Sciences Appliquées, École Polytechnique de Montréal, Montreal, Quebec, Canada, June 1990 (in French).

[17]NACA TM-83556, Jan. 1984.

[18]Hoblit, F. M., "Gust Loads on Aircraft: Concept and Applications," AIAA Education Series, AIAA, Washington, DC, 1988.

[19]Woodfield, A. A., "Wind Shear and Its Effects on Aircraft," AGARD-LS-197, Paper 5, 1994.

[20]Mulgund, S. S., and Stengel, R. F., "Target Pitch Angle for the Microburst Escape Maneuver," *Journal of Aircraft,* Vol. 30, No. 6, 1993, pp. 826–832.

[21]Dunham, R. E., Jr., "Heavy Rain Effects," AGARD-LS-197, Paper 6, 1994.

[22]Erickson, G. E., and Gilbert, W. P., "Investigation of Forebody and Wing Leading-Edge Vortex Interact Experimental Ions at High Angles of Attack," AGARD-CP-342, Paper 11, April 1983.

[23]Abbott, I. H., and Von Doenhoff, A. E., *Theory of Wing Sections,* 2nd Ed., Dover, New York, 1959.

Index

Texts Published in the AIAA Education Series

Texts Published in the AIAA Education Series (continued)

Boundary Layers
A. D. Young   *1989*

Aircraft Design: A Conceptual Approach
Daniel P. Raymer   *1989*

Gust Loads on Aircraft: Concepts and
Applications
Frederic M. Hoblit   *1988*

Aircraft Landing Gear Design:
Principles and Practices
Norman S. Currey   *1988*

Mechanical Reliability: Theory, Models
and Applications
B. S. Dhillon   *1988*

Re-Entry Aerodynamics
Wilbur L. Hankey   *1988*

Aerothermodynamics of Gas Turbine
and Rocket Propulsion,
  Revised and Enlarged
Gordon C. Oates   *1988*

Advanced Classical Thermodynamics
George Emanuel   *1988*

Radar Electronic Warfare
August Golden Jr.   *1988*

An Introduction to the Mathematics
and Methods of Astrodynamics
Richard H. Battin   *1987*

Aircraft Engine Design
Jack D. Mattingly, William H.   *1987*
Heiser, and Daniel H. Daley

Gasdynamics: Theory and Applications
George Emanuel   *1986*

Composite Materials for Aircraft
Structures
Brian C. Hoskins and   *1986*
Alan A. Baker, Editors

Intake Aerodynamics
J. Seddon and E. L. Goldsmith   *1985*

Fundamentals of Aircraft Combat
Survivability Analysis and Design
Robert E. Ball   *1985*

Aerothermodynamics of Aircraft
Engine Components
Gordon C. Oates, Editor   *1985*

Aerothermodynamics of Gas Turbine
and Rocket Propulsion
Gordon C. Oates   *1984*

Re-Entry Vehicle Dynamics
Frank J. Regan   *1984*

Published by
American Institute of Aeronautics
and Astronautics, Inc.
Washington, DC